SOMATOSTATIN ANALOGUES

SOMATOSTATIN ANALOGUES

From Research to Clinical Practice

Edited by

ALICJA HUBALEWSKA-DYDEJCZYK
ALBERTO SIGNORE
MARION DE JONG
RUDI A. DIERCKX
JOHN BUSCOMBE
CHRISTOPHE VAN DE WIELE

Published by John Wiley & Sons, Inc., Hoboken, New Jersey
Published simultaneously in Canada

For general information on our other products and services or for technical support, please contact our Customer Care Department within the United States at (800) 762-2974, outside the United States at (317) 572-3993 or fax (317) 572-4002.

Wiley also publishes its books in a variety of electronic formats. Some content that appears in print may not be available in electronic formats. For more information about Wiley products, visit our web site at www.wiley.com.

Library of Congress Cataloging-in-Publication Data

Somatostatin analogues : from research to clinical practice / edited by Alicja Hubalewska-Dydejczyk, Alberto Signore, Marion de Jong, Rudi A. Dierckx, John Buscombe, Christophe Van de Wiele.
　　p. ; cm.
　Includes bibliographical references and index.
　ISBN 978-1-118-52153-3 (cloth)
I. Hubalewska-Dydejczyk, Alicja, editor.
[DNLM:　1. Somatostatin–analogs & derivatives.　2. Receptors, Somatostatin–therapeutic use.
3. Somatostatin–therapeutic use. WK 515]
　QP572.S59
　612.405–dc23
2014043035

Printed in the United States of America

10　9　8　7　6　5　4　3　2　1

1　2015

CONTENTS

CONTRIBUTORS

Manuela Albertelli, Endocrinology Unit, Department of Internal Medicine and Center of Excellence for Biomedical Research, University of Genova, Genova, Italy

Richard P. Baum, THERANOSTICS Center for Molecular Radiotherapy and Molecular Imaging, ENETS Center of Excellence, Zentralklinik Bad Berka, Germany

Lisa Bodei, Division of Nuclear Medicine, European Institute of Oncology, Milan, Italy

Adrienne H. Brouwers, Department of Nuclear Medicine and Molecular Imaging, University of Groningen, University Medical Center Groningen, Groningen, The Netherlands

John Buscombe, Addenbrookes Hospital, Cambridge, UK

Adele Cassenti, Department of Oncology, University of Turin, Orbassano, Turin, Italy

Marco Chianelli, Endocrinology Unit, "Regina Apostolorum" Hospital, Albano (Rome), Italy

Rudi A. Dierckx, Department of Nuclear Medicine and Molecular Imaging, University of Groningen, and University Medical Center of Groningen, Groningen, The Netherlands

Philip H. Elsinga, Department of Nuclear Medicine and Molecular Imaging, University of Groningen, and University Medical Center of Groningen, Groningen, The Netherlands

Melpomeni Fani, Clinic of Radiology and Nuclear Medicine, University of Basel Hospital, Basel, Switzerland

Diego Ferone, Endocrinology Unit, Department of Internal Medicine and Center of Excellence for Biomedical Research, University of Genova, Genova, Italy

Helle-Brit Fiebrich, Department of Medical Oncology, University of Groningen, and University Medical Center of Groningen, Groningen, The Netherlands

Luz Kelly Anzola Fuentes, Nuclear Medicine, ClinicaColsanitas, Bogotà, Colombia

Aleksandra Gilis-Januszewska, Department of Endocrinology with Nuclear Medicine Unit, Medical College, Jagiellonian University, Krakow, Poland

Wouter W. de Herder, Department of Internal Medicine, Erasmus MC, Sector of Endocrinology, Rotterdam, The Netherlands

Anouk N.A. van der Horst-Schrivers, Departments of Medical Endocrinology, University of Groningen, University Medical Center Groningen, Groningen, The Netherlands

Alicja Hubalewska-Dydejczyk, Department of Endocrinology with Nuclear Medicine Unit, Medical College, Jagiellonian University, Krakow, Poland

Agata Jabrocka-Hybel, Department of Endocrinology, University Hospital in Krakow; Department of Endocrinology, Medical College, Jagiellonian University, Krakow, Poland

Werner Jaschke, Department of Radiology, Innsbruck Medical University, Innsbruck, Austria

Marion de Jong, Department of Nuclear Medicine and Radiology, Erasmus MC, University Medical Center, Rotterdam, The Netherlands

Boen L.R. Kam, Department of Nuclear Medicine and Radiology, Erasmus MC, University Medical Center, Rotterdam, The Netherlands

Ido P. Kema, Department of Laboratory Center, University of Groningen, and University Medical Center of Groningen, Groningen, The Netherlands

Mark Konijnenberg, Erasmus MC, Rotterdam, The Netherlands

Klaas Pieter Koopmans, Department of Radiology and Nuclear Medicine, Martini Hospital Groningen, Groningen, The Netherlands

Eric P. Krenning, Department of Nuclear Medicine and Radiology, Erasmus MC, University Medical Center, Rotterdam, The Netherlands

Harshad R. Kulkarni, THERANOSTICS Center for Molecular Radiotherapy and Molecular Imaging, ENETS Center of Excellence, Zentralklinik Bad Berka, Germany

Dik J. Kwekkeboom, Department of Nuclear Medicine and Radiology, Erasmus MC, University Medical Center, Rotterdam, The Netherlands

Thera P. Links, Department of Endocrinology, University of Groningen, and University Medical Center Groningen, Groningen, The Netherlands

Helmut R. Maecke, Department of Nuclear Medicine, University Hospital Freiburg, Freiburg, Germany

Theodosia Maina, Molecular Radiopharmacy, INRASTES, NCSR "Demokritos," Athens, Greece

Frédérique Maire, Service de Gastroentérologie-Pancréatologie, Hôpital Beaujon, Clichy and Université Paris Denis-Diderot, Paris, France

Renata Mikołajczak, Radioisotope Centre Polatom, National Centre for Nuclear Research, Otwock, Poland

Berthold A. Nock, Molecular Radiopharmacy, INRASTES, NCSR "Demokritos," Athens, Greece

Dorota Pach, Department of Endocrinology, University Hospital in Krakow; Department of Endocrinology with Nuclear Medicine Unit, Medical College, Jagiellonian University, Krakow, Poland

Giovanni Paganelli, Division of Nuclear Medicine, European Institute of Oncology, Milan, Italy

Mauro Papotti, Department of Oncology, University of Turin, Orbassano, Turin, Italy

Daniel Putzer, Department of Radiology, Innsbruck Medical University, Innsbruck, Austria

Ida Rapa, Department of Oncology, University of Turin, Orbassano, Turin, Italy

Luisella Righi, Department of Oncology, University of Turin, Orbassano, Turin, Italy

Philippe Ruszniewski, Service de Gastroentérologie-Pancréatologie, Hôpital Beaujon, Clichy and Université Paris Denis-Diderot, Paris, France

Alberto Signore, Nuclear Medicine Unit, Department of Medical-Surgical Sciences and of Translational Medicine, Faculty of Medicine and Psychology, "Sapienza" University of Rome, Rome, Italy

Anna Sowa-Staszczak, Department of Endocrinology, University Hospital in Krakow, Krakow, Poland

Agnieszka Stefańska, Department of Endocrinology, University Hospital in Krakow, Krakow, Poland

Aikaterini Tatsi, Molecular Radiopharmacy, INRASTES, NCSR "Demokritos," Athens, Greece

Jaap J.M. Teunissen, Department of Nuclear Medicine and Radiology, Erasmus MC, University Medical Center, Rotterdam, The Netherlands

Marily Theodoropoulou, Department of Endocrinology, Max Planck Institute of Psychiatry, München, Germany

Henri J.L.M. Timmers, Department of Endocrinology, Radboud University Nijmegen Medical Centre, Nijmegen, The Netherlands

Malgorzata Trofimiuk-Müldner, Department of Endocrinology with Nuclear Medicine Unit, Medical College, Jagiellonian University, Krakow, Poland

Christophe Van de Wiele, Department of Nuclear Medicine, Ghent University Hospital, Belgium

Irene J. Virgolini, Department of Nuclear Medicine, Innsbruck Medical University, Innsbruck, Austria

Esther I. van Vliet, Department of Nuclear Medicine and Radiology, Erasmus MC, University Medical Center, Rotterdam, The Netherlands

Marco Volante, Department of Oncology, University of Turin, Orbassano, Turin, Italy

Elisabeth G.E. de Vries, Department of Medical Oncology, University of Groningen, and University Medical Center of Groningen, Groningen, The Netherlands

Dietmar Waitz, Department of Nuclear Medicine, Innsbruck Medical University, Innsbruck, Austria

Annemiek M.E. Walekamp, Department of Medical Oncology, University of Groningen, and University Medical Center of Groningen, Groningen, The Netherlands

Gerlig Widmann, Department of Radiology, Innsbruck Medical University, Innsbruck, Austria

PREFACE

Huge progress has been made in recent years in modern medicine owing to, inter alia, the development of molecular biology. Better understanding of the nature of the disease is a continuous challenge to look for more effective forms of diagnostics and therapy resulting in the improvement of the quality of life of our patients.

The model example of such progress is "somatostatin story". Somatostatin isolation over 40 years ago not only resulted in the Nobel Prize for its discoverers but has also greatly impacted the current clinical practice. The hormone, which at beginning was known only as regulating factor, has now become a potent drug and imaging medium. It has changed the fate of many acromegalic patients and has been applied in other oncological and non-oncological diseases.

Somatostatin was the first peptide to be obtained by bacterial recombination. Although its first therapeutic administration took place before the exact mechanisms of its action were elucidated, it was the discovery of somatostatin receptors and their subtypes, which gave rise to interdisciplinary research leading to the use of somatostatin analogues in routine clinical practice.

The development of radiolabeled somatostatins has to some degree defined the development of nuclear medicine over the past 20 years. During this journey, much has been learned about the nature of cancer with particular reference to those tumors originating from neuroendocrine tissue. What has been unique about this process is the key role played by nuclear medicine scientists both clinical and preclinical. Initial work was supported by industry with imaging agents such as In-111 pentetreotide and Tc-99m depreotide becoming licensed products. However, for the past 10 years, every new advance has been led by academia, and not industry. Nuclear medicine has been able to care for the patient is a holistic way, imaging for diagnosis, staging and re-staging, and treatment to palliate symptoms and extend life. To aid in this nuclear medicine, physicians have interacted with a wide range of clinicians and built multi-disciplinary clinics, a pattern followed in other cancer types.

The nuclear medicine community continues to innovate using radiolabeled somatostatin analogues: developing Lu-177 as a therapeutic isotope offering efficacy with reduced toxicity, using Ga-68 DOTATATE in imaging which differentiated thyroid cancer, and the administrating Y-90 DOTATOC intra-arterially for treating gliomas. All these innovations depended on the imagination, careful science, and dedication of a range of scientists and clinicians around the world.

The best examples of somatostatin research importance in clinical practice are neuroendocrine tumors, particularly originating from the gastroenteropancreatic system. Whilst it is true that neuroendocrine tumors are rare. The slow progression of many tumors has resulted in a prevalence that is much higher than the incidence and at any time 10% of patients visiting gastroenterological oncology clinics may have neuroendocrine tumors. Firstly, somatostatin and its analogues were applied to control the clinical symptoms in syndromic patients, particularly carcinoid ones. Then somatostatin receptor scintigraphy in its various forms became the imaging of choice in gastroenteropancreatic neuroendocrine tumor (GEP-NET) patients. Development of radioguided surgery improved the surgical outcomes. The diagnostic application of the radiolabeled somatostatin analogues led to the therapeutic approach with In-111, replaced by 90-Y and 177-Lu labeled compounds. Those centers worldwide that offer radio-peptide therapy for neuroendocrine tumors have together treated more patients than the centers that use licensed radiolabeled products to treat much more common lymphomas. But the last word has not been said yet. Locoregional therapies for liver metastases with alpha-particles emitting isotopes are being tested. Although the direct cytostatic effect of "old" non-labeled somatostatin analogues in neuroendocrine tumors has been confirmed, the new ones seem to be even more promising. Currently there are attempts to organize and unify the GEP-NET classification to establish important prognostic factors, which would allow the prediction of the disease outcome with a great probability, and to choose optimal diagnostic and therapeutic options for each individual patient. Those methods have still not been optimized and remain a huge challenge in this field of oncology.

The book presented to you is a compendium of current knowledge on the use of somatostatin analogues in diagnostics and therapy, and it also shows the directions of further research in this field. The authors of the book chapters are experts of various scientific disciplines involved in work on somatostatin analogues as well as well-known authorities interested in management of patients with different neoplasms, especially neuroendocrine tumors. The book has been greatly supported by, among others, people actively working in the European Neuroendocrine Tumor Society (ENETs) and those who managed COST Actions (European Cooperation in the field of Scientific and Technical Research) devoted to the development of targeted therapy based on radiolabeled somatostatin compounds. Last but not least involved has been the International Research group in Immuno-Scintigraphy and Therapy (IRIST) and the editors of this book have all been or are Presidents of IRIST. This group has been intimately involved in the development of radiolabeled somatostatins for diagnosis and therapy and as such, is ideally placed to share this knowledge with a wider medical audience.

We should believe the words said by Orioson Swett Marden: *There is no medicine like hope, no incentive so great and no tonic so powerful as expectations of something better than tomorrow*, and that every day of our work helps our patients. Hence, there still is a field for development of research to find out the new compounds with superior efficacy to current treatments, or labeled molecules to be used in imaging diagnostics.

We hope that this book will become a guide for all those who deal with the issue presented herein.

ALICJA HUBALEWSKA-DYDEJCZYK, ALBERTO SIGNORE,
MARION DE JONG, RUDI A. DIERCKX,
JOHN BUSCOMBE, AND CHRISTOPHE VAN DE WIELE

ACKNOWLEDGMENTS

We would like to acknowledge all people managing and actively participating in COST BM0607 Action for providing the idea of this book and constant encouragement. We would like to thank all the contributors for their work and for enabling us to complete the book in timely fashion. The input of the coworkers from our departments should be stressed for helping us with obtaining good quality artwork for the book. And last but not least, a special note of appreciation is to be given to our families for constant support and endless patience.

ALICJA HUBALEWSKA-DYDEJCZYK, ALBERTO SIGNORE,
MARION DE JONG, RUDI A. DIERCKX,
JOHN BUSCOMBE, AND CHRISTOPHE VAN DE WIELE

1

SOMATOSTATIN: THE HISTORY OF DISCOVERY

MALGORZATA TROFIMIUK-MÜLDNER AND ALICJA HUBALEWSKA-DYDEJCZYK

Department of Endocrinology with Nuclear Medicine Unit, Medical College, Jagiellonian University, Krakow, Poland

ABBREVIATIONS

FDA   the Food and Drug Administration
GIF   growth hormone-inhibiting factor
PET   positron emission tomography
SPECT   single photon emission computed tomography
SRIF   somatotropin release-inhibiting factor

Now, here, you can see, it takes all the running you can do, to keep in the same place.
If you want to get somewhere else, you must run at least twice as fast as that!
Lewis Carroll, Through the Looking Glass

The beginning of the second half of the twentieth century, the great era of discovery of factors regulating anterior pituitary hormones synthesis and release, resulted also in isolation and characterization of somatostatin. The history started with search for growth hormone-releasing factor. In 1968, Krulich and colleagues noted that extracts from different parts of rat hypothalamus either stimulated or inhibited release of pituitary growth hormone [1]. The inhibiting substance was named growth hormone-inhibiting factor (GIF). The group of Roger Guillemin developed highly sensitive assay for rat growth hormone, which enabled the confirmation of negative linear

Somatostatin Analogues: From Research to Clinical Practice, First Edition. Edited by Alicja Hubalewska-Dydejczyk, Alberto Signore, Marion de Jong, Rudi A. Dierckx, John Buscombe, and Christophe Van de Wiele.

relationship between the production of the growth hormone by anterior pituitary cell culture and amount of hypothalamic extract added [2]. About 500,000 sheep hypothalami later Brazeau and Guillemin isolated the substance responsible for inhibiting effect—somatotropin release-inhibiting factor—SRIF. The structure of 14-aminoacid peptide was then sequenced, the sequence of the residues confirmed, and the molecule was resynthesized. The synthetic molecule inhibiting properties were confirmed in both *in vivo* and *in vitro* experiments. The result of the discovery was paper published in *Science* in 1973 [3]. Roger Guillemin renamed the hormone—since 1973 it has been known as the somatostatin [4]. The new hormone was extracted also from hypothalami of other species.

Those times were also regarded the gut hormones era. In 1969, Hellman and Lernmark announced the inhibiting effect of extract from alfa-1 cells of pigeon pancreas on insulin secretion from pancreatic islets derived from obese, hyperglycemic mice [5]. In 1974, group of C. Gale from Seattle noticed the lowering of fasting insulin and glucagon levels in baboons as well as tampering of arginine-stimulated insulin release by somatostatin—directly and in dose-dependent manner [6]. This finding was confirmed also in other animal models and humans shortly after. As the hypothalamic somatostatin seemed to act locally, the search for local, pancreatic source of the hormone started. The antibodies against somatostatin proved to be useful tool. The presence of somatostatin in delta (D) cells of the pancreas (formerly alfa-1 cells) was proved by immunofluorescence [7, 8]. In 1979, somatostatin was isolated from the pigeon pancreas, and next from other species [9]. The somatostatin-reactive cells were also found in gastrointestinal mucosa, and then in other tissues, including tumors. Concurrently, the multiple groups worked on the somatostatin action and its pan-inhibiting properties were gradually characterized. In 1977, Roger Guillemin and Andrew Schally were awarded the Nobel Prize in medicine and physiology for their work on somatostatin and other regulating hormones. Of interest, somatostatin-like peptides were also discovered in plants [10].

Other somatostatin forms, somatostatin-28 particularly, and somatostatin precursor—preprosomatostatin—were characterized in late 1970s/early 1980s. Human cDNA coding preprosomatostatin was isolated and cloned in 1982 [11, 12].

The possible pathological implications and potential therapeutic use of somatostatin were postulated early in the somatostatin discovery era. The clinical description of somatostatin-producing pancreatic tumor in human came from Larsson and colleagues in 1977 [13]. Somatostatin administration to block the growth hormone secretion in acromegalic patients was reported as early as in 1974 [14]. The potency of the hormone to block carcinoid flush was also observed in late 1970s and early 1980s [15, 16]. Somatostatin was the first human peptide to be produced by bacterial recombination. In 1977 Itakura, Riggs and Boyer group synthesized gene for somatostatin-14, fused it with *Escherichia coli* beta-galactosidase gene on the plasmid and transformed the *E. coli* bacteria with chimeric plasmid DNA. As the result, they obtained the functional human polypeptide [17]. The synthesis of recombinant human somatostatin led to the commercial human recombinant insulin production.

Although it was possible to produce somatostatin in large quantities, the short half-life of the hormone was one of the reasons why the native hormone was not feasible for routine clinical practice. The search for more stable yet functional hormone analogue started in 1974. The search was focused on the peptide analogues. The somatostatin receptor agonists were first to be used in clinical practice. In 1980–1982, octapeptide SMS 201–995 was synthetized and proved to be more resistant to degradation and more potent than native hormone in inhibiting growth hormone synthesis [18]. The drug, currently known as octreotide, was the first Food and Drug Administration (FDA)-approved somatostatin analogue. It was followed by other analogues, such as lanreotide (BIM 23014), and by the long-acting formulas. High selective affinity of octreotide and lanreotide for somato-statin receptor type 2 (lesser to the receptor types 3 and 5) was one of the triggers for further research. In 2005 vapreotide (RC160), somatostatin analogue with additional affinity to receptor type 4, was initially accepted for treatment of acute oesophageal vari-ceal bleeding and granted the orphan drug status in 2008 in the United States (although final FDA approval has not been granted). Lately, promising results of large phase III studies on "universal" multitargeted somatostatin analogue, cyclohexapeptide SOM-230 pasireotide, in acromegaly and Cushing's disease, have been published [19, 20]. The drug has been granted the European Medicines Agency and the FDA approval for pituitary adrenocorticotropic hormone (ACTH)-producing adenomas treatment. The research on first nonpeptide receptor subtype selective agonists was published in 1998; however, none of tested compounds have been introduced to clinical practice [21]. The studies on somatostatin receptors antagonists have been conducted since 1990s.

The other areas for research were somatostatin receptors. The high affinity-binding sites for somatostatin were found on pancreatic cells and in brain surface by group of J.C. Reubi in 1981–1982. The different pharmacological properties of the receptors were noted early. At the beginning two types of somatostatin receptors, with high and low affinity for octreotide, were characterized [22, 23]. In 1990s, all five subtypes of somatostatin receptors were cloned and their function was discovered. The other important step was the discovery of the somatostatin receptors overexpression in tumor cells, particularly of neuroendocrine origin [24]. This led to the first successful trials on diagnostic use of radioisotope-labeled hormones. The iodinated octreotide was used in localization of the neuroendocrine tumors in 1989–1990 [25, 26]. The Iodine-123 was replaced by the Indium-111, and later on by the Technetium 99 m [27–29]. The first Gallium-68 labeled somatostatin analogues for positron emission tomography (PET) studies were proposed in 1993 [30]. Feasibility of labeled somatostatin receptor antagonist for single photon emission computed tomography (SPECT) or PET tumor imaging has been reported in 2011 [31]. Together with diagnostics, the concept of therapeutic use radioisotope labeled somatostatin analogues has evolved. The first pep-tides for therapy were those labeled with Indium-111 [32]. In 1997, the Yttrium-90 labeled analogues, followed by Lutetium-177 labeled ones, were introduced in pallia-tive treatment of neuroendocrine disseminated tumors [33, 34].

The co-expression of somatostatin and dopamine receptors, as well as discovery of receptor heterodimerization, led to the search for chimeric somatostatin-dopamine molecules, dopastatins [35]. Other area of recent research is cortistatin, a member of somatostatin peptides family, with somatostatin receptors affinity but also with dis-tinct properties [36].

Summing up the multicenter research on somatostatin led to the discovery of the hormone probably second only to the insulin in its clinical use.

REFERENCES

[1] Krulich, L.; Dhariwal, A. P.; McCann, S. M. Endocrinology 1968, 83, 783–790.

[2] Rodger, N. W.; Beck, J. C.; Burgus, R.; Guillemin, R. Endocrinology 1969, 84, 1373–1383.

[3] Brazeau, P.; Vale, W.; Burgus, R.; et al. Science 1973, 179, 77–79.

[4] Burgus, R.; Ling, N.; Butcher, M.; Guillemin, R. Proceedings of the National Academy of Sciences of the U S A 1973, 70, 684–688.

[5] Hellman, B.; Lernmark, A. Diabetologia 1969, 5, 22–24.

[6] Koerker, D. J.; Ruch, W.; Chideckel, E.; et al. Science 1974, 184, 482–484.

[7] Polak, J. M.; Pearse, A. G.; Grimelius, L.; Bloom, S. R. Lancet 1975, 31, 1220–1222.

[8] Luft, R.; Efendic, S.; Hökfelt, T.; et al. Medical Biology 1974, 52, 428–430.

[9] Spiess, J.; Rivier, J. E.; Rodkey, J. A.; et al. Proceedings of the National Academy of Sciences of the U S A 1979, 76, 2974–2978.

[10] Werner, H.; Fridkin, M.; Aviv, D.; Koch, Y. Peptides, 1985, 6, 797–802.

[11] Böhlen, P.; Brazeau, P.; Benoit, R.; et al. Biochemical and Biophysical Research Communications 1980, 96, 725–734.

[12] Shen, L. P.; Pictet, R. L.; Rutter, W. J. Proceedings of the National Academy of Sciences of the U S A 1982, 79, 4575–4579.

[13] Larsson, L. I.; Hirsch, M. A.; Holst, J. J.; et al. Lancet 1977, 26(8013), 666–668.

[14] Yen, S. S.; Siler, T. M.; DeVane, G. W. New England Journal of Medicine 1974, 290, 935–938.

[15] Thulin, L.; Samnegård, H.; Tydén, G.; et al. Lancet 1978, 2, 43.

[16] Frölich, J. C.; Bloomgarden, Z. T.; Oates, J. A.; et al. New England Journal of Medicine 1978, 299, 1055–1057.

[17] Itakura, K.; Hirose, T.; Crea, R.; et al. Science, 1977, 198, 1056–1063.

[18] Bauer, W.; Briner, U.; Doepfner, W.; et al. Life Sciences 1982, 31, 1133–1140.

[19] Colao, A.; Petersenn, S.; Newell-Price, J.; et al. New England Journal of Medicine 2012, 366, 914–924. Erratum in: New England Journal of Medicine 2012, 367, 780.

[20] Petersenn, S.; Schopohl, J.; Barkan, A.; et al. Journal of Clinical Endocrinology and Metabolism 2010, 95, 2781–2789.

[21] Yang, L.; Guo, L.; Pasternak, A.; et al. Journal of Medicinal Chemistry 1998, 41, 2175–2179.

[22] Reubi, J. C.; Perrin, M. H.; Rivier, J. E.; Vale, W. Life Sciences 1981, 28, 2191–2198.

[23] Reubi, J. C.; Rivier, J.; Perrin, M.; et al. Endocrinology 1982, 110, 1049–1051.

[24] Reubi, J. C.; Maurer, R.; von Werder, K.; et al. Cancer Research 1987, 47, 551–558.

[25] Krenning, E. P.; Bakker, W. H.; Breeman, W. A.; et al. Lancet 1989, 1(8632), 242–244.

[26] Bakker, W. H.; Krenning, E. P.; Breeman, W. A.; et al. Journal of Nuclear Medicine 1990, 31, 1501–1509.

[27] Bakker, W. H.; Krenning, E. P.; Reubi, J. C.; et al. Life Sciences 1991, 49, 1593–1601.

[28] Decristoforo, C.; Mather, S. J. European Journal of Nuclear Medicine 1999, 26, 869–876.

[29] Bangard, M.; Béhé, M.; Guhlke, S.; et al. European Journal of Nuclear Medicine 2000, 27, 628–663.

[30] Mäcke, H. R.; Smith-Jones, P.; Maina, T.; et al. Hormone and Metabolism Research. Supplement Series 1993, 27, 12–17.

[31] Wild, D.; Fani, M.; Behe, M.; et al. Journal of Nuclear Medicine 2011, 52, 1412–1417.

[32] Krenning, E. P.; Kooij, P. P.; Bakker, W. H.; et al. Annals of New York Academy of Sciences 1994, 733, 496–506.

[33] de Jong, M.; Bakker, W. H.; Krenning, E. P.; et al. European Journal of Nuclear Medicine 1997, 24, 368–371.

[34] Kwekkeboom, D. J.; Bakker, W. H; Kooij, P. P.; et al. European Journal of Nuclear Medicine 2001, 28, 1319–1325.

[35] Jaquet, P.; Gunz, G.; Saveanu, A.; et al. Journal of Endocrinological Investigation 2005, 28(11 Suppl International), 21–27.

[36] Fukusumi, S.; Kitada, C.; Takekawa, S.; et al. Biochemical and Biophysical Research Communications 1997, 232, 157–163.

2

PHYSIOLOGY OF ENDOGENOUS SOMATOSTATIN FAMILY: SOMATOSTATIN RECEPTOR SUBTYPES, SECRETION, FUNCTION AND REGULATION, AND ORGAN SPECIFIC DISTRIBUTION

MARILY THEODOROPOULOU

Department of Endocrinology, Max Planck Institute of Psychiatry, München, Germany

ABBREVIATIONS

AIP	aryl hydrocarbon receptor interacting protein
Akt	AKT8 virus oncogene cellular homolog
Bax	Bcl-2-associated X
cAMP/cGMP	cyclic adenosine/guanosine monophosphate
Cdk	cyclin-dependent kinase
DAG	diacylglycerol
DR4	death receptor 4
GH	growth hormone
GI	gastrointestinal track
GPCR	G-protein-coupled receptors
grb	growth factor receptor-bound protein
GSK3	glycogen synthase kinase 3

Somatostatin Analogues: From Research to Clinical Practice, First Edition. Edited by
Alicja Hubalewska-Dydejczyk, Alberto Signore, Marion de Jong, Rudi A. Dierckx,
John Buscombe, and Christophe Van de Wiele.

IGF-I	insulin-like growth factor I
IP3	inositol 1,4,5-triphosphate
JNK	c-Jun NH(2)-terminal kinase
MAPK	mitogen-activated protein kinase
mTOR	mammalian target of rapamycin
NOS	nitric oxide synthase
PI3K	phosphatidyl inositol 3-kinase
PIP2	phosphatidylinositol 4,5-bisphosphate
PKC	protein kinase C
PLA	phospholipase A
PLC	phospholipase C
PTP	protein tyrosine phosphatase
Raf	rapidly accelerated fibrosarcoma
Rb	retinoblastoma
SH2	src homology 2
SHP	SH2-containing phosphatase
Sos	son of sevenless
Src	Rous sarcoma oncogene cellular homolog
SSTR	somatostatin receptors
TNFR1	tumor necrosis factor receptor 1
TSC2	tumor sclerosis complex 2
TSP-1	thrombospondin-1

INTRODUCTION

Somatostatin mediates its action upon binding to somatostatin receptors (SSTR) which belong to the seven-transmembrane domain, G-protein-coupled receptors (GPCRs) superfamily and are mainly coupled to the Gi protein and therefore inhibit adenylate cyclase and cAMP accumulation [1]. There are five somatostatin receptors SSTR1-5. The genes encoding human SSTR1-5 are located in chromosome 14q13, 17q24, 22q13.1, 20p11.2 and 16p13.3. The gene encoding for SSTR2 has an intron and the transcribed mRNA can be spliced to encode SSTR2A and B isoforms [2]. SSTR5 also exists as truncated isoforms with four or five transmembrane domains (sst5TDM4 and sst5TDM5; [3]) generated by cryptic splice sites in the coding sequence and the 3′ untranslated region of the *SSTR5* gene. All SSTR are Gi coupled and inhibit adenylate cyclase. However, as it will be described more extensively later, they also trigger several signaling cascades that may be pertussis toxin (i.e., Gi) dependent or independent.

SECRETION

Somatostatin was initially identified as a hypothalamic peptide able to inhibit growth hormone (GH) secretion from the pituitary [4]. Two biological forms of somatostatin exist, somatostatin-14 and -28, which are derived from a 92 aminoacid pro-somatostatin precursor [5, 6]. Somatostatin is a neurotransmitter and can be regarded as a secretory pan-inhibitor; it suppresses GH, prolactin, thyroid-stimulating hormone [7, 8] and adrenocorticotropic hormone (ACTH) [9] secretion from the anterior pituitary; cholecystokinin, gastrin, secretin, vasoactive intestinal peptide, motilin, gastric inhibitory polypeptide from the gastrointestinal track (GI); glucagon, insulin, and pancreatic polypeptide from the endocrine pancreas [10]; triiodothyronine, thyroxin, and calcitonin from the thyroid; and renin and aldosterone from the kidney and the adrenals [11]. In addition to its endocrine action, it also suppresses exocrine secretion (e.g., gastric acid from intestinal mucosa, bicarbonate, and digestive enzymes from exocrine pancreas). In the GI, it also inhibits bile flow from the gallbladder, bowel motility and gastric emptying, smooth muscle contraction and nutrient absorption from the intestine. Somatostatin also inhibits cytokine and growth factors production from immune and various tumor cells.

Somatostatin suppresses GH and TSH through SSTR2 and SSTR5, and prolactin predominantly through SSTR5 [12, 13]. GH secretion is also inhibited by SSTR1 [14]. *Sstr2* knockout mice have elevated ACTH levels, indicating a regulatory role for SSTR2 [15]. Both SSTR2 and SSTR5 decrease ACTH synthesis [16], with SSTR5 displaying a more potent suppressive action on ACTH release [17]. Insulin secretion is primarily inhibited by SSTR5, while glucagon secretion is primarily inhibited by SSTR2 [18]. Gastric acid and pancreatic amylase release is inhibited by SSTR2 and SSTR4, while other GI hormones are inhibited by SSTR1, 2, and 5 [19, 20].

Somatostatin exerts its antisecretory action mainly by inhibiting exocytosis. This is mediated by its inhibitory action on adenylate cyclase and subsequent decrease in cyclic adenosine monophosphate (cAMP) production [21–24]. The effect is pertussis toxin-dependent indicating the involvement of the Gi protein [25]. In addition to cAMP suppression, somatostatin activates potassium (K^+) channels (delayed rectifying, inward rectifying and ATP sensitive) and induces membrane hyperpolarization that inhibits depolarization-induced Ca^{2+} influx via voltage-sensitive Ca^{2+} channels. This reduces intracellular Ca^{2+} and inhibits exocytosis [26–30]. The inhibitory action of somatostatin on Ca^{2+} is mediated through the Gi and Go protein subtypes [31, 32]. In addition, an alternative pathway involving a cGMP-dependent protein kinase was identified behind the inhibitory action of somatostatin on neuronal calcium channels [33]. All SSTRs, except SSTR3, couple to voltage-gated K^+ channels, but SSTR2 and 4 are more potent in increasing K^+ currents [34]. SSTR1, 2, 4 and 5 couple to N- and L-type voltage-sensitive Ca^{2+} channels indicating a direct effect [35–38]. In addition, somatostatin has a distal to secondary messengers effect on exocytosis by activating the Ca^{2+}-dependent phosphatase calcineurin [39, 40].

Regarding the effect of somatostatin on hormone transcription, initial studies did not find changes in GH mRNA levels after somatostatin administration, supporting the hypothesis that somatostatin suppresses GH secretion by blocking exocytosis rather than transcription [41–43]. However, studies *in vitro* and in tumors from patients with acromegaly who were preoperatively treated with somatostatin analogs revealed reduced GH transcript levels after somatostatin treatment [44–47]. Somatostatin suppressed GH-releasing hormone-induced GH promoter activity in a pertussis toxin-sensitive manner [48]. SSTR2 overexpression in human somatotropinomas and prolactinomas in primary cell cultures suppressed GH and PRL transcripts, indicating a role for this receptor in somatostatin's suppressive action on GH [49]. Interestingly, somatostatin was shown to stimulate GH secretion at low doses (below 10^{-13}M), an effect that was mediated by SSTR5 [50, 51]. By contrast, SSTR5 agonists suppress PRL secretion, but not transcription *in vitro* [52]. Somatostatin analogs suppress POMC promoter activity, an effect that is abolished by SSTR2 knockdown [53].

ANTIPROLIFERATIVE SIGNALING

Somatostatin limits cell growth through cytostatic or apoptotic mechanisms depending on the SSTR [54, 55]. One of the first described mechanisms behind the antiproliferative action of SSTR was the inhibitory action on growth factor receptor signaling [56–58]. Protein tyrosine phosphatases (PTPs) were shown to play a central role in this process by de-phosphorylating the growth factor bound tyrosine kinase receptors [59]. PTP activity was found to be increased after somatostatin treatment in many cell systems [60–63] and in human tumors in primary cell culture [64, 65]. PTP were shown to be activated by Gαi [59] and Gαi/o [66]. SSTR associate with the cytosolic src homology 2 (SH2) domain containing PTP, SHP-1 (PTP1C) and SHP-2 (PTP1D), and the membrane anchored PTPη (DEP1) [67–74]. Through PTPs, somatostatin blocks cell cycle progression by arresting cells at the G1/S (SSTR1, 2, 4 and 5) or the G2/M (SSTR3) boundary [75, 76]. In addition, SSTR2 and SSTR3 were shown to induce apoptosis [77–79]. SSTRs also induce acidification, which results in apoptosis via a SHP-1-dependent mechanism [80], while SSTR1, 3 and 4 inhibit the Na^+/H^+ exchanger NHE1, leading to increased intracellular acidification [81, 82]. Finally, SSTR1, 2, 3 and 5 block nitric oxide synthase (NOS), revealing an additional regulation point in the antiproliferative action of somatostatin [83, 84].

SSTR have common and individual signaling aspects, which are covered in more detail further (Fig. 2.1).

SSTR1

SSTR1 couples to Gαi3 and Gαi1/2 [85–87] and inhibits adenylate cyclase when overexpressed in Chinese hamster ovarian (CHO) cells [88]. SSTR1 also increases PTP activity [60, 69, 89]. In fact, it uses SHP-2 to activate the serine/threonine mitogen-activated protein kinase (MAPK) concomitantly with its antiproliferative action in these cells [64]. The MAPK pathway usually mediates the mitogenic action of growth factors, cytokines and hormones. However, depending on the cell system and extracellular milieu, the MAPK pathway can also halt cell growth in order to promote cell differentiation. Typically, the pathway starts with activation of the tyrosine kinase domain of the growth factor receptors and the association through special adaptors to Sos which enhances the GTP-binding activity of the GTP-ase Ras. GTP-bound activated Ras associate with, brings to the membrane and activates the Raf family of kinases (A-Raf, B-Raf, and c-Raf/Raf-1). Raf kinases (MAPK kinase

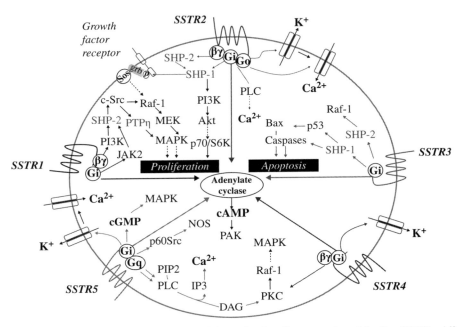

FIGURE 2.1 Schematic presentation of the main signaling cascades of the five SSTRs. All SSTRs are coupled to Gi, inhibit adenylate cyclase and lower cAMP. SSTR1, 2, and 3 transduce their antiproliferative action by stimulating one or more PTP which in turn affects the mitogenic MAPK and the survival PI3K pathways. By contrast, SSTR5 mediates its antiproliferative action through PTP-independent pathways. Open arrowheads: stimulatory effect; blunt arrowheads: inhibitory effect; interrupted lines: indirect effect.

kinases) phosphorylate and activate the MAPK kinases MEK1/2 which then phosphorylate and activate the p44 and p42 MAPK. Raf-1 can also be activated by the src family of tyrosine kinases. SSTR1 activated SHP-2 dephosphorylates c-src at an inhibitory site (Tyr529) which enables its phosphorylation at the stimulatory Tyr418. This enables c-src to phosphorylate and activate Raf-1, which in turn phosphorylates and activates MEK/MAPK leading to upregulation of the cell cycle inhibitor p21/Cip1. This pathway is inhibited by the Gi inhibitor pertussis toxin and is mediated by the βγ subunits of the Gi protein. It also involves an active phosphatidyl inositol 3 kinase (PI3K) although the exact mechanism is not clear [64].

Somatostatin treatment induces a long-lasting PTP activity that cannot be explained by the rapidly activated SHP-2. This PTP is the membrane anchored PTPη, which was described as a tumor suppressor in several tumor types [90, 91]. The importance of PTPη in mediating the antiproliferative action of SSTR1 was demonstrated in the PC CI3 clonal thyroid cells, which loose their ability to respond to somatostatin after oncogene-induced cellular transformation that suppresses PTPη; re-introducing PTPη restores their response to the antiproliferative action of somatostatin [73]. SSTR1 inhibits MAPK through PTPη in glioma and neuroblastoma cells

[92]. SSTR1 activates Jak2, in a pertussis toxin-sensitive manner, which then phosphorylates and activates SHP-2 leading to c-src dephosphorylation and activation, and eventually to PTPη phosphorylation [93].

SSTR2

SSTR2 is the best-studied mediator of somatostatin's antiproliferative action. In fact, SSTR2 is considered as a tumor suppressor in pancreatic cancer since its expression is lost in these tumors [94, 95].

SSTR2A and B inhibit adenylate cyclase, and this effect was found to depend on the G protein subunits available in each cell type [86, 96, 97]. In pituitary tumor GH4C1 cells, the ability of SSTR2A to inhibit adenylate cyclase and subsequently cAMP production resulted in decreased protein kinase A (PKA) activity [98]. The antiproliferative action of SSTR2 also begins with PTP activation. The PTP associated with SSTR2 is the cytosolic SH2 domain containing SHP-1, which associates with the receptor constitutively through Gαi3 [70–89]. Somatostatin treatment leads to SHP-1 dissociation from the receptor and activation resulting in the dephosphorylation of tyrosine kinase receptors (e.g., insulin receptor) and its substrates (e.g., insulin receptor substrate-1, IRS-1) [99]. Another mechanism leading to SHP-1 activation is through SHP-2, which also associates with SSTR2 [100]. Upon receptor activation, the βγ subunits of the Gi proteins activate src, probably by binding src to β-arrestin, which then phosphorylates SHP-2 and subsequently activates SHP-1 [101]. Finally, SSTR2 activates SHP-1 through the α subunit of the Gi protein and the receptor-bound tyrosine kinase JAK2 and inhibits fibroblast growth factor (FGF)-2 isoform of 210 amino acids (HMW FGF-2)-induced pancreatic tumor cell growth [102]. This was a novel finding since JAK2 is traditionally considered to associate with the cytokine receptor family.

SSTR2 was shown to inhibit growth factor induced MAPK phosphorylation and activation [103, 104], but also to activate MAPK, which together with the activated p38-MAPK leads to decreased cell proliferation [105]. In this setting, the SSTR2-induced MAPK activation was mediated by Ras and B-Raf, but also by Rap1 that is another member of the Ras subfamily of small GTP-ases [106]. SSTR2 also activates the survival PI3K signaling, in a mechanism involving Gβγ and SHP-2 [106, 107]. By contrast, activation of overexpressed or endogenous SSTR2 inhibits the PI3K pathway in tumor cell systems [108, 109]. SSTR2 binds directly p85 and this is a unique feature of SSTR2 not shared by another member of the SSTR family. SSTR2 activation disrupts its association with p85 by associating filamin A, resulting in PI3K inhibition [110]. In pituitary tumor cells, p85 physically associates with SHP-1 and SSTR2 activation with octreotide leads to decreased p85 tyrosine phosphorylation, which was SHP-1 dependent. Although the effect of octreotide was pertussis toxin sensitive, indicating involvement of the Gi, it was not depending on Gβγ showing that Gi-linked GPCR could interact with and inhibit PI3K through the Gi α-subunit. This way SSTR2 inhibits the serine/threonine kinase Akt that mediates the antiapoptotic and cell survival effects of several growth factors. This is done in part by phosphorylating and subsequently inhibiting glycogen synthase kinase-3 (GSK3β) which halts cell cycle progression. Cell cycle progression starts with the activation of D-type cyclins and their associated cyclin-dependent kinases Cdk4 and 6 [111]. The G1 to S transition is primarily governed by cyclin E and its associated kinase Cdk2, which hyperphosphorylates retinoblastoma (Rb) [112]. Phosphorylated Rb dissociates from E2F transcription factors resulting in the transcription of genes that will bring the cell to the S phase of the cell cycle [113]. Cyclin/CDK complexes are inhibited by cyclin kinase inhibitors such as p21/Cip1 and p27/Kip1. p27/Kip1 is the primary regulator of cyclin E/CDK2 complex, since by sequestering Cdk2 it prevents the complex formation. GSK3β phosphorylates and marks for proteolytic degradation the cyclins E and D1 and activates p27/Kip1. SSTR2 upregulates p21/Cip1 after stimulating both ERK1/2 and p38-MAPK [105] and p27/Kip1 in a mechanism involving SHP-1 [72, 83].

Although p27/Kip1 is an important downstream target of somatostatin's antiproliferative signaling, cells like the rat pituitary tumor GH3 that do not express p27/Kip1 also respond to SSTR2 activation by decreasing cell proliferation [114]. In these cells, SSTR2 was shown to induce the expression of the tumor suppressor Zac1, in a mechanism involving Gαi, SHP-1, GSK3β, and the Zac1 activator p53 [109]. Zac1 (gene name *Plagl1*) is a zinc finger protein able to induce apoptosis and cell cycle arrest that is frequently downregulated/lost in several solid cancers [115]. RNA interference experiments in pituitary tumor cells revealed that Zac1 is essential for octreotide's antiproliferative action. A retrospective immunohistochemical analysis on archival paraffin embedded tumoral tissue from acromegalic patients treated with somatostatin analogs pre-operatively revealed a strong positive correlation between treatment response and ZAC1 immunoreactivity, with strong ZAC1 immunoreactivity positively correlating with IGF-I normalization and tumor shrinkage after treatment [116]. Interestingly, in GH3 cells ZAC1 gene expression was suppressed after knocking down the aryl hydrocarbon receptor interacting protein (AIP), which is triggered by octreotide treatment [117]. The gene encoding for AIP was found to have germline mutations in patients with familial and sporadic acromegaly and *AIP* mutations predict an unfavorable response to somatostatin analogs [118, 119].

In addition to its action on cell cycle proteins, GSK3β also activates the tumor suppressor tuberin (TSC2), which inhibits the mammalian target of rapamycin (mTOR) controlling cap-dependent translation and subsequently cell growth in terms of cell size rather than cell proliferation. SSTR2 by inhibiting Akt decreased GSK3β phosphorylation and increased its activity leading to decreased phosphorylation of the mTOR effectors p70/S6K and 4E-BP1 [120]. Suppression of the mTOR pathway may explain the observations reporting tumor shrinkage in acromegalic patients treated with SSTR2 agonists not due to apoptosis but rather due to decrease in cell volume [121, 122].

There is increasing evidence that SSTR2 is not only cytostatic but also able to induce apoptosis by upregulating the death receptor 4 (DR4) and tumor necrosis factor receptor 1 (TNFR1) and downregulating the antiapoptotic Bcl2 [123].

SSTR3

SSTR3 inhibits adenylate cyclase activity in a pertussis toxin sensitive pathway by coupling to $G\alpha i1$ [96]. Similar to SSTR1 and 2, SSTR3 is also able to activate a PTP; overexpressed SSTR3 was found to activate SHP-2 and subsequently inactivate Raf-1 [63, 71]. Nevertheless, SSTR3 was initially described as the only SSTR able to induce apoptosis, since its activation in cells selectively expressing SSTR3 led to apoptosis but not to cell cycle arrest [77, 124]. This effect is mediated by upregulating p53 and the proapoptotic protein Bax. In addition, an involvement of SHP-1 and activated caspase 8 was described in the somatostatin-induced cell acidification and apoptosis in SSTR3-expressing cells [80, 125]. SSTR3 is also characterized by a unique antiproliferative action in endothelial cells, constituting it as the primary apoptotic and antiangiogenic SSTR [126, 127].

SSTR4

This receptor type is the less studied in the family. The original studies failed to demonstrate a coupling of SSTR4 to Gi and adenylate cyclase; but eventually, it was shown to suppress cAMP production similar to the other members of the family [128]. Furthermore, SSTR4 was found to activate MAPK in a pertussis toxin sensitive manner by activating phospholipase A (PLA)-2 and arachidonate production. In fact, this is the only SSTR that is reported to stimulate cell proliferation. SSTR4 is also coupled to K^+ channels (delayed rectifier) leading to decreased Ca^{2+} influx. SSTR4 displays an unusually long lasting effect and is hypothesized to mediate the antiepileptic properties of somatostatin [129, 130]. Interestingly, this receptor was also shown to mediate the anti-inflammatory properties of somatostatin [131].

SSTR5

SSTR5, together with SSTR2, is the main SSTR inhibiting hormone release. SSTR5 (initially termed "SSTR4") was cloned as an adenylate cyclase coupled SSTR with high affinity to somatostatin-28 [132]. Similar to the other SSTRs it is able to inhibit adenylate cyclase in a pertussis toxin sensitive mechanism. SSTR5 induces K^+ leading to cell hyperpolarization which subsequently closes the L-type voltage-sensitive Ca^{2+} channels resulting in decreased Ca^{2+} influx [133]. SSTR5 also affects phospholipase C (PLC) in a mechanism only partially involving Gi and requiring the $G\alpha q$ [134]. PLC cleaves phosphatidylinositol 4,5-bisphosphate (PIP2) into diacyl glycerol (DAG) and inositol 1,4,5-triphosphate (IP3), which gets released into the cytosol where it binds to Ca^{2+} channels and increases Ca^{2+} influx into the cytosol. DAG is membrane bound and together with Ca^{2+} functions in recruiting and activating protein kinase C (PKC). Overexpressed SSTR5 was reported to increase IP3 and subsequent Ca^{2+} increase [135]. By contrast, it was found to inhibit cholecystokinin (CCK)-induced Ca^{2+} influx by inhibiting PLC and IP3 generation [89]. Contrary to what is the case for the other SSTRs, no PTP is required for SSTR5 antiproliferative effect [89]. Instead, SSTR5 acts by inhibiting CCK-induced cyclic GMP (cGMP), which can activate specific kinases (G kinases) able to upregulate *c-fos* and subsequently cell proliferation [136]. In this model, SSTR5 by decreasing cGMP inhibits MAPK. In addition, SSTR5 activation in human pancreatic carcinoid cells increases the receptor association with the src-like tyrosine kinase p60src, which phosphorylates and inactivates neuronal nitric oxide synthase (nNOS), and therefore suppresses tumor cell proliferation [137]. These data show that SSTR5 employs completely different cascades to induce its antiproliferative effect compared to the other SSTR.

INDIRECT ANTIPROLIFERATIVE ACTION OF SSTRS

SSTR do not abolish the mitogenic action of growth factors only by inhibiting their signaling cascades, but also by downregulating the synthesis of the growth factors themselves. The founding example of somatostatin-induced growth factor downregulation is IGF-I, which is primarily regulated by GH. Somatostatin analogs used in the treatment of acromegaly decrease circulating IGF-I levels by inhibiting GH synthesis. In addition a direct action on hepatocyte IGF-I production was shown with the activation of hepatic SSTR2 and 3 inhibiting GH-induced IGF-I by dephosphorylating STAT5b, an important transcription factor for IGF-I promoter activation, in a pertussis toxin sensitive mechanism involving a PTP [138].

The ability of SSTRs to suppress growth factor synthesis is also responsible for their antiangiogenic action. Angiogenesis is regulated by the vascular endothelial growth factor (VEGF), which drives the development of new vessels under the trigger of hypoxia in the growing tumor. Somatostatin treatment in an *in vivo* model of Kaposi sarcoma inhibited tumor growth despite the complete lack of SSTR in these cells, an effect that was attributed to the antiangiogenic action of somatostatin [127]. SSTR1 is highly expressed in vessels where it inhibits endothelial proliferation, migration and neovascularization [139, 140]. Endothelial SSTR3 downregulates VEGF and endothelial NOS (eNOS) transcription [126]. The ability of SSTR3 to decrease eNOS activity is also shared by SSTR1 and SSTR2 [84, 126]. More recently, SSTR2 activation was found to block angiogenesis by upregulating the secretion of antiangiogenic factor thrombospondin-1 (TSP-1) from pancreatic cancer cells bringing another twist in the antiangiogenic action of somatostatin [141].

ORGAN SPECIFIC DISTRIBUTION

All SSTR are expressed in the brain: SSTR1 in the cortex, hippocampus, hypothalamus, midbrain, and cerebellum; SSTR2 in the cortex, basal ganglia, and hypothalamus; SSTR3 in the cortex, hypothalamus (arcuate and ventromedial nuclei), and basal ganglia; SSTR5 (and SSTR4 in less extent) in the hypothalamus in the arcuate/ventromedial and arcuate/median eminence, respectively; and SSTR4 mainly in the hippocampus [142]. SSTR2 and SSTR5 are the main receptors found in the adenohypophysis with SSTR1 and SSTR3 being expressed at lower levels [143]. All SSTRs are found in parts of the GI and the spleen [144]. SSTR1 is expressed in jejunum and stomach and SSTR2 in kidney [145]. In the pancreas, alpha cells express mainly SSTR2, beta cells SSTR1 and SSTR5, and delta cells SSTR5. The adrenals express SSTR2 and SSTR5. In the immune system, lymphocytes express SSTR3 and thymus SSTR1, SSTR2, and SSTR3. Liver expresses SSTR1 and SSTR2. SSTR4 is present in the lung, pancreas, and heart. It has to be considered that most of these data were obtained by *in situ* hybridization and autoradiography techniques. The development of specific SSTR antibodies will enable a thorough mapping of SSTR expression in normal tissues.

CONCLUSION

SSTR expression pattern and complex signaling is what makes somatostatin such an extraordinary neurotransmitter and hormone. Their ability to trigger common but also unique pathways fine-tune somatostatin's action depending on the cell type, receptor types expressed, and physiological circumstances. The potent inhibitory action of SSTR on cellular processes as diverse as secretion, proliferation, and apoptosis is what makes somatostatin an invaluable target for drug development.

REFERENCES

[1] Patel, Y. C. Frontiers in Neuroendocrinology 1999, 20, 157–198.

[2] Patel, Y. C.; Greenwood, M.; Kent, G.; et al. Biochemical and Biophysical Research Communications 1993, 192, 288–294.

[3] Duran-Prado, M.; Gahete, M. D.; Martinez-Fuentes, A. J.; et al. Journal of Clinical Endocrinology & Metabolism 2009, 94, 2634–2643.

[4] Brazeau, P.; Vale, W.; Burgus, R.; et al. Science 1973, 179, 77–79.

[5] Schally, A. V.; Huang, W. Y.; Chang, R. C. C.; et al. Proceedings of the National Academy of Sciences of the United States of America-Biological Sciences 1980, 77, 4489–4493.

[6] Brown, M.; Rivier, J.; Vale, W. Endocrinology 1981, 108, 2391–2393.

[7] Siler, T. M.; Yen, S. C.; Vale, W.; Guillemin, R. Journal of Clinical Endocrinology & Metabolism 1974, 38, 742–745.

[8] Drouin, J.; Delean, A.; Rainville, D.; et al. Endocrinology 1976, 98, 514–521.

[9] Tyrrell, J. B.; Lorenzi, M.; Gerich, J. E.; Forsham, P. H. Journal of Clinical Endocrinology & Metabolism 1975, 40, 1125–1127.

[10] Koerker, D. J.; Ruch, W.; Chideckel, E.; et al. Science 1974, 184, 482–484.

[11] Gomez-Pan, A.; Snow, M. H.; Piercy, D. A.; et al. Journal of Clinical Endocrinology & Metabolism 1976, 43, 240–243.

[12] Shimon, I.; Taylor, J. E.; Dong, J. Z.; et al. Journal of Clinical Investigation 1997, 99, 789–798.

[13] Shimon, I.; Yan, X. M.; Taylor, J. E.; et al. Journal of Clinical Investigation 1997, 100, 2386–2392.

[14] Kreienkamp, H. J.; Akgun, E.; Baumeister, H.; et al. FEBS Letters 1999, 462, 464–466.

[15] Viollet, C.; Vaillend, C.; Videau, C.; et al. European Journal of Neuroscience 2000, 12, 3761–3770.

[16] Strowski, M. Z.; Dashkevicz, M. P.; Parmar, R. M.; et al. Neuroendocrinology 2002, 75, 339–346.

[17] Hofland, L. J.; van der Hoek, J.; Feelders, R.; et al. European Journal of Endocrinology 2005, 152, 645–654.

[18] Strowski, M. Z.; Parmar, R. M.; Blake, A. D.; Schaeffer, J. M. Endocrinology 2000, 141, 111–117.

[19] Rossowski, W. J.; Gu, Z. F.; Akarca, U. S.; et al. Peptides 1994, 15, 1421–1424.

[20] den Bosch, J. V.; Adriaensen, D.; Van Nassauw, L.; Timmermans, J. P. Regulatory Peptides 2009, 156, 1–8.

[21] Jakobs, K. H.; Aktories, K.; Schultz, G. Nature 1983, 303, 177–178.

[22] Reisine, T.; Axelrod, J. Endocrinology 1983, 113, 811–813.

[23] Koch, B. D.; Schonbrunn, A. Endocrinology 1984, 114, 1784–1790.

[24] Heisler, S.; Reisine, T. D.; Hook, V. Y. H.; Axelrod, J. Proceedings of the National Academy of Sciences of the United States of America-Biological Sciences 1982, 79, 6502–6506.

[25] Hildebrandt, J. D.; Sekura, R. D.; Codina, J.; et al. Nature 1983, 302, 706–709.

[26] Pace, C. S.; Tarvin, J. T. Diabetes 1981, 30, 836–842.

[27] Yamashita, N.; Shibuya, N.; Ogata, E. Proceedings of the National Academy of Sciences of the United States of America 1986, 83, 6198–6202.

[28] Yatani, A.; Codina, J.; Sekura, R. D.; et al. Molecular Endocrinology 1987, 1, 283–289.

[29] Koch, B. D.; Blalock, J. B.; Schonbrunn, A. Journal of Biological Chemistry 1988, 263, 216–225.

[30] Sims, S. M.; Lussier, B. T.; Kraicer, J. Journal of Physiology 1991, 441, 615–637.

[31] Schlegel, W.; Wuarin, F.; Zbaren, C.; et al. FEBS Letters 1985, 189, 27–32.

[32] Kleuss, C.; Hescheler, J.; Ewel, C.; et al. Nature 1991, 353, 43–48.

[33] Meriney, S. D.; Gray, D. B.; Pilar, G. R. Nature 1994, 369, 336–339.

[34] Yang, S. K.; Parkington, H. C.; Blake, A. D.; et al. Endocrinology 2005, 146, 4975–4984.

[35] Fujii, Y.; Gonoi, T.; Yamada, Y.; et al. FEBS Letters 1994, 355, 117–120.

[36] Tallent, M.; Liapakis, G.; O'Carroll, A. M.; et al. Neuroscience 1996, 71, 1073–1081.

[37] Hou, C. F.; Gilbert, R. L.; Barber, D. L. Journal of Biological Chemistry 1994, 269, 10357–10362.

[38] Smith, P. A. Endocrinology 2009, 150, 741–748.

[39] Renstrom, E.; Ding, W. G.; Bokvist, K.; Rorsman, P. Neuron 1996, 17, 513–522.

[40] Gromada, J.; Hoy, M.; Buschard, K.; et al. Journal of Physiology 2001, 535, 519–532.

[41] Simard, J.; Labrie, F.; Gossard, F. DNA 1986, 5, 263–270.

[42] Levy, A.; Lightman, S. L. Journal of Molecular Endocrinology 1988, 1, 19–26.

[43] Namba, H.; Morita, S.; Melmed, S. Endocrinology 1989, 124, 1794–1799.

[44] Wood, D. F.; Docherty, K.; Ramsden, D. B.; Sheppard, M. C. Molecular and Cellular Endocrinology 1987, 52, 257–261.

[45] Tanner, J. W.; Davis, S. K.; Mcarthur, N. H.; et al. Journal of Endocrinology 1990, 125, 109–115.

[46] Sugihara, H.; Minami, S.; Okada, K.; et al. Endocrinology 1993, 132, 1225–1229.

[47] Tsukamoto, N.; Nagaya, T.; Kuwayama, A.; et al. Endocrine Journal 1994, 41, 437–444.

[48] Morishita, M.; Iwasaki, Y.; Onishi, A.; et al. Journal of Molecular Endocrinology 2003, 31, 441–448.

[49] Acunzo, J.; Thirion, S.; Roche, C.; et al. Cancer Research 2008, 68, 10163–10170.

[50] Luque, R. M.; Duran-Prado, M.; Garcia-Navarro, S.; et al. Endocrinology 2006, 147, 2902–2908.

[51] Cordoba-Chacon, J.; Gahete, M. D.; Culler, M. D.; et al. Journal of Neuroendocrinology 2012, 24, 453–463.

[52] Gruszka, A.; Culler, M. D.; Melmed, S. Molecular and Cellular Endocrinology 2012, 362, 104–109.

[53] Castillo, V.; Theodoropoulou, M.; Stalla, J.; et al. Neuroendocrinology 2011, 94, 124–136.

[54] Reubi, J. C.; Laissue, J. A. Trends in Pharmacological Sciences 1995, 16, 110–115.

[55] Susini, C.; Buscail, L. Annals of Oncology 2006, 17, 1733–1742.

[56] Tsuzaki, S.; Moses, A. C. Endocrinology 1990, 126, 3131–3138.

[57] Lee, M. T.; Liebow, C.; Kamer, A. R.; Schally, A. V. Proceedings of the National Academy of Sciences of the United States of America 1991, 88, 1656–1660.

[58] Florio, T.; Schettini, G. Journal of Molecular Endocrinology 1996, 17, 89–100.

[59] Pan, M. G.; Florio, T.; Stork, P. J. S. Science 1992, 256, 1215–1217.

[60] Buscail, L.; Delesque, N.; Esteve, J. P.; et al. Proceedings of the National Academy of Sciences of the United States of America 1994, 91, 2315–2319.

[61] Florio, T.; Rim, C.; Hershberger, R. E.; et al. Molecular Endocrinology 1994, 8, 1289–1297.

[62] Florio, T.; Scorziello, A.; Fattore, M.; et al. Journal of Biological Chemistry 1996, 271, 6129–6136.

[63] Reardon, D. B.; Wood, S.L.; Brautigan, D. L.; et al. Biochemical Journal 1996, 314, 401–404.

[64] Florio, T.; Thellung, S.; Arena, S.; et al. European Journal of Endocrinology 1999, 141, 396–408.

[65] Florio, T.; Thellung, S.; Corsaro, A.; et al. Clinical Endocrinology 2003, 59, 115–128.

[66] Dent, P.; Reardon, D. B.; Wood, S. L.; et al. Journal of Biological Chemistry 1996, 271, 3119–3123.

[67] Delesque, N.; Buscail, L.; Esteve, J. P.; et al. Somatostatin and Its Receptors 1995, 190, 187–196.

[68] Srikant, C. B.; Shen, S. H. Endocrinology 1996, 137, 3461–3468.

[69] Lopez, F.; Esteve, J. P.; Buscail, L.; et al. Metabolism-Clinical and Experimental 1996, 45, 14–16.

[70] Lopez, F.; Esteve, J. P.; Buscail, L.; et al. Journal of Biological Chemistry 1997, 272, 24448–24454.

[71] Reardon, D. B.; Dent, P.; Wood, S. L.; et al. Molecular Endocrinology 1997, 11, 1062–1069.

[72] Pages, P.; Benali, N.; Saint-Laurent, N.; et al. Journal of Biological Chemistry 1999, 274, 15186–15193.

[73] Florio, T.; Arena, S.; Thellung, S.; et al. Molecular Endocrinology 2001, 15, 1838–1852.

[74] Massa, A.; Barbieri, F.; Aiello, C.; et al. Journal of Biological Chemistry 2004, 279, 29004–29012.

[75] Cheung, N. W.; Boyages, S. C. Endocrinology 1995, 136, 4174–4181.

[76] Srikant, C. B. Biochemical and Biophysical Research Communications 1995, 209, 400–406.

[77] Sharma, K.; Patel, Y. C.; Srikant, C. B. Molecular Endocrinology 1996, 10, 1688–1696.

[78] Teijeiro, R.; Rios, R.; Costoya, J. A.; et al. Cellular Physiology and Biochemistry 2002, 12, 31–38.

[79] Ferrante, E.; Pellegrini, C.; Bondioni, S.; et al. Endocrine-Related Cancer 2006, 13, 955–962.

[80] Thangaraju, M.; Sharma, K.; Leber, B.; et al. Journal of Biological Chemistry 1999, 274, 29549–29557.

[81] Lin, X.; Voyno-Yasenetskaya, T. A.; Hooley, R.; et al. Journal of Biological Chemistry 1996, 271, 22604–22610.

[82] Lin, C. Y.; Varma, M. G.; Joubel, A.; et al. Journal of Biological Chemistry 2003, 278, 15128–15135.

[83] Lopez, F.; Ferjoux, G.; Cordelier, P.; et al. FASEB Journal 2001, 15, 2300–2302.

[84] Arena, S.; Pattarozzi, A.; Corsaro, A.; et al. Molecular Endocrinology 2005, 19, 255–267.

[85] Kubota, A.; Yamada, Y.; Kagimoto, S.; et al. Biochemical and Biophysical Research Communications 1994, 204, 176–186.

[86] Kubota, A.; Yamada, Y.; Kagimoto, S.; et al. Metabolism 1996, 45, 42–45.

[87] Hadcock, J. R.; Strnad, J.; Eppler, C. M. Molecular Pharmacology 1994, 45, 410–416.

[88] Hershberger, R. E.; Newman, B. L.; Florio, T.; et al. Endocrinology 1994, 134, 1277–1285.

[89] Buscail, L.; Esteve, J. P.; SaintLaurent, N.; et al. Proceedings of the National Academy of Sciences of the United States of America 1995, 92, 1580–1584.

[90] Keane, M. M.; Lowrey, G. A.; Ettenberg, S. A.; et al. Cancer Research 1996, 56, 4236–4243.

[91] Iuliano, R.; Trapasso, F.; Le Pera, I.; et al. Cancer Research 2003, 63, 882–886.

[92] Barbieri, F.; Pattarozzi, A.; Gatti, M.; et al. Endocrinology 2008, 149, 4736–4746.

[93] Arena, S.; Pattarozzi, A.; Massa, A.; et al. Molecular Endocrinology 2007, 21, 229–246.

[94] Delesque, N.; Buscail, L.; Esteve, J. P.; et al. Cancer Research 1997, 57, 956–962.

[95] Benali, N.; Cordelier, P.; Calise, D.; et al. Proceedings of the National Academy of Sciences of the United States of America 2000, 97, 9180–9185.

[96] Law, S. F.; Yasuda, K.; Bell, G. I.; Reisine, T. Journal of Biological Chemistry 1993, 268, 10721–10727.

[97] Kagimoto, S.; Yamada, Y.; Kubota, A.; et al. Biochemical and Biophysical Research Communications 1994, 202, 1188–1195.

[98] Tentler J. J.; Hadcock, J. R.; Gutierrez-Hartmann, A. Molecular Endocrinology 1997, 11, 859–866.

[99] Bousquet, C.; Delesque, N.; Lopez, F.; et al. Journal of Biological Chemistry 1998, 273, 7099–7106.

[100] Ferjoux, G.; Lopez, F.; Esteve, J. P.; et al. Molecular Biology of the Cell 2003, 14, 3911–3928.

[101] Luttrell, L. M.; Lefkowitz, R. J. Journal of Cell Science 2002, 115, 455–465.

[102] Hortala, M.; Ferjoux, G.; Estival, A.; et al. Journal of Biological Chemistry 2003, 278, 20574–20581.

[103] Cattaneo, M. G.; Taylor, J. E.; Culler, M. D.; et al. FEBS Letters 2000, 481, 271–276.

[104] Held-Feindt, J.; Krisch, B.; Forstreuter, F.; Mentlein, R. Journal of Physiology-Paris 2000, 94, 251–258.

[105] Sellers, L. A.; Alderton, F.; Carruthers, A. M.; et al. Molecular and Cellular Biology 2000, 20, 5974–5985.

[106] Lahlou, H.; Saint-Laurent, N.; Esteve, J. P.; et al. Journal of Biological Chemistry 2003, 278, 39356–39371.

[107] Stetak, A.; Csermely, P.; Ullrich, A.; Keri, G. Biochemical and Biophysical Research Communications 2001, 288, 564–572.

[108] Bousquet, C.; Guillermet-Guibert, J.; Saint-Laurent, N.; et al. EMBO Journal 2006, 25, 3943–3954.

[109] Theodoropoulou, M.; Zhang, J.; Laupheimer, S.; et al. Cancer Research 2006, 66, 1576–1582.

[110] Najib, S.; Saint-Laurent, N.; Esteve, J. P.; et al. Molecular and Cellular Biology 2012, 32, 1004–1016.

[111] Sherr, C. J. Cancer Research 2000, 60, 3689–3695.

[112] Koff, A.; Giordano, A.; Desai, D.; et al. Science 1992, 257, 1689–1694.

[113] Schwarz, J. K.; Devoto, S. H.; Smith, E. J.; et al. EMBO Journal 1993, 12, 1013–1020.

[114] Qian, X.; Jin, L.; Grande, J. P.; Lloyd, R. V. Endocrinology 1996, 137, 3051–3060.

[115] Theodoropoulou, M.; Stalla, G. K.; Spengler, D. Molecular and Cellular Endocrinology 2010, 326, 60–65.

[116] Theodoropoulou, M.; Tichomirowa, M. A.; Sievers, C.; et al. International Journal of Cancer 2009, 125, 2122–2126.

[117] Chahal, H. S.; Trivellin, G.; Leontiou, C. A.; et al. Journal of Clinical Endocrinology & Metabolism 2012, 97, E1411–E1420.

[118] Daly, A. F.; Tichomirowa, M. A.; Petrossians, P.; et al. Journal of Clinical Endocrinology & Metabolism 2010, 95, E373–E383.

[119] Tichomirowa, M. A.; Barlier, A.; Daly, A. F.; et al. European Journal of Endocrinology 2011, 165, 509–515.

[120] Azar, R.; Najib, S.; Lahlou, H.; et al. Cellular and Molecular Life Sciences 2008, 65, 3110–3117.

[121] Bevan, J. S. Journal of Clinical Endocrinology and Metabolism 2005, 90, 1856–1863.

[122] Melmed, S.; Sternberg, R.; Cook, D.; et al. Journal of Clinical Endocrinology and Metabolism 2005, 90, 4405–4410.

[123] Guillermet, J.; Saint-Laurent, N.; Rochaix, P.; et al. Proceedings of the National Academy of Sciences of the United States of America 2003, 100, 155–160.

[124] Sharma, K.; Srikant, C. B. International Journal of Cancer 1998, 76, 259–266.

[125] Liu, D.; Martino, G.; Thangaraju, M.; et al. Journal of Biological Chemistry 2000, 275, 9244–9250.

[126] Florio, T.; Morini, M.; Villa, V.; et al. Endocrinology 2003, 144,1574–1584.

[127] Albini, A.; Florio, T.; Giunciuglio, D.; et al. FASEB Journal 1999, 13, 647–655.

[128] Patel, Y. C.; Greenwood, M. T.; Panetta, R.; et al. Life Sciences 1995, 57, 1249–1265.

[129] Kreienkamp, H. J.; Roth, A.; Richter, D. DNA and Cell Biology 1998, 17, 869–878.

[130] Qiu, C.; Zeyda, T.; Johnson, B.; et al. Journal of Neuroscience 2008, 28, 3567–3576.

[131] den Bosch, J. V.; Torfs, P.; De Winter, B. Y.; et al. Journal of Cellular and Molecular Medicine 2009, 13, 3283–3295.

[132] O'Carroll, A. M.; Lolait, S. J.; Konig, M.; Mahan, L. C. Molecular Pharmacology 1992, 42, 939–946.

[133] Lussier, B. T.; Wood, D. A.; French, M. B.; et al. Endocrinology 1991, 128, 583–591.

[134] Akbar, M.; Okajima, F., Tomura, H.; et al. FEBS Letters 1994, 348, 192–196.

[135] Wilkinson, G. F.; Feniuk, W.; Humphrey, P. P. A. European Journal of Pharmacology 1997, 340, 277–285.

[136] Cordelier, P.; Esteve, J. P.; Bousquet, C.; et al. Proceedings of the National Academy of Sciences of the United States of America 1997, 94, 9343–9348.

[137] Cordelier, P.; Esteve, J. P.; Najib, S.; et al. Journal of Biological Chemistry 2006, 281, 19156–19171.

[138] Murray, R. D.; Kim, K., Ren, S. G.; et al. Journal of Clinical Investigation 2004, 114, 349–356.

[139] Curtis, S. B.; Hewitt, J.; Yakubovitz, S.; et al. American Journal of Physiology-Heart and Circulatory Physiology 2000, 278, H1815–H1822.

[140] Bocci, G.; Culler, M. D.; Fioravanti, A.; et al. European Journal of Clinical Investigation 2007, 37, 700–708.

[141] Laklai, H.; Laval, S.; Dumartin, L.; et al. Proceedings of the National Academy of Sciences of the United States of America 2009, 106, 17769–17774.

[142] Florio, T.; Thellung, S.; Schettini, G. Pharmacological Research 1996, 33, 297–305.

[143] Day, R.; Dong, W. J.; Panetta, R.; et al. Endocrinology 1995, 136, 5232–5235.

[144] Hunyady, B.; Hipkin, R. W.; Schonbrunn, A.; Mezey, E. Endocrinology 1997, 138, 2632–2635.

[145] Yamada, Y.; Post, S. R.; Wang, K.; et al. Proceedings of the National Academy of Sciences of the United States of America 1992, 89, 251–255.

3

SOMATOSTATIN RECEPTORS IN MALIGNANCIES AND OTHER PATHOLOGIES

Marco Volante, Adele Cassenti, Ida Rapa, Luisella Righi, and Mauro Papotti

Department of Oncology, University of Turin, Orbassano, Turin, Italy

ABBREVIATIONS

RT-PCR   reverse transcriptase-polymerase chain reaction
SSTR   somatostatin receptors

PREMISE

Somatostatin (SS) is an acidic polypeptide, originally described by Krulich and coworkers [1] in hypothalamic extracts, that exerts several biological functions through the interaction with specific transmembrane receptors, the somatostatin receptor (SSTR) family. The physiological and pharmacological properties, as well as the distribution in normal cells and tissues, of SSTR have been described previously in this book. The present chapter will discuss the different methods used to localize SSTR at the tissue level and the data available on the distribution of SSTR in human tumors, with special reference to neuroendocrine ones, and other diseases. These informations will be crucial to understand the role in the clinical practice of targeting SSTR with specific analogues, both in terms of developing diagnostic tools

Somatostatin Analogues: From Research to Clinical Practice, First Edition. Edited by
Alicja Hubalewska-Dydejczyk, Alberto Signore, Marion de Jong, Rudi A. Dierckx,
John Buscombe, and Christophe Van de Wiele.

and of manipulating SSTR signaling as a therapeutic strategy, that will be discussed extensively in the following sections of this book.

METHODS TO IDENTIFY SSTR IN TISSUES

Before discussing the current knowledge on SSTR expression in tumors and other pathologies, a short reappraisal of the different methods available and reported to determine SSTR at the tissue level is mandatory to critically consider the expression data currently available.

Several methods have been used to determine the expression of SSTR in tissue samples. All of them have intrinsically limitations and advantages, and therefore, the most powerful data have been generated by the combination of two or more of them. A comparison of the three most relevant methods is represented in Table 3.1.

SSTR tissue localization had originally been demonstrated by means of binding assays of radiolabeled SS analogues [2–4]. However, this method that has the unique property of tracing the presence of "functionally active" receptors is affected by a scarce reproducibility, is applicable to high-quality frozen material only, and does not recognize the SSTR subtype expressed, unless using highly subtype-specific analogues [5].

Since the cloning of the five genes, between 1992 and 1994, several studies detected the specific mRNA expression of SSTR in normal tissue and tumors by means of alternative techniques such as the Northern blot [6], *in situ* hybridization [7, 8], or

TABLE 3.1   Comparison between different methods of tissue identification of SSTR

Method	Pros	Cons	Applicability
Gene expression (PCR)	• Sensitive • Identifies SSTR subtypes	• Lack of tissue localization Potentially aspecific (normal contaminating SSTR+cells) • Needs frozen material	Low
Autoradiography	• Tissue localization • Identifies "functional" SSTR	• Does not identify SSTR subtype • Scarce reproducibility • Needs frozen material	Very low
Immunohistochemistry	• Tissue localization • Cost effective Applicable to archival tissues (including biopsies) • Identifies SSTR subtypes	• Needs standardization • Lack of "robust" commercially available antibodies	High

reverse transcriptase-polymerase chain reaction (RT-PCR), either qualitative or quantitative [9–11]. However, in general terms, mRNA expression from tissue extracts is variably affected by signals determined by SSTR-expressing cells different from those that represent the target, and the quality of the tissue sample is again a relevant clue for obtaining adequate results.

In parallel with mRNA determination, several groups aimed at the development of SSTR-specific antibodies. The vast majority of them are polyclonal antisera and have been determined either against the N- or the C-terminus of the protein [12–17]. Most of these antibodies are working nicely on paraffin-embedded tissue, are raised against all or most of SSTR receptor subtypes (although with variable reliability), are commercially available, and, therefore, allowed extensive investigations on large archival case series. Moreover, some monoclonal antibodies have also been developed, with special reference to SSTR subtype 2A [18–21], although not commercially available yet at the time of this manuscript preparation.

Besides research purposes, the advantages of immunohistochemical SSTR detection in the clinical practice include a high cost/benefit ratio, high reproducibility in pathology laboratories worldwide, possibility of tissue localization of tumor cells, recognition of the different SSTR subtypes, and, last but not least, applicability on retrospective archival material. In addition, immunohistochemical methods may be applied to preoperative fine-needle aspiration of cytological or biopsy material [22], allowing SSTR demonstration in inoperable tumors or offering specific information to diagnostic or therapeutic decisions before surgery (Fig. 3.1). Several studies have also documented that immunohistochemistry correlates with other methods in a

(a)   (b)

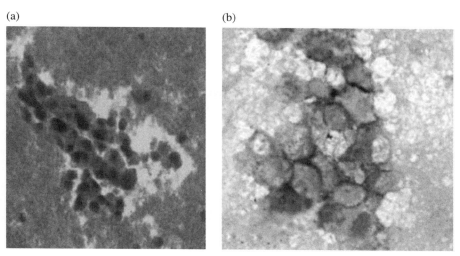

FIGURE 3.1 SSTR type 2A determination in a cellblock preparation from a liver metastasis of a well-differentiated neuroendocrine carcinoma of unknown primary. (a) Hematoxylin and eosin; (b) immunoperoxidase; (a and b) original magnification 400×. (*See insert for color representation of the figure.*)

significant proportion of cases, with only minor discrepancies [10, 13, 18, 21–23]. However, a major caveat of immunohistochemistry for SSTR, with special reference to its use as a diagnostic tool, is represented by the lack of standardization, although some scoring systems have been proposed [24, 25]. In general terms, a fine membranous staining is considered the most specific [26], although a cytoplasmic staining may be a result of receptor internalization, especially in the case of tumors coexpressing both SSTR and their natural ligand or in the cases of patients treated with SS analogues [27].

SSTR EXPRESSION IN HUMAN MALIGNANCIES

A range of different tumors overexpress SSTR, as compared to nontransformed cells. The underlying stimuli that induce this overexpression as well as the functional meaning for the biology of tumor cells have not been conclusively explained. It is possible that the upregulation of SSTR serves as homeostatic growth inhibitory auto-crine/paracrine response to the deregulated tumor cell proliferation [28]. Nevertheless, with all the limitations related to the complexity of the role of SSTR in cancer biology, it can be asserted that SSTR and their intracellular signaling pathways should generally be considered as tumor suppressive.

SSTR EXPRESSION IN NEUROENDOCRINE NEOPLASMS

Neuroendocrine neoplasms originate from a normal cell population that is physio-logically a target of SS, thus expressing SSTR. Therefore, the generally high level of SSTR expression in these groups of tumors is not surprising.

A tentative list of all data on SSTR-positive neuroendocrine neoplasms would greatly fail to be comprehensive. Wide literature data (see for review [29–32]) show that SSTR are highly expressed in pituitary adenomas; neuroendocrine tumors of the pancreas, gastrointestinal tract, and lung; paragangliomas/pheochromocytomas; Merkel cell carcinomas; neuroblastomas; and medullary thyroid carcinomas. Hundreds of such tumors have been analyzed by means of various techniques, including binding assays, immunohistochemistry, and mRNA analysis. A wide heterogeneity in SSTR subtype expression in the different tumor types, in different cases of the same tumor type, and even in different cell populations within individual lesions has been reported. Such high heterogeneity of SSTR distribution partially explains some discrepancies in the clinical features and response to SS analogue therapy observed in neuroendocrine tumors from various sites. According to such literature data, in most cases, individual tumors coexpress different SSTR subtypes, SSTR type 2 being the most frequently represented in all locations, followed by types 3, 5, and 1. Subtype 4 is generally poorly expressed. Higher SSTR expression is generally observed in "well-differentiated" tumors although a considerable proportion of positive cases might be observed also in poorly differentiated—highly aggressive—lesions, such as small cell lung cancer [33]. Although poorly elucidated from a biological point of view, hormonal secretion by the tumor is also associated with different patterns of expression of SSTR subtypes, such as in the case of pancreatic insulin-producing tumors that show a low SSTR type 2 expression [34].

Apart from detailing a prevalence of expression, SSTR determination at the tissue level is potentially a complementary approach to better define the diagnostic and therapeutic strategies in patients affected by neuroendocrine tumors. A recent literature focused on the comparison between SSTR tissue determination (mainly by means of immunohistochemistry) and *in vivo* imaging using different SS analogue-based methodologies [24, 25, 35–37], with in general a good correlation. In a previous paper by our group, a relatively high correlation was observed with the response to SS analogue treatment [24]. Moreover, some studies claimed a prognostic value of SSTR type 2 determination in neuroendocrine tumors of the gastrointestinal tract and pancreas [38–40].

Future studies are therefore needed to validate SSTR testing as a relevant marker of clinical usefulness in neuroendocrine neoplasms.

SSTR EXPRESSION IN NONNEUROENDOCRINE MALIGNANCIES

A wide spectrum of solid or hematological malignancies has been demonstrated to variably express SSTR [41, 42]. A variety of carcinomas showed *in vivo* and tissue localization-based evidence of SSTR expression: such tumors include, among others, cancers from the breast [43–45], lung [18, 22], kidney [45], pancreatobiliary tract [46], stomach [47], liver [48], colorectum [49], ovary [50], thyroid follicular cells [51], and prostate [52–55] (Fig. 3.2). Unpublished data from our group onto cell lines (Table 3.2) are consistent with what was reported on tissues earlier. In some cases, an antiproliferative activity of SS analogues could be demonstrated

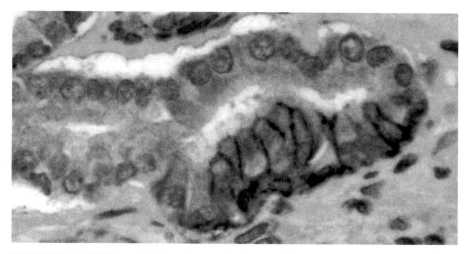

FIGURE 3.2 SSTR type 2A clonal expression in prostatic adenocarcinoma (immunoperoxidase; original magnification 400×). (*See insert for color representation of the figure.*)

TABLE 3.2   SSTR expression at MRNA level in nonneuroendocrine cell lines[a]

Cell line	Derivation	SSTR mRNA expressed
MONOMAC	Monoblastic leukemia	4
MCF7	Breast cancer	2, 5
T47D	Breast cancer, apocrine	2, 5
MDAMB231	Breast cancer	2, 4
CALU-1	Lung cancer, squamous	3
KATO III	Gastric cancer	1, 2, 5
HT29	Colon cancer	1, 2, 5
H716	Colon cancer (with neuroendocrine features)	1, 2, 3, 5
MOG UVW	Glioblastoma	2

[a]Volante, M., unpublished.

in vitro in hormone independent prostate cancer models [56]. SSTR subtypes have also been detected in meningiomas, medulloblastomas, and gliomas, in soft tissue sarcomas, and in malignant melanomas; this distribution correlated with either scintigraphic imaging or *in vitro* tests on SS analogue response [57–59]. Promising clinical applications of SS analogues have also been reported in lymphohematological malignancies [60] and in thymomas [61].

SSTR IN NONNEOPLASTIC DISEASES

There is strong evidence that selected nontumoral lesion may also express SSTR. For instance, active granulomas in sarcoidosis express SSTR on epithelioid cells [62], whereas inactive or successfully treated fibrosing granulomas devoid of epithelioid cells lack SSTR. Inflamed joints in active rheumatoid arthritis express SSTR, preferentially located in the proliferating synovial vessel [63].

Furthermore, inflammatory bowel disease is characterized by an overexpression of SSTR in the vascular system [64] of the altered parts of the gastrointestinal tract. Concerning SSTR presence and possible applications of SS analogues in nonneoplastic diseases, SS actions in modulating the immunological response and angiogenesis together with the high density of expression of SSTR (viz., type 1 and 2) in the retina represent the baseline of very promising applications for therapy in various retinal diseases, from macular edema or macular degeneration to thyrotoxic orbitopathy, retinal ischemic damage, and proliferative diabetic retinopathy [65, 66].

UNMET CLUES

Despite the wide body of evidence on the presence and function of SSTR in several neoplastic and nonneoplastic human diseases, several issues still deserve further investigation and elucidation.

The correct methodology for SSTR determination that could be applied in the clinical practice is not well established, so far. Immunohistochemistry seems the most promising but is still limited by the scarce availability of reliable and clinically validated reagents (mostly related to SSTR subtype 2), as well as by the lack of standardization in the interpretation. Moreover, novel molecules with a wider spectrum of affinity to different SSTR subtypes than those currently available [67] claim a reinterpretation of the data available with a better understanding of coexpression modalities of the different SSTR subtypes, also taking into consideration the capability of different subtype to form functionally active homo- or heterodimers that modify significantly the activation of intracellular signaling pathways also due to altered agonist-induced desensitization [68]. Moreover, the SS/SSTR axis is more complex, due to both other peptides with selective affinity to SSTR such as cortistatin [31] and due to the capability of SSTR to heterodimerize with other receptors such as dopamine receptors [69]. In this respect, recent studies claim that the coexpression of SSTR and dopamine receptors (type 2) might open to novel therapeutic strategies with chimeric molecules [70–72].

REFERENCES

[1] Krulich, L.; Dhariwal, A. P.; McCann, S. M.; et al. Endocrinology 1968, 83, 783–790.

[2] Lamberts, S. W.; Hofland, L. J.; van Koetsveld, P. M.; et al. Journal of Clinical Endocrinology and Metabolism 1990, 71, 566–574.

[3] Papotti, M., Macrí, L.; Bussolati, G.; Reubi, J. C. International Journal of Cancer 1989, 43, 365–369.

[4] Reubi, J. C.; Maurer, R.; von Werder, K.; et al. Cancer Research 1987, 47, 551–558.

[5] Reubi, J. C.; Waser, B.; Schaer, J. C.; Laissue, J. A. European Journal of Nuclear Medicine 2001, 28, 836–846.

[6] Kong, H.; DePaoli, A. M.; Breder, C. D.; et al. Neuroscience 1994, 59, 175–184.

[7] Reubi, J. C.; Schaer, J. C.; Waser, B.; Mengod, G. Cancer Research 1994, 54, 3455–3459.

[8] Janson, E. T.; Gobl, A.; Kalkner, K. M.; Oberg, K. Cancer Research 1996, 56, 2561–2565.

[9] Sestini, R.; Orlando, C.; Peri, A.; et al. Clinical Cancer Research 1996, 2, 1757–1765.

[10] Papotti, M.; Bongiovanni, M.; Volante, M.; et al. Virchows Archiv 2002, 440, 461–475.

[11] Nakayama, Y.; Wada, R.; Yajima, N.; et al. Pancreas 2010, 39, 1147–1154.

[12] Helboe, L.; Møller, M.; Nørregaard, L.; et al. Brain Research Molecular Brain Research 1997, 49, 82–88.

[13] Hofland, L. J.; Liu, Q.; Van Koetsveld, P. M.; et al. Journal of Clinical Endocrinology and Metabolism 1999, 84, 775–780.

[14] Janson, E. T. Stridsberg, M.; Gobl, A.; et al. Cancer Research 1998, 58, 2375–2378.

[15] Kumar, U.; Sasi, R.; Suresh, S.; et al. Diabetes 1999, 48, 77–85.

[16] Reubi, J. C.; Kappeler, A.; Waser, B.; et al. American Journal of Pathology 1998, 153, 233–245.

[17] Schindler, M.; Sellers, L. A.; Humphrey, P. P.; Emson, P. C. Neuroscience 1997, 76, 225–240.

[18] Papotti, M.; Croce, S.; Macri, L.; et al. Diagnostic Molecular Pathology 2000, 9, 47–57.

[19] Kuan, C. T.; Wikstrand, C. J.; McLendon, R. E.; et al. Hybridoma (Larchmt) 2009, 28, 389–403.

[20] Fischer, T.; Doll, C.; Jacobs, S.; et al. Journal of Clinical Endocrinology and Metabolism 2008, 93, 4519–4524.

[21] Mundschenk, J.; Unger, N.; Schulz, S.; et al. Journal of Clinical Endocrinology and Metabolism 2003, 88, 5150–5157.

[22] Papotti, M.; Croce, S.; Bello, M.; et al. Virchows Archiv 2001, 439, 787–797.

[23] Körner, M.; Eltschinger, V.; Waser, B.; et al. American Journal of Surgical Pathology 2005, 29, 1642–1651.

[24] Volante, M.; Brizzi, M. P.; Faggiano, A.; et al. Modern Pathology 2007, 2, 1172–1182.

[25] Miederer, M.; Seidl, S.; Buck, A.; et al. European Journal of Nuclear Medicine and Molecular Imaging 2009, 36, 48–52.

[26] Reubi, J. C.; Waser, B.; Liu, Q.; et al. Journal of Clinical Endocrinology and Metabolism 2000, 85, 3882–3891.

[27] Reubi, J. C.; Waser, B.; Cescato, R.; et al. Journal of Clinical Endocrinology and Metabolism 2010, 95, 2343–2350.

[28] Msaouel, P.; Galanis, E.; Koutsilieris, M. Expert Opinion on Investigational Drugs 2009, 18, 1297–1316.

[29] de Herder, W. W.; Hofland, L. J.; van der Lely, A. J.; Lamberts, S. W. Endocrine Related Cancer 2003, 10, 451–458.

[30] Volante, M.; Bozzalla-Cassione, F.; Papotti M. Endocrine Pathology 2004, 15, 275–291.

[31] Volante, M.; Rosas, R.; Allìa, E.; et al. Molecular and Cellular Endocrinology 2008, 286, 219–229.

[32] Pelosi, G.; Volante, M.; Papotti, M.; et al. Quarterly Journal of Nuclear Medicine and Molecular Imaging 2006, 50, 272–287.

[33] Righi, L.; Volante, M.; Tavaglione, V.; et al. Annals of Oncology 2010, 21, 548–555.

[34] Vezzosi, D.; Bennet, A.; Rochaix, P.; et al. European Journal of Endocrinology 2005, 152, 757–767.

[35] Müssig, K.; Oksüz, M. O., Dudziak, K.; et al. Hormone and Metabolic Research 2010, 42, 599–606.

[36] Kaemmerer, D.; Peter, L.; Lupp, A.; et al. European Journal of Nuclear Medicine and Molecular Imaging 2011, 38, 1659–1668.

[37] Ferone, D.; Pivonello, R.; Kwekkeboom, D. J.; et al. Journal of Endocrinological Investigations 2012, 35, 528–534.

[38] Asnacios, A.; Courbon, F.; Rochaix, P.; et al. Journal of Clinical Oncology 2008, 26, 963–970.

[39] Kim, H. S.; Lee, H. S.; Kim, W. H. Cancer Research and Treatment 2011, 43, 181–188.

[40] Corleto, V. D.; Falconi, M.; Panzuto, F.; et al. Neuroendocrinology 2009, 89, 223–230.

[41] Patel, Y. C. Journal of Endocrinological Investigations 1997, 20, 348–367.

[42] Hofland, L. J.; Lamberts, S. W. Endocrine Reviews 2003, 24, 28–47.

[43] Reubi, C.; Gugger, M.; Waser, B. European Journal of Nuclear Medicine and Molecular Imaging 2002, 29, 855–862.

[44] Kumar, U.; Grigorakis, S. I.; Watt, H. L.; et al. Breast Cancer Research and Treatment 2005, 92, 175–186.

[45] Vikić-Topić, S.; Raisch, K. P.; Kvols, L. K.; Vuk-Pavlović, S. Journal of Clinical Endocrinology and Metabolism 1995, 80, 2974–2979.

[46] Zhao, B., Zhao, H.; Zhao, N.; et al. Journal of Hepato-Biliary-Pancreatic Surgery 2002, 9, 497–502.

[47] Miller, G. V.; Farmery, S. M.; Woodhouse, L. F.; Primrose, J. N. British Journal of Cancer 1992, 66, 391–395.

[48] Verhoef, C.; van Dekken, H.; Hofland, L. J.; et al. Digestive Surgery 2008, 25, 21–26.

[49] Qiu, C. Z.; Wang, C.; Huang, Z. X.; et al. World Journal of Gastroenterology 2006, 12, 2011–2015.

[50] Hall, G. H.; Turnbull, L. W.; Richmond, I.; et al. British Journal of Cancer 2002, 87, 86–90.

[51] Klagge, A.; Krause, K.; Schierle, K.; et al. Hormone and Metabolic Research 2010, 42, 237–240.

[52] Alonso, O.; Gambini, J. P.; Lago, G.; et al. Clinical Nuclear Medicine 2011, 36, 1063–1064.

[53] Dizeyi, N.; Konrad, L.; Bjartell, A.; et al. Urologic Oncology 2002, 7, 91–98.

[54] Hansson, J.; Bjartell, A.; Gadaleanu, V.; et al. Prostate 2002, 53, 50–59.

[55] Halmos, G.; Schally, A. V.; Sun, B.; et al. Journal of Clinical Endocrinology and Metabolism 2000, 85, 2564–2571.

[56] Plonowski, A.; Schally, A. V.; Nagy, A.; et al. International Journal of Oncology 2002, 20, 397–402.

[57] Frühwald, M. C.; Rickert, C. H.; O'Dorisio, M. S.; et al. Clinical Cancer Research 2004, 10, 2997–3006.

[58] Florio, T.; Montella, L.; Corsaro, A.; et al. Anticancer Research 2003, 23, 2465–2471.

[59] Arena, S.; Barbieri, F.; Thellung, S.; et al. Journal of Neuro-Oncology 2004, 66, 155–166.

[60] Ferone, D.; Resmini, E.; Boschetti, M.; et al. Journal of Endocrinological Investigations 2005, 28, S111–117.

[61] Ferone, D.; Montella, L.; De Chiara, A.; et al. Frontiers in Bioscience 2009, 14, 3304–3309.

[62] ten Bokum, A. M.; Hofland, L. J.; de Jong, G.; et al. European Journal of Clinical Investigations 1999, 29, 630–636.

[63] Paran, D.; Paran, H. Current Opinion in Investigational Drugs 2003, 4, 578–582.

[64] Reubi, J. C.; Laissue, J.; Waser, B.; et al. Annals of New York Academy of Sciences 1994, 733, 122–137.

[65] Davis, M. I.; Wilson, S. H.; Grant, M. B. Hormone and Metabolic Research 2001, 33, 295–299.

[66] Hernaez-Ortega, M. C.; Soto-Pedre, E.; Piniés, J. A. Diabetes Research and Clinical Practice 2008, 80, e8–10.

[67] Schmid, H. A. Molecular and Cellular Endocrinology 2008, 286, 69–74.

[68] Pfeiffer, M.; Koch, T.; Schröder, H.; et al. Journal of Biological Chemistry 2001, 276, 14027–14036.

[69] Ferone, D.; Gatto, F.; Arvigo, M.; et al. Journal of Molecular Endocrinology 2009, 42, 361–370.

[70] Srirajaskanthan, R.; Watkins, J.; Marelli, L.; et al. Neuroendocrinology 2009, 89, 308–314.

[71] Diakatou, E.; Kaltsas, G.; Tzivras, M.; et al. Endocrine Pathology 2011, 22, 24–30.

[72] Saveanu, A.; Muresan, M.; De Micco, C.; et al. Endocrine Related Cancer 2011, 18, 287–300.

4

THE USE OF RADIOLABELED SOMATOSTATIN ANALOGUE IN MEDICAL DIAGNOSIS: INTRODUCTION

ALBERTO SIGNORE

Nuclear Medicine Unit, Department of Medical-Surgical Sciences and of Translational Medicine, Faculty of Medicine and Psychology, "Sapienza" University of Rome, Rome, Italy

ABBREVIATIONS

NET   neuroendocrine tumor
SRS   somatostatin receptor scintigraphy
SST   somatostatin
SSTR somatostatin receptor

Nuclear medicine has always been a powerful tool for the study of biological functions and, more recently, for the molecular and histological characterization of tissues under physiological and pathological conditions. The growing knowledge of the biology of normal and pathological tissues leads to the discovery of several biologically active peptides that mediate their function by binding to external membrane-bound receptors with high affinity. This property made peptide/receptor complex a potential target to be used for molecular characterization of tissues *in vivo*.

Peptides have good characteristics as radiopharmaceuticals: they are small in size and have a fast renal clearance and easily penetrate into tissues with consequent low

Somatostatin Analogues: From Research to Clinical Practice, First Edition. Edited by
Alicja Hubalewska-Dydejczyk, Alberto Signore, Marion de Jong, Rudi A. Dierckx,
John Buscombe, and Christophe Van de Wiele.

background blood pool activity after few minutes from injection. They can be synthetic or of natural human recombinant origin and do not usually elicit allergic reactions.

Somatostatin (SST) is a peptide with a broad distribution in the nervous system and acts as a neurotransmitter in several organs, having a wide range of mainly inhibiting effects, such as the suppression of growth hormone release, as well as the inhibition of pancreatic and gastrointestinal hormone release [1, 2]. Five SST receptor (SSTR) subtypes have been cloned, of which SSTR1 and SSTR4 are grouped into one family and SSTR2, SSTR3, and SSTR5 into another. They all are G-protein-coupled receptors located at the cell membrane [3], which recognize the ligand and generate a transmembrane signal. The resulting hormone–receptor complexes have the ability to be internalized. Once internalized, these vesicles fuse with lysosomes, resulting in hormone degradation or receptor recycling [4]. SSTR2 appears to be the most frequently represented receptor subtype over the surface of neuroendocrine tumor (NET) cells [5] and activated lymphocytes providing the molecular basis for many clinical applications of radiolabeled SST analogues [6]. Toward the end of the 1980s, the *in vivo* demonstration of SSTRs on the surface of some tumors raised interest in receptor imaging [7], and indeed, the peptide receptor overexpression on tumors cells, as compared to normal tissues [8, 9], constituted the basis for molecular imaging of these tumors.

Both natural SST-14 and SST-28 bind with high affinity to all five SSTRs but have a short plasma half-life (~3 min) owing to rapid enzymatic degradation by endogenous peptidases. First attempts to label recombinant SST-28 were made by Signore and coworkers at the "Sapienza" University of Rome, Italy, in 1982. The SST was labeled with ^{123}I and used for imaging pituitary tumors. Unfortunately, the plasma half-life of this hormone was extremely rapid, thus not allowing good imaging with the available technology. In an early study by Amartey et al. [10], SST was labeled with ^{99m}Tc. As could be expected in view of its short half-life, no specific accumulation in receptor-rich tissue was observed. In the following years, the availability of SSTR analogues with a longer half-life (of 1.5–2 h) and preserved receptor binding has allowed significant improvements for diagnosis and therapy of NET. Furthermore, SST analogues have no major side effects, and their use is very safe, particularly when used as radiopharmaceuticals [11].

Molecular imaging by radiolabeled SST analogue highlights the presence of pathological tissues overexpressing SSTR. It can be used to characterize lesions with respect to the expression of SSTR, to contribute to differential diagnosis, for staging diseases by detecting or excluding sites of metastases, and for grading the disease since in NET the loss of the expression of SSTR is a sign of dedifferentiation and has a poor prognostic value.

Somatostatin receptor scintigraphy (SRS), however, is not disease specific but receptor specific, and to be able to interpret the results of the scan, one needs to understand the underlying tumor biology. A negative scan may indicate absence of tumor, tumor regression, or tumor dedifferentiation. Correlation with anatomical imaging is always mandatory.

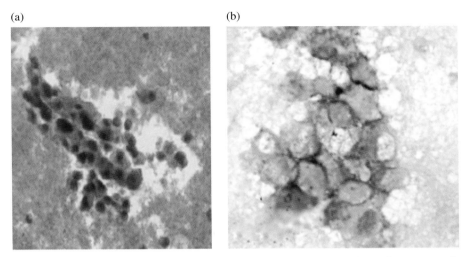

FIGURE 3.1 SSTR type 2A determination in a cellblock preparation from a liver metastasis of a well-differentiated neuroendocrine carcinoma of unknown primary. (a) Hematoxylin and eosin; (b) immunoperoxidase; (a and b) original magnification 400×.

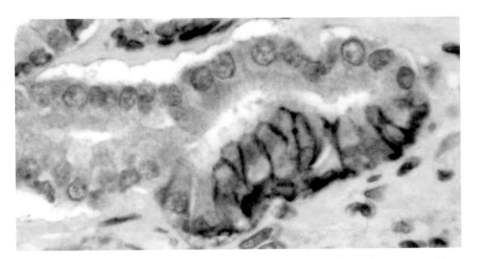

FIGURE 3.2 SSTR type 2A clonal expression in prostatic adenocarcinoma (immunoperoxidase; original magnification 400×).

Somatostatin Analogues: From Research to Clinical Practice, First Edition. Edited by Alicja Hubalewska-Dydejczyk, Alberto Signore, Marion de Jong, Rudi A. Dierckx, John Buscombe, and Christophe Van de Wiele.
© 2015 John Wiley & Sons, Inc. Published 2015 by John Wiley & Sons, Inc.

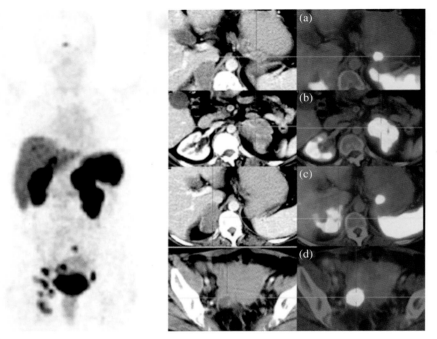

FIGURE 4.2.1 ⁶⁸Ga-DOTATATE PET/CT (maximum intensity projection image on extreme left) shows multiple primary neuroendocrine tumors in pancreatic tail with an SUVmax of 15.4 (a); in both adrenals, massive enlargement and partial necrosis in the left adrenal with an SUVmax of 29.7 (b) and comparatively smaller and less dense SSTR expression in the right adrenal with an SUVmax of 10.3 (c); and right ovary with an SUVmax of 11.6 (d). There was also widespread metastasis in the liver, multiple abdominal/extra-abdominal lymph nodes, and possibly the pituitary gland (SUVmax of 5.4) (a, b, c, d—images in transverse view; CT on the left and fused PET/CT on the right).

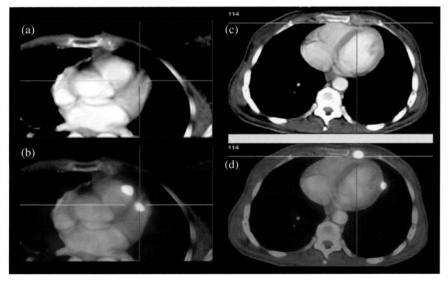

FIGURE 4.2.2 Identification of rare metastases of neuroendocrine tumors with high-sensitivity on ⁶⁸Ga-SSTR PET/CT: not only were myocardial metastases, which were otherwise difficult to appreciate on CT (a), localized on ⁶⁸Ga-DOTA-SSTR PET/CT (b) but also pericardial metastases (c, CT; d, fused PET/CT).

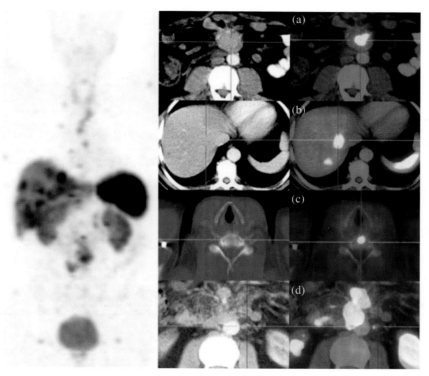

FIGURE 4.2.3 In this case of CUP syndrome, ^{68}Ga-DOTATOC PET/CT (MIP image on extreme left) revealed somatostatin receptor-positive primary tumor in the jejunum with SUVmax of 13.8 (a), along with multiple metastases in the liver (b), bone (c), lymph nodes (d), and a metastasis in the left adrenal. (a, b, c, d—images in transverse view; CT on the left and fused PET/CT on the right.)

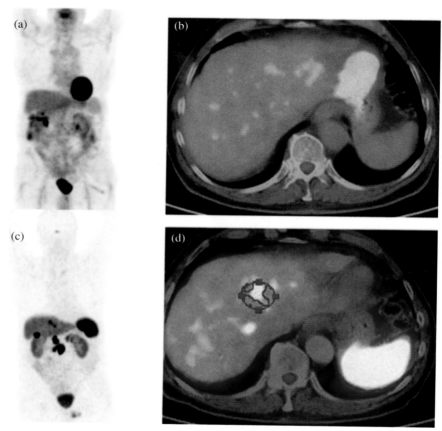

FIGURE 4.2.4 Nonfunctional, well-differentiated neuroendocrine neoplasm of the pancreas head with hepatic and abdominal lymph node metastases and proliferation rate (Ki-67) of 7%. ^{18}F-FDG PET/CT (a, b) demonstrated no increased glucose metabolism in the metastases. ^{68}Ga-DOTATOC PET/CT (c, d), on the other hand, demonstrated intense expression of somatostatin receptors. This mismatch is a sign of good prognosis and an indication for PRRT (a, c, MIP images; b, d, fused PET/CT images in transverse view, demonstrating hepatic metastases).

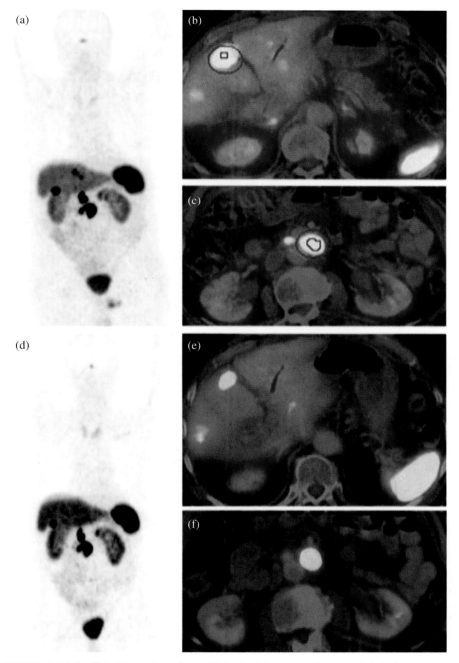

FIGURE 4.2.5 ^{68}Ga-DOTATATE PET/CT (a, MIP image pretherapy; d, MIP image post-therapy) demonstrated significant therapeutic response in the same patient (described in Fig. 4.2.4) (partial response according to molecular imaging criteria) after PRRT with 2700 MBq ^{90}Y-DOTATOC. Hepatic metastasis in segment S5 inferoventral segment of the liver (b, fused PET/CT image in transverse view pretherapy; e, fused PET/CT image in transverse view posttherapy) showed an SUV fall of 68% from 61.1 to 19.4 posttherapy. Preaortic lymph node metastasis (c, fused PET/CT image in transverse view pretherapy; f, fused PET/CT image in transverse view posttherapy) showed an SUV fall of 29% from 55.5 to 39.5 posttherapy.

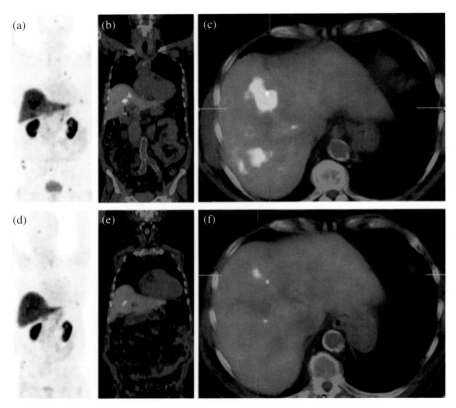

FIGURE 4.2.6 Neuroendocrine neoplasm of the pancreatic tail with hepatic and skeletal metastases: ^{68}Ga-DOTATATE PET/CT (pretherapy images—a, MIP; b, coronal PET/CT fused; c, transverse PET/CT fused) demonstrated excellent response in the liver metastases after two cycles of PRRT with a cumulative administered activity of 8000 MBq of [177]Lu (posttherapy images—d, MIP; e, coronal PET/CT fused; f, transverse PET/CT fused).

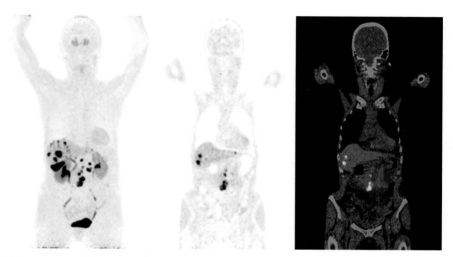

FIGURE 4.3.2 ^{18}F-DOPA PET/CT. Images of a 73-year-old female with a serotonin-producing metastasized well-differentiated NET of the small bowel (carcinoid). On the left, a maximum intensity projection (MIP) image is depicted; in the middle, a coronal slice; and on the right, a coronal fusion slice of PET and low-dose CT. Physiological uptake of ^{18}F-DOPA is visible in the striata and excretion via kidneys with physiological activity in the ureters and urinary bladder. Metastases are visible in the liver and mesenteric tissue/lymph nodes. ^{18}FDOPA, 6-^{18}F-L-3,4-dihydroxyphenylalanine; PET/CT, positron emission tomography/computed tomography.

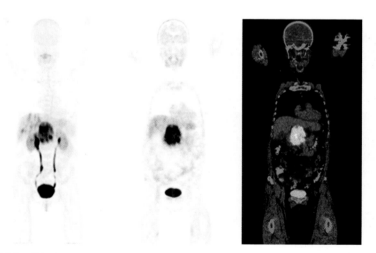

FIGURE 4.3.3 ^{11}C-5-HTP PET/CT. Images of a 62-year-old female with a gastrinoma. In contrary to ^{18}F-DOPA PET, the striata are not visualized. On the left, an MIP is depicted; in the middle, a coronal slice; and on the right, a coronal fusion slice of PET and low-dose CT. Physiological excretion of ^{11}C-5-HTP metabolites is via kidneys with physiological activity in the ureters and urinary bladder. Tumor is visible in the head of the pancreas and metastases in the liver. ^{11}C-5-HTP, β-[^{11}C]-5-hydroxy-l-tryptophan; ^{18}F-DOPA, 6-^{18}F-L-3,4-dihydroxyphenylalanine; PET/CT, positron emission tomography/computed tomography.

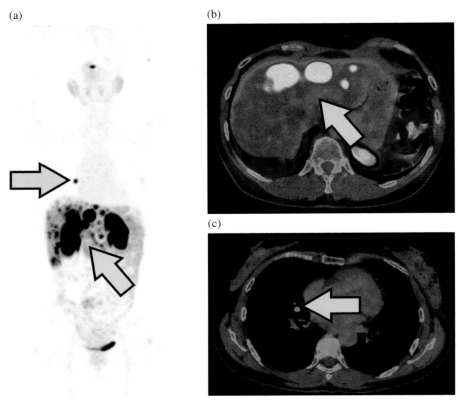

FIGURE 4.4.1.4 PET/CT with radiolabeled somatostatin analogue (^{68}Ga-DOTATOC) in patient with disseminated pancreatic neuroendocrine tumor. Arrows indicate liver (a, b) and lymph node metastases (a, c).

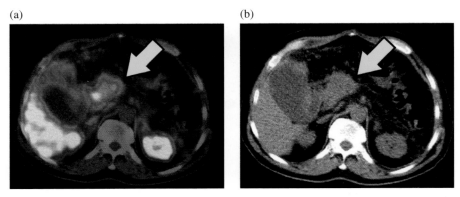

FIGURE 4.4.1.7 A 60-year-old man with recurrence of pancreatic glucagonoma. Somatostatin receptor scintigraphy with ^{99m}Tc-EDDA/HYNIC-TOC showing a recurrence of the disease in the head of the pancreas. Arrows indicate the lesion in the head of the pancreas (a - SPECT/CT image; b - CT image).

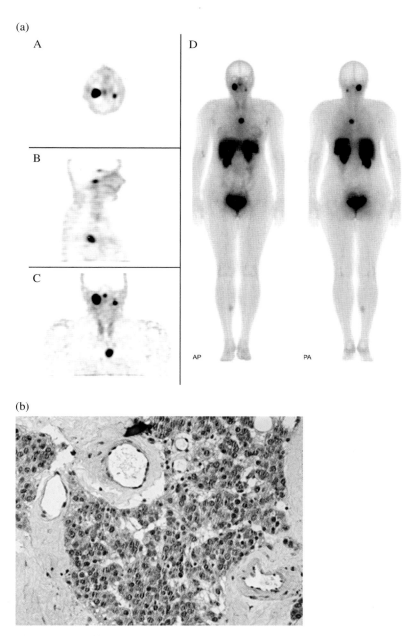

FIGURE 4.4.2.3 (a) ^{99m}Tc-EDDA/HYNIC-TOC SRS in a 37-year-old woman with multiple paragangliomas with prevalent expression of SSTR4 (80% of cells). (A) Axial, (B) sagittal, (C) coronal views, (D) whole body scan. (b) Immunohistochemical staining for SSTR subtype 4 (magn. ×400). Courtesy of Prof. R. Tomaszewska, Dept. of Pathology, Jagiellonian University, Medical College, Krakow.

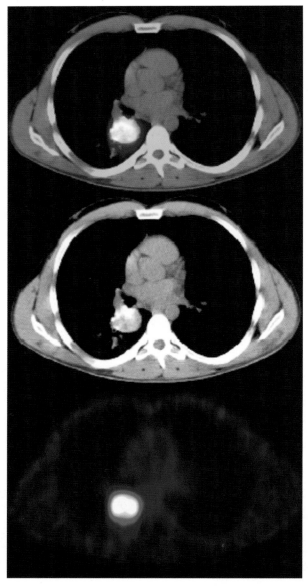

FIGURE 4.4.4.1 ^{99m}Tc-tectrotide scintigraphy in bronchial carcinoid (SPECT/CT) (Nuclear Medicine Unit, Department of Endocrinology, University Hospital in Krakow, Poland).

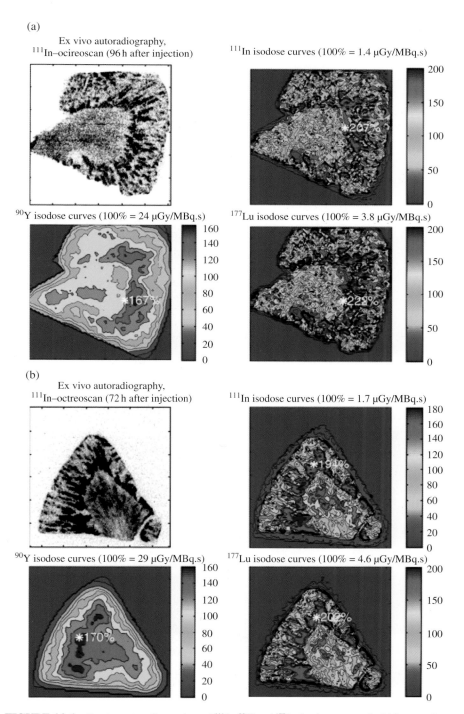

FIGURE 6.2.6 *Ex vivo* autoradiographs and ^{111}In, ^{90}Y, and ^{177}Lu isodose curves for kidney sections from 3 patients (a, b, and c) at different time intervals after administration of ^{111}In-DTPA-octreotide (Figure reproduced with permission from M. Konijnenberg et al. [59]).

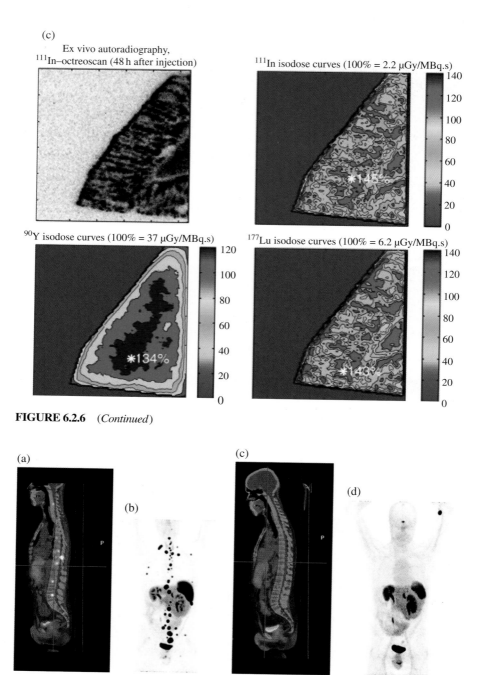

(c)

Ex vivo autoradiography,
^{111}In–octreoscan (48 h after injection)

^{111}In isodose curves (100% = 2.2 µGy/MBq.s)

$*14$ %

^{90}Y isodose curves (100% = 37 µGy/MBq.s)

$*134\%$

^{177}Lu isodose curves (100% = 6.2 µGy/MBq.s)

$*143\%$

FIGURE 6.2.6 (*Continued*)

(a)

(c)

(b)

(d)

FIGURE 6.3.1 Morphologic response in a patient affected by bone marrow metastases and previous liver metastases from a well-differentiated neuroendocrine carcinoma, Ki-67 10% with an unknown primary. The patient underwent a right liver hepatectomy with a homolateral adrenalectomy. After surgery, the patient was treated with ^{177}Lu-DOTATATE (24.6 GBq cumulative activity). The evaluation of the status at baseline by means of ^{68}Ga-DOTATOC PET/CT (a, fusion sagittal image; b, MIP image) shows diffuse bone marrow involvement, particularly in the axial skeleton. The evaluation performed after the end of PRRT by means of ^{68}Ga-DOTATOC PET/CT (c, fusion sagittal image; d, MIP image) shows the almost complete disappearance of the lesions. Please note the hypertrophy of the left liver lobe, as a consequence of the right lobe resection.

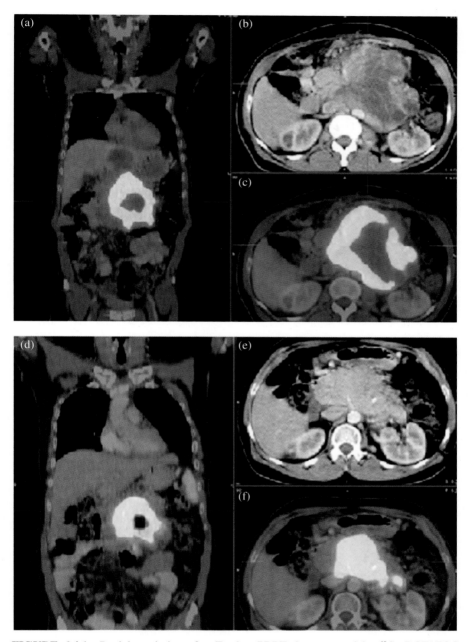

FIGURE 6.4.1 Partial remission after Tandem-PRRT demonstrated by ⁶⁸Ga-DOTATOC PET/CT (a, b, c, before; d, e, f, after Tandem-PRRT using 4000 MBq ¹⁷⁷Lu-DOTATOC and 4500 MBq ⁹⁰Y-DOTATOC) of a very large, partially necrotic nonfunctional well-differentiated neuroendocrine neoplasm in the pancreas tail and body, with intense somatostatin receptor expression. The tumor was infiltrating the neighboring structures and caused life-threatening recurrent bleedings which stopped after PRRT. SUVmax dropped from 39.9 to 35.3, molecular tumor volume dropped from 542 to 214 ml, and the size overall diminished (a and d, coronal fused PET/CT images; b and e, CT images in transverse view; c and f, fused PET/CT images in transverse view).

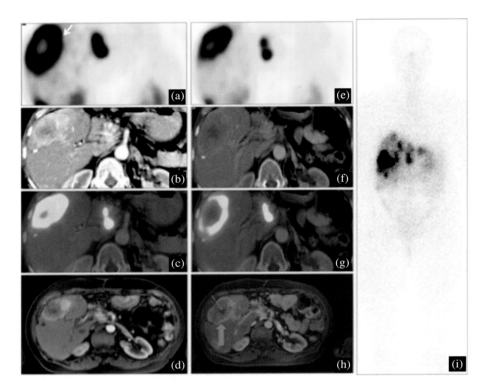

FIGURE 6.4.2 A 52-year-old patient had nonfunctioning, well-differentiated neuroendocrine tumor of the pancreatic tail with local extension into the stomach and neighboring organs, lymph node metastases, as well as extensive bilobar liver metastases and few bone metastases and underwent Tandem-PRRT with 5.5 GBq of ^{177}Lu and 3.5 GBq of ^{90}Y. ^{68}Ga-DOTATOC PET showed good molecular response in the lesion in segment S8 of the liver (white arrow). There was 47% fall in the SUV from 70.8 to 37.7. T1-weighted MRI with contrast showed posttherapy increase in the size of the lesion but with progressive necrosis (long arrow) and hypervascularity in the rim (short bold arrow) indicating a favorable therapeutic response. The left panel (a, b, c, d) shows pretherapy and middle panel (e, f, g, h) shows posttherapy images in transverse view (a and e, PET; b and f, CT; c and g, PET/CT fused; d and h, MRI). The whole body planar scan (i) acquired 44 h posttherapy (^{177}Lu gamma-energy window) showed good uptake and long retention in the metastases.

(a)   (b)

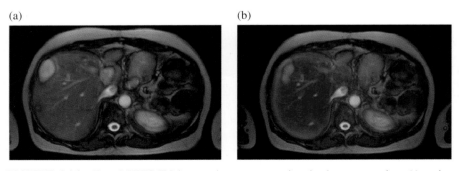

FIGURE 6.4.3 Fused PET/MRI images in transverse view in the same patient (A, using ^{68}Ga-DOTATOC; B, using ^{18}F-FDG) show mismatch of the uptake of the 2 tracers (^{68}Ga-DOTATOC $>^{18}$F-FDG) in the lesion in segment S8 of the liver, demonstrating good differentiation and high somatostatin receptor density.

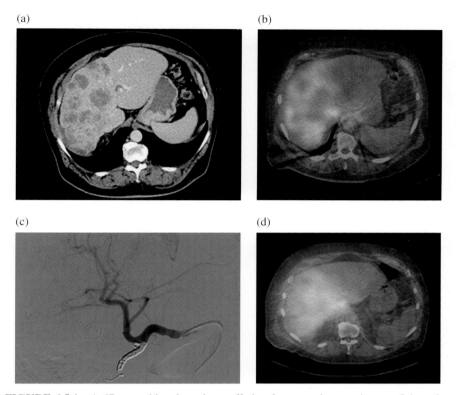

(a)   (b)

(c)   (d)

FIGURE 6.5.1 A 67-year-old male patient suffering from an adenocarcinoma of the colon sigmoideum, stage pT3 N1b M1 R0 (K-ras mutation not present), and UICC stage 4. The patient had undergone resection of the primary in 10/2010. He then underwent multiple palliative chemotherapeutic treatments until February 2013, when CT showed progressive liver metastases to the right liver lobe (a). Pulmonary shunt scintigraphy using 150 MBq Tc-99m-MAA on 8/4/2013 revealed faint uptake over both lungs, reaching a pulmonary shunt rate of 14% in the semiquantitative analysis. The scan showed increased liver uptake in the right liver lobe (b). Embolization of the gastroduodenal and right gastric artery was performed using coils (c). Application of 1400 MBq Y-90-SIR-spheres was performed on 18/4/2013. 0.3 GBq was applied in segment IV and 1.11 GBq was applied in the right liver lobe, selectively. Posttherapeutic bremsstrahlung scan revealed increased uptake in the right liver lobe on SPECT/CT imaging, showing no extrahepatic uptake (d).

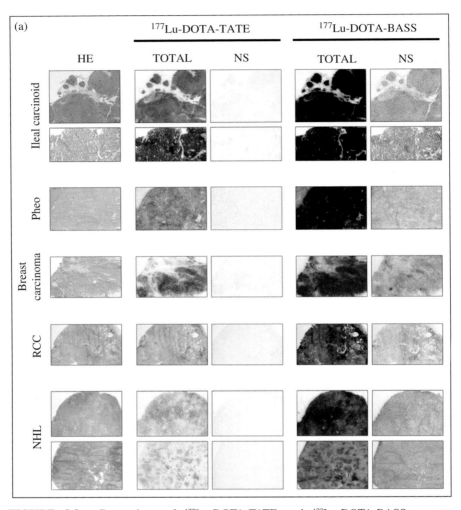

FIGURE 8.3a Comparison of ^{177}Lu-DOTA-TATE and ^{177}Lu-DOTA-BASS receptor autoradiographic binding in successive sections of various types of human cancers (ileal carcinoid, pheochromocytoma, breast carcinoma, renal cell carcinoma, and non-Hodgkin lymphoma) expressing sst2. Columns from left to right represent hematoxylin and eosin staining, total and nonspecific binding of ^{177}Lu-DOTA-TATE, and total and nonspecific binding of ^{177}Lu-DOTABASS. Binding is markedly stronger with the antagonist ^{177}Lu-DOTA-BASS. HE, hematoxylin and eosin; NHL, non-Hodgkin lymphoma; NS, nonspecific; Pheo, pheochromocytoma; RCC, renal cell carcinoma.

The application of ^{111}In-labeled octreotide analogue to target SSTRs on tumor cells still represents a paradigm in the field of peptide radiopharmaceuticals. The SST analogue ^{111}In-labeled octreotide (OctreoScan) was officially introduced in 1994, and its use to visualize various SSTR-positive tumors and tissues is widely accepted. Many tumors (most of them neuroendocrine related) may express a combination of the five receptor subtypes (SSTR1–SSTR5) in different percentages [12, 13].

Since that time, a very large "panel" of tumors and diseases were studied by OctreoScan scintigraphy, and extensive clinical studies have been performed mainly in NET [14–20] but also in other tumors like brain tumors [21], melanomas [22], and lung [23] and breast [24] cancer. In order to overcome the limitations of the use of OctreoScan like the high cost of [111]In (a cyclotron-produced radionuclide) and the nonoptimal physical features of this radioisotope, many SST analogues have been labeled with [99m]Tc. Depreotide, a synthetically produced ten-amino-acid peptide with affinity for SSTR2, SSTR3, and SSTR5, has been labeled with [99m]Tc [25] and successfully used to characterize malignancy in solitary pulmonary nodules [26]. Another interesting peptide is vapreotide, which binds SSTR2 and SSTR5 with high affinity and moderately to SSTR3 and SSTR4 and has been used in animals [27]. Another promising radiopharmaceutical, the ^{99m}Tc-HYNIC-tyr(3)-octreotide, has been labeled and successfully used in humans [28–31].

A further evolution in the field of SST analogues is represented by the development of macrocyclic chelators that exhibit the property to bind also beta particle emitters like ^{90}Y and [177]Lu. These chelators include DOTA, DOTAOC, DOTATOC, DOTAVAP, DOTATATE, and lanreotide DOTALAN. These radiolabeled compounds including these chelators seem to show favorable binding and biodistribution characteristics with high uptake and retention in target organs, thus being promising candidates for peptide receptor radionuclide therapy. It is well known that at least one of the first generation of ^{90}Y-labeled octreotide analogues, ^{90}Y-DOTATOC, has real therapeutic activity in the treatment of NET, which have proved resistant to other forms of treatment [32, 33].

REFERENCES

[1] Brazeau, P.; Vale, W.; Burnus, R.; et al. Science 1973, 179, 77–79.

[2] Plewe, G.; Beyer, J.; Krause, U.; Neufeld, M.; del Pozo, E. Lancet 1984, 2, 782–784.

[3] Patel, Y. C.; Greenwood, M. T.; Warszynska, A., Panetta, R.; Srikant, C. B. Biochemical & Biophysical Research Communications 1994, 198, 605–612.

[4] Hofland, L. J.; Lamberts, S. W. Endocrine Reviews 2003, 24, 28–47.

[5] Reubi, J. C. Endocrine Reviews 2003, 24, 389–427.

[6] Rufini, V.; Calcagni, M. L.; Baum, R. P. Seminars in Nuclear Medicine 2006, 36, 228–247.

[7] Krenning, E. P.; Bakker, W. H.; Breeman, W. A.; et al. Lancet 1989, 1, 242–244.

[8] Virgolini, I.; Yang, Q.; Li, S.; et al. Cancer Research 1994, 54, 690–700.

[9] Reubi, J. C.; Schaer, J. C.; Waser, B.; Mengod, G. Cancer Research 1994, 54, 3455–3459.

[10] Amartey, J. K. Nuclear Medicine and Biology 1993, 20, 539–543.

[11] Pepe, G.; Moncayo, R.; Bombardieri, E.; Chiti, A. European Journal of Nuclear Medicine and Molecular Imaging 2012, 39 (Suppl 1), S41–S51.

[12] Patel, Y. C.; Greenwood, M. T.; Panetta, R.; Demchyshyn, L.; Niznik, H.; Srikant, C. B. Life Sciences 1995, 57, 1249–1265.

[13] Reubi, J. C.; Schaer, J. C.; Laissue, J. A.; Waser, B. Metabolism 1996, 45, 39–41.

[14] Angeletti, S.; Corleto, V. D.; Schillaci, O.; et al. Gut 1998, 42, 792–794.

[15] Leners, N.; Jamar, F.; Fiasse, R.; Ferrant, A.; Pauwels, S. Journal of Nuclear Medicine 1996, 37, 916–922.

[16] Krenning, E. P.; Kwekkeboom, D. J.; Oei, H. Y.; et al. Digestion 1994, 55 (Suppl 3), 54–59.

[17] Krenning, E. P.; Kwekkeboom, D. J.; Oei, H. Y.; et al. Annals of New York Academy of Sciences 1994, 15, 733, 416–424.

[18] Kwekkeboom, D. J.; Krenning, E. P. World Journal of Surgery 1996, 20, 157–161.

[19] Oppizzi, G.; Cozzi, R.; Dallabonzana, D.; et al. Journal of Endocrinological Investigations 1998, 21, 512–519.

[20] Signore, A.; Procaccini, E.; Chianelli, M.; et al. Quarterly Journal of Nuclear Medicine 1995, 39 (suppl 1), 111–112.

[21] Maini, C. L.; Sciuto, R., Tofani, A.; et al. Nuclear Medicine Communications 1995, 16, 756–766.

[22] Fletcher, W. S.; Lum, S. S.; Nance, R. W.; Pommier, R. F.; O'Dorisio, M. S. Yale Journal of Biology & Medicine 1997, 70, 561–563.

[23] Bombardieri, E.; Chiti, A.; Crippa, F.; et al. Quarterly Journal of Nuclear Medicine 1995, 39 (Suppl 1), 104–107.

[24] van Eijck, C. H.; Kwekkeboom, D. J.; Krenning, E. P. Quarterly Journal of Nuclear Medicine 1998, 42, 18–25.

[25] Blum, J. E.; Handmaker, H.; Rinne, N. A. Chest 1999, 115, 224–232.

[26] Blum, J.; Handmaker, H.; Lister-James, J.; Rinne, N. Chest 2000, 117, 1232–1238.

[27] Thakur, M. L.; Kolan, H.; Li, J.; et al. Nuclear Medicine and Biology 1997, 24, 105–113.

[28] Decristoforo, C.; Cholewinski, W.; Donnemiller, E.; Riccabona, G.; Moncayo, R.; Mather, S. J. European Journal of Nuclear Medicine 2000, 27, 1580.

[29] Decristoforo, C.; Mather, S. J.; Cholewinski, W.; Donnemiller, E.; Riccabona, G.; Moncayo, R. European Journal of Nuclear Medicine 2000, 27, 1318–1325.

[30] Decristoforo, C.; Melendez-Alafort, L.; Sosabowski, J. K.; Mather, S. J. Journal of Nuclear Medicine 2000, 41, 1114–1119.

[31] Bangard, M.; Behe, M.; Guhlke, S.; et al. European Journal of Nuclear Medicine 2000, 27, 628–637.

[32] Waldherr, C.; Haldemann, A.; Maecke, H. R.; Crazzolara, A.; Mueller-Brand, J. Clinical Oncology (Royal College of Radiologists) 2000, 12, 121–123.

[33] Paganelli, G.; Zoboli, S.; Cremonesi, M.; Macke, H. R.; Chinol, M. Cancer Biotherapy & Radiopharmaceuticals 1999, 14, 477–483.

4.1

SOMATOSTATIN RECEPTOR SCINTIGRAPHY-SPECT

RENATA MIKOŁAJCZAK[1] AND ALBERTO SIGNORE[2]

[1]*Radioisotope Centre Polatom, National Centre for Nuclear Research, Otwock, Poland*
[2]*Nuclear Medicine Unit, Department of Medical-Surgical Sciences and of Translational Medicine, Faculty of Medicine and Psychology, "Sapienza" University of Rome, Rome, Italy*

ABBREVIATIONS

BCA	bifunctional metal chelating agent
COST	European Cooperation in Science and Technology
CT	computed tomography
DFO	desferrioxamine
DOTA	1,4,7,10-tetraazacyclododecane-1,4,710-tetraacetic acid
DTPA	diethylenetriaminepentaacetic acid
EDDA	ethylenediamine-N, N'-diacetic acid
GEP-NET	gastroenteropancreatic neuroendocrine tumor
HYNIC	hydrazinonicotinamide
IAEA	International Atomic Energy Agency
MEAP	medium-energy all-purpose collimators
MEN-1	multiple endocrine neoplasia type 1
MIBG	metaiodobenzylguanidine
MRI	magnetic resonance imaging
NET	neuroendocrine tumor
NHS-HYNIC	N-hydroxysuccinimidyl hydrazinonicotinamide

Somatostatin Analogues: From Research to Clinical Practice, First Edition. Edited by Alicja Hubalewska-Dydejczyk, Alberto Signore, Marion de Jong, Rudi A. Dierckx, John Buscombe, and Christophe Van de Wiele.
© 2015 John Wiley & Sons, Inc. Published 2015 by John Wiley & Sons, Inc.

PET   positron emission tomography
SPECT   single-photon emission computed tomography
SPN   solitary pulmonary nodule
SRS   somatostatin receptor scintigraphy
Sst   somatostatin receptor

RADIOPHARMACEUTICALS FOR SRS-SPECT

The development of radiolabeled peptides for successful receptor targeting requires consideration of several factors, such as the accumulation in the target and nontarget tissues, the clearance from the body, the excretory pathway, and the *in vivo* stability of the radiopeptide. The radiolabeled peptides that successfully went through all tests, including toxicological studies, and with well-established preparation method may enter clinical studies in humans. The issues related to the peptide-based radiopharmaceutical design and their development have been well described in several excellent reviews [1–3].

Particularly for the well-characterized somatostatin receptors, the design of a peptide and its synthetic pathway was possible in order to produce metabolically stabilized peptide analogues, which preserved most of the biological activity of the original molecule and high affinity for the corresponding receptor. They could be labeled with various radionuclides for both diagnosis and therapy, while the choice of radiolabeling approach depended on the radionuclide properties and characteristics of the chelator. As a common feature, it is required that the labeling protocols allow very high labeling yield, radiochemical purity, and specific activity and the peptide retains the affinity for the receptor.

There are certain protocols established for *in vitro* characterization of radioligand affinity for the receptors expressed on the tumor cell membrane, their internalization rate, dissociation from the tumor cells, etc., which are helpful in selecting the most promising radiopeptides during preclinical investigations. Biodistribution and imaging techniques are used with suitable animal models to evaluate *in vivo* the pharmacological behavior and pharmacokinetics of the radiopeptides. However, it is a long way from the design of a new peptide until its use in the clinic, both due to the radiopharmaceutical development and to the regulatory constraints. As a result, from a large number of newly developed radiopeptides, only very few found their way into routine clinical application.

Historically, the development of agents used for imaging of somatostatin receptors reflected the above considerations and followed the increasing knowledge of the role of somatostatin and its analogues in the diagnosis and treatment of tumors. It has been shown that sst-expressing tumors can be treated with somatostatin or synthetic analogues to either reduce hypersecretion of hormones or inhibit tumor growth [4]. However, because somatostatin undergoes rapid *in vivo* enzymatic degradation, somatostatin analogues that are more resistant to *in vivo* degradation have been developed [5–9]. The molecule was modified in various ways resulting in improved biological characteristics, but mostly in increased affinity for sst_2 and to some extent for sst_5. Introduction of D-amino acids and shortening of the molecule to the bioactive core sequence resulted in eight amino acid-containing somatostatin analogues such as octreotide (OC) (Sandostatin, SMS 201-995), lanreotide (BIM23014), and vapreotide (RC-160). Lanreotide and OC are widely used for the symptomatic treatment of neuroendocrine-active tumors, such as growth hormone-producing pituitary adenomas and gastroenteropancreatic tumors [10].

Nowadays, new somatostatin-based agents labeled with gamma emitters found their way to the clinic offering improved imaging characteristics. Three radiopharmaceuticals for somatostatin receptor scintigraphy (SRS)-SPECT, which were granted marketing authorization, are briefly discussed.

^{111}In-Pentetreotide

The evidence of the overexpression of somatostatin receptors by primary and metastatic malignant disease, mainly of neuroendocrine origin, has prompted a worldwide search for radiolabeled somatostatin analogues for use in SRS [4, 11]. First, the successful visualization of somatostatin receptor-positive neoplastic lesions with a radioiodinated synthetic somatostatin analogue (^{123}I-Tyr^3-octreotide) was reported [12–14]. Soon, an improved octreotide-based radioligand labeled with indium-111 was introduced [15, 16]. Both radiolabeled octreotide derivatives, ^{123}I-Tyr^3-octreotide and ^{111}In-DTPA-D-Phe^1-octreotide, were shown to be very useful in detecting small neuroendocrine tumors and metastases not detected by conventional means and for identifying tumors that respond to therapeutic doses of "cold" octreotide. The limitations of ^{123}I-Tyr^3-octreotide, however, were the high cost of ^{123}I, the short half-life, and the unfavorable clearance via bile ducts, which did not allow imaging of tumors in the abdominal region. The improved imaging properties of ^{111}In-DTPA-D-Phe^1-octreotide, the first somatostatin analogue developed for indirect labeling with ^{111}In, resulted in its wider application in medical diagnosis [4, 17]. This imaging agent was developed by Mallinckrodt Medical, Inc., in conjunction with the University Hospital of Rotterdam in the Netherlands and Sandoz Pharma Ltd., Basel, Switzerland. It has undergone clinical trials in Europe and in the United States and has been granted marketing authorization (^{111}In-pentetreotide, OctreoScan®) (see Fig. 4.1.1).

SRS with ^{111}In-DTPA-D-Phe^1-octreotide became a "gold standard" in the localization, staging, and therapy follow-up in patients with neuroendocrine tumors. Though a very powerful noninvasive imaging technique, the application of ^{111}In-DTPA-D-Phe^1-octreotide in diagnostic oncology is restricted by the increased cost and limited availability of the cyclotron-produced ^{111}In and its suboptimal nuclear characteristics, such as a long half-life ($T_{1/2} = 67$ h) and the two medium-energy photons (171 keV, 245 keV), which lead to a poor image resolution and a high radiation dose to the patient.

^{99m}Tc-Depreotide

Two new somatostatin analogues, ^{99m}Tc-P587 and ^{99m}Tc-P829, were synthesized and evaluated preclinically in comparison to ^{111}In-DTPA-octreotide [18]. Both P587 (with the sequence of –Gly-Gly-Cys- as a triamide-thiol chelator) and P829 (with the monoamine, bisamide, and monothiol chelating sequence of –(β-Dap)-Lys-Cys- appended to the homocysteine side chain) were labeled with ^{99m}Tc by ligand exchange from ^{99m}Tc-glucoheptonate with specific activity higher than 2.2 TBq/mmol. Tumor/blood and tumor/muscle ratios at 90 min post injection to Lewis rats bearing CA20948 rat pancreatic tumors were 6 and 33 for ^{99m}Tc-P587, 21 and 68 for ^{99m}Tc-P829, and 22 and 64 for ^{111}In-DTPA-octreotide. In addition, the uptake of labeled peptides was shown to be specific and saturable (diminished up to 80–90% by increasing doses of coinjected parent peptide up to the dose of 4 mg/kg). ^{99m}Tc-P829 has been selected for clinical studies due to its high tumor uptake and low gastrointestinal uptake. It has been studied in patients with various tumors and showed good results in the identification of solitary pulmonary nodules (SPN) in patients with non-small

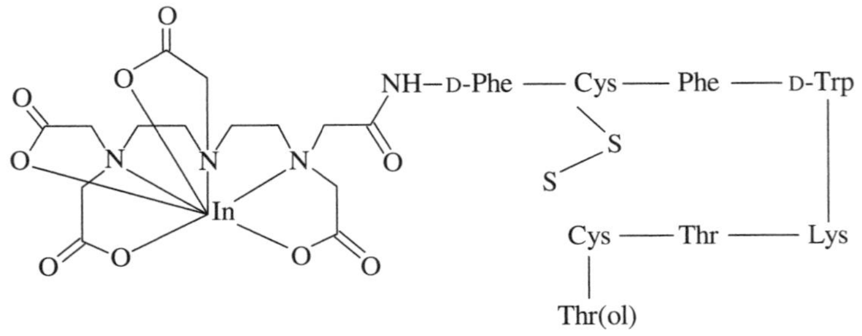

FIGURE 4.1.1   Structure of ^{111}In-DTPA-D-Phe^1-octreotide.

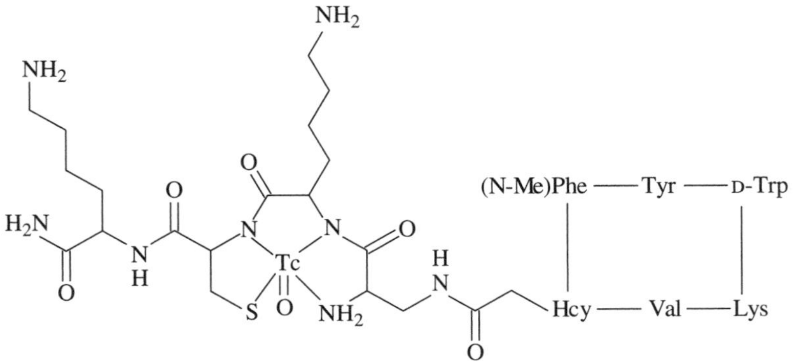

FIGURE 4.1.2   Structure of ^{99m}Tc-depreotide.

cell lung cancer [19]. The agent was approved for human use (^{99m}Tc-depreotide, NeoSpect; GE Healthcare: Amersham Health). Its chemical structure is presented in Figure 4.1.2.

In patients with endocrine tumors, the detection rate using ^{99m}Tc-depreotide scintigraphy was lower than that of ^{111}In-pentetreotide scintigraphy, which appeared to be more sensitive, especially for liver metastases, because of high liver uptake of ^{99m}Tc-depreotide [20]. Currently, the product registration is discontinued.

^{99m}Tc-EDDA/HYNIC-Tyr[3]-Octreotide

The first report on the diagnostic usefulness of ^{99m}Tc-tricine/HYNIC-Tyr[3]-octreotide (TOC) [^{99m}Tc-HYNIC-TOC] compared to ^{111}In-pentetreotide was published in 2000 [21]. Further tracer development resulted in the new radiopharmaceutical ^{99m}Tc-EDDA/HYNIC-TOC that was then compared to ^{111}In-pentetreotide in various neuroendocrine tumors and confirmed the superior imaging features of ^{99m}Tc-labeled tracer [22]. Figure 4.1.3 shows the structure of this complex.

In the process of clinical validation of the ^{99m}Tc-EDDA/HYNIC-TOC kit (Tektrotyd, POLATOM, Poland), the first results showing its diagnostic efficacy were obtained in collaboration within the European Cooperation in Science and Technology (COST) and International Atomic Energy Agency (IAEA) programs. A pilot study showed that ^{99m}Tc-EDDA/HYNIC-TOC can be effectively utilized for the diagnosis of neuroendocrine tumors and that it is useful in imaging of primary tumors and metastatic lesions [23]. In the course of further investigations, the attention was focused on the good imaging features of this agent in detecting non-small cell lung cancer [24–28]. In a direct comparison, it has been shown that ^{99m}Tc-EDDA/HYNIC-TOC is equivalent to ^{99m}Tc-depreotide in the detection of SPN [29]. Currently, the tracer is granted marketing authorization in some European countries.

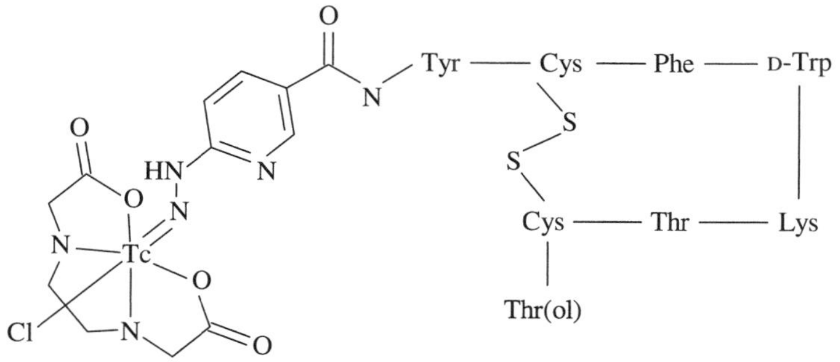

FIGURE 4.1.3 Structure of ^{99m}Tc-EDDA/HYNIC-Tyr[3]-octreotide.

RADIOLABELING OF PEPTIDES FOR SRS-SPECT

Only a few radionuclides are available, which can be used for radiolabeling of peptides for scintigraphy. Their physical properties are summarized in Table 4.1.1. Depending on the chemical properties of the radioactive element, the strategies for peptide radiolabeling can be developed.

Iodine-123 is a useful gamma emitter for SPECT with ligands such as metaiodo-benzylguanidine (MIBG). Small peptides can be radioiodinated by electrophilic substitution of an aromatic proton by electrophilic radioiodine ($*I^-$), and this reaction can take place at an amino acid residue of the peptide, which contains aromatic rings, for example, tyrosine or histidine. To enable iodination with ^{123}I, octreotide was modified by replacing Phe^3 in the amino acid chain by Tyr^3 and further electrophilic reaction on the hydroxyl group present in the aromatic ring of tyrosine and evaluated for imaging of neuroendocrine tumors [12–14]. The labeling of the peptide with iodine radioisotopes via electrophilic substitution makes the obtained bond susceptible to *in vivo* enzymatic attack resulting in their reduced stability, which is a limitation of this method. Another approach for iodination is the acylation reactions via prelabeled prosthetic groups; however, the attachment of a bulky prosthetic group in a small peptide often significantly influences the binding affinity for the receptor and the *in vivo* pharmacokinetics of the labeled peptide [2].

^{111}In, ^{67}Ga, and ^{99m}Tc are radiometals and to incorporate a radiometal into the peptide structure, a chelator is required. Usually, the bifunctional metal chelating agent (BCA) is coupled with the peptide, and the radionuclide is coordinated to the peptide–chelator compound. It is important that the chelator is at sufficient distance from the binding sites of the peptide to avoid adverse interaction of these two entities. This may necessitate the addition of a spacer between them to separate the active regions of both components. During the conjugation process, the chelator reacts with a free terminal amine group of the peptide. Therefore, the chelator must have carboxylic acid groups for this reaction. A wide range of suitable chelators ready for coupling with peptides are available [30–32]. The choice on an appropriate chelator for a given radiometal is crucial both for the efficiency of radiometallation and for the *in vivo* performance of the radiometallated peptide. ^{111}In-pentetreotide has only moderate binding affinity to the sst$_2$, and acyclic diethylenetriaminepenta-acetic acid (DTPA) is not a suitable chelator for β-emitters, such as ^{90}Y and ^{177}Lu. For these radiometals, DTPA has been replaced by 1,4,7,10-tetraazacyclododec-ane-1,4,710-tetraacetic acid (DOTA), which forms thermodynamically and

TABLE 4.1.1　Physical properties of radionuclides used in SRS-SPECT

Radionuclide	Half-life	γ-Energy (keV)	Decay mode	Production mode
^{123}I	13.2 h	159 (83%)	EC	Cyclotron
^{67}Ga	78.3 h	93 (10%), 185 (24%), 296 (22%)	EC	Cyclotron
^{99m}Tc	6.02 h	141 (89%)	IT	Generator
^{111}In	2.83 days	171 (88%), 247 (94%)	EC (100) Auger	Cyclotron

kinetically stable complexes with +3 cations of radiometals. This was of high importance, since patients with disseminated neoplasms and positive result after SRS were referred for therapy with another somatostatin analogue labeled with β-emitter, such as ^{90}Y or ^{177}Lu [33, 34]. DOTA can be used for radiolabeling with ^{111}In and also with $^{67/68}$Ga. Earlier chelator modifications designed to stably bind gallium radioisotopes were based on desferrioxamine (DFO) conjugated to octreotide via a succinyl linker to form a stable conjugate (DFO-β-succinyl-DPhe^1-octreotide). Although this ligand demonstrated specific binding *in vivo*, its receptor affinity was found reduced [35, 36].

It has been shown that not only the amino acid sequence but also the radiolabeling method affects the biological behavior of radiopeptides due to the small number of sites available for labeling and the likelihood of modifying amino acid residues that are essential for biological activity. Due to the high potency of many peptides and, on the other hand, the low tissue concentration of their receptors, specific activity is often critical [37, 38]. This is more relevant to therapeutic than diagnostic applications of radiolabeled peptides, where usually the administered dose of peptide is lower. In addition, the radiolabeled BCA–peptide conjugate must be thermodynamically stable and kinetically inert to survive physiological conditions.

From the point of view of SRS-SPECT technique, ^{99m}Tc-labeled sst-binding radiotracers were of main interest. ^{99m}Tc is called a "perfect radioisotope," because of its very low radiotoxicity (group IV of radiotoxicity; $k = 0.001$; k-estimate reduction of biologic effect of radiation with the use of radioisotopes of the same activity), short physical half-life ($T_{1/2} = 6.02$ h), and optimum radiation energy for detection by gamma camera (141 keV). ^{111}In belongs to group III of radiotoxicity ($k = 0.01$). It has a longer physical half-life ($T_{1/2} = 67$ h) and two energy photons (171 keV with abundance 90% and 245 keV with abundance 94%) requiring the use of medium-energy all-purpose collimators (MEAP), a 3-day examination protocol, and eventually resulting in images of inferior spatial resolution. Most importantly, patient's and staff's exposure to radiation is considerably smaller when using ^{99m}Tc-labeled compounds. The wide availability and cost-effectiveness of ^{99m}Tc are of major importance for routine clinical applications. As a consequence, the search for a ^{99m}Tc-based somatostatin analogue has been intense [39, 40].

^{99m}Tc can be easily obtained from ^{99}Mo/^{99m}Tc generators, in the chemical form of pertechnetate ion ^{99m}TcO$_4^-$, Tc(VII), and needs to be reduced to a lower oxidation state to be effectively bound to biomolecules. The rich redox chemistry of ^{99m}Tc makes it difficult to control the oxidation state and solution stability of ^{99m}Tc chelates, but on the other hand, it provides opportunities to modify the structures and properties of technetium complexes by the choice of chelators.

The main strategies for labeling peptides with radionuclides are generally similar to those used for labeling proteins [41, 42]. The direct labeling approach used for labeling antibodies after converting the cysteine disulfide bridges into free thiols by a reducing agent, which in turn are free to bind ^{99m}Tc in a very efficient way, can't be used for ^{99m}Tc labeling of somatostatin analogues. The forming ^{99m}Tc complexes are unstable, and there is poor control on the labeling site [43]. For small disulfide bond-containing peptides, these bonds are often critical for biological function, and even slight alterations in the ring structure can result in dramatic alterations in biological activity. The reducing agent (usually stannous ion) used in ^{99m}Tc labeling can reduce (open) the disulfide bond with consequent considerable loss of receptor-binding affinity [44].

Therefore, usually, the indirect labeling approach is the method of choice whereby the BCA is attached to the peptide to form a BCA–peptide conjugate and the conjugate is then labeled with the radiometal. In the case of ^{99m}Tc, the labeling proceeds either directly by reduction of ^{99m}TcO$_4^-$ or indirectly by ligand exchange via an intermediate ^{99m}Tc complex (such as ^{99m}Tc-glucoheptonate, ^{99m}Tc-diphosphonate, or ^{99m}Tc-tricine). In general, this approach is easy to carry out and has a well-defined chemistry.

The attractiveness of the indirect approach was growing after the solid-phase peptide synthesis was introduced in the development of peptide-based ligands. The technique consists of two major steps: first, the peptide chain with protected amino acid lateral chains is assembled on a polymeric support (resin), and second, the peptide is released from the resin, yielding the crude product. During peptide synthesis, functional groups not involved in peptide bond formation need to be reversibly protected to prevent unwanted side reactions. Side-chain protecting groups are either fully retained, partially removed, or completely removed, depending on the requirements for further workup and derivatization. Compared to synthesis in solution, solid-phase peptide synthesis has major advantages: first, the ease of reagent removal and purification of the intermediate peptides by simple washing of the resin and, second, its high yield. For the somatostatin analogues, the introduction of the chelator by simple elongation of the octapeptide is the main advantage [45].

As a result, over the past few years, several somatostatin analogues have emerged carrying a variety of chelators utilized for efficient ^{99m}Tc labeling of biomolecules including small peptides [46]. Examples of such ligands are presented in Figure 4.1.4. Among them were tetradentate chelators, such as N$_2$S$_2$ diamidedithiols, N$_3$S triamidethiols, N$_2$S$_2$ diaminedithiols [18, 47–49], propylene amine oxime [50], or open-chain or cyclic tetraamines [51, 52]. ^{99m}Tc forms mainly penta- or hexacoordinated complexes containing TcO$_3^+$ (N$_2$S$_2$ and N$_3$S) or TcO$_2^+$ (N$_4$) core. Another labeling approach is based on organometallic ^{99m}Tc carbonyl complexes, which are characterized by high stability and can be formed in high specific activity due to the d^6 electron configuration of the Tc(I) metal center. The active species in this process is the aquoion ^{99m}Tc(CO)$_3$(H$_2$O)$_3^+$, which exchanges water molecules with mono-, di-, and tridentate chelators to form stable complexes. The easy access to the [^{99m}Tc(CO)$_3$]$^+$ core has been reported [53], after synthesis of Tc(I) and Re(I) complexes [M(H$_2$O)$_3$(CO)$_3$]$^+$ (M = ^{99m}Tc and ^{188}Re) by direct reduction of ^{99m}Tc-pertechnetate with sodium borohydride in aqueous solution. Usually, histidine is of particular interest as the labeling site, since it is a natural amino acid, or Nα-His moiety [54]. However, many of these analogues did not find practical application due to complicated labeling procedures and/or unfavorable pharmacokinetics.

The new quality in the investigation of possibilities for ^{99m}Tc labeling of small peptides was found in the application of HYNIC core with *N*-hydroxysuccinimidyl

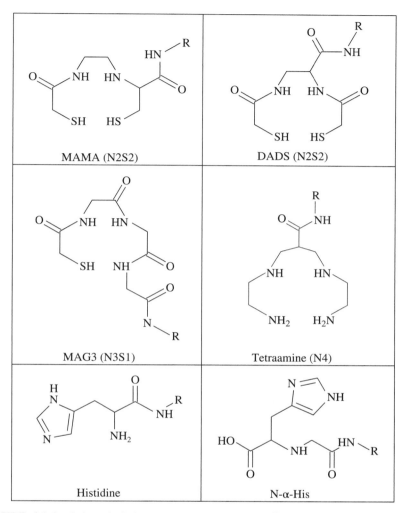

FIGURE 4.1.4 Selected chelators used in the indirect ^{99m}Tc labeling of somatostatin analogues for SRS-SPECT.

hydrazinonicotinamide (NHS-HYNIC, HYNIC) as a BCA precursor [55]. It has initially been developed for radiolabeling of polyclonal immunoglobulin [56] and was then recommended for preparation of hydrazino-modified proteins and synthesis of ^{99m}Tc–protein conjugates [57] and chemotactic peptides [58].

Initially, tricine (*N*-[tris(hydroxymethyl)methyl]glycine) was used as coligand for ^{99m}Tc in the HYNIC core [59]. It was assumed that the ^{99m}Tc species is coordinated by two tricine molecules and the terminal N-atom of the hydrazine group of HYNIC in the resulting ^{99m}Tc-HYNIC–protein complex [60]. Detailed HPLC analysis indicated that the complex can reversibly adopt various forms, dependent on the temperature, reaction time, and pH. Replacement of tricine by other coligands such as ethylenedi-amine-N,N'-diacetic acid (EDDA) resulted in more stable complexes and lower number of isomers [61, 62]. Study of potential structures by LC-MS confirmed that HYNIC may function as a monodentate or a bidentate chelator [63, 64]. Therefore, ^{99m}Tc labeling is performed in the presence of one or more coligands, which saturate the hexacoordinate coordination sphere of the Tc(V) core with donor groups such as amine, carboxylate, or hydroxyl [40]. The HYNIC core has become one of the most popular and effective BCA used for ^{99m}Tc labeling of somatostatin analogues.

The coligands studied for ^{99m}Tc labeling of somatostatin analogues were EDDA, tricine, nicotinic acid, and combinations thereof. Changing the coligand can signifi-cantly affect the lipophilicity of the complex and allows to modify its biodistribution. Several studies have been published on the ^{99m}Tc labeling of octreotide via HYNIC in combination with different coligands [65, 66].[99m]Tc-HYNIC-TOC after labeling with ^{99m}Tc using tricine and EDDA as coligands retained its receptor affinity as deter-mined *in vitro* in rat brain cortex membranes and showed favorable biodistribution *in vivo* in tumor-bearing animals [67, 68]. In animal models, the tracer accumulation ratio in the tumor compared to the kidneys and liver was higher than in the case of ^{111}In-DTPA-octreotide [69].

KITS FOR ^{99m}Tc-LABELED SOMATOSTATIN ANALOGUES

There are considerable advantages of having the somatostatin analogues delivered to the hospital in the form of dry kits allowing their labeling with ^{99m}Tc directly before administration to the patient. The kits offer high and reproducible labeling yields combined with the long shelf life. They contain all chemical components, including the reducing agent, the ligand-functionalized peptide, as well as potential supporting ligands, buffers, and other excipients, in lyophilized form, and the labeling reaction is initiated by adding the generator eluate ([99m]TcO$_4^-$ in saline) to the kit vial.

The direct ^{99m}Tc labeling of HYNIC-derivatized peptides in presence of EDDA resulted in low labeling yields [49, 60, 70] unsuitable for a routine clinical use, while the labeling yield in presence of tricine (N-[tris(hydroxymethyl)methyl]glycine) as coligand was very high at room temperature and could be achieved in a short time. To overcome these limitations, the single-step and two-step kit approaches were developed [71], both demanding incubation at elevated temperature. Based on the labeling approach utilizing ligand exchange from tricine to EDDA at the elevated temperature with almost quantitative yields in a short time [61], the kit formulation has been developed after optimization of pH of freeze-drying solution and the content of stannous chloride. Mannitol as the bulking agent was added to improve the features of freeze-dried pellet and ensure rapid dissolution of all reagents in the labeling process [72]. Manufacturing conditions leading to the optimized freeze-dried composition of a kit for ^{99m}Tc-EDDA/HYNIC-TOC radiopharmaceutical were also described [73], leading to the radiopharmaceutical of good imaging properties.

Similarly to other kits for ^{99m}Tc radiopharmaceuticals, it is generally assumed that the peptide-based radiopharmaceutical can be obtained in the ^{99m}Tc labeling procedure lasting 30–60 min. Most of the labeling reactions are performed at 80–100°C (with the exception of N_4-derivatized peptides that are labeled at room temperature). The radiolabeled product is sterile and can be administered to patients without any further purification steps. The administration of the radiopharmaceutical should not induce pharmacologic effects.

MODIFICATIONS OF OCTREOTIDE FOR SPECT-SRS

The pharmacophore of octreotide consists of the sequence -Phe-(D)Trp-Lys-Thr- (see Fig. 4.1.5), which is forced into a β-turn conformation by a disulfide bridge formed between the two cysteinyl residues close to the N- and C-terminus, respectively, thus mimicking the spatial disposition of the corresponding amino acids in native somatostatin-14. In order to maintain high sst-binding affinity, the pharmacophore of an sst ligand needs to be conformationally constrained [74].

Also, the presence of the ε-amino group of Lys^5 is a prerequisite for the biological activity and especially for receptor recognition. Hence, modification of one of the phenylalanine residues through replacement by tyrosine or derivatization of the α-amino group of (D)-Phe^1 was the most reasonable approach. The hydroxy group present in tyrosine allows an electrophilic reaction of the aromatic ring with iodine isotopes, while coupling of an appropriate chelator at the N-terminus allows stable binding of certain metallic radionuclides.

Modifications of the amino acid sequence of octreotide analogues, that is, the replacement of Phe^3 by Tyr^3 that was useful for radioiodination, appeared to also improve sst_2 affinity, while the C-terminal substitution of Thr(ol) by Thr resulted in a further improvement of sst_2 affinity, to a higher rate of internalization, and to a higher tumor uptake in animal models [11]. The preclinical comparison of [111]In- and ^{99m}Tc-labeled somatostatin analogues—OC, TOC, and [Tyr^3, Thr^8]-octreotide (TATE) with either DOTA or HYNIC as chelators [71]—showed that all of them were characterized by specific internalization in AR42J rat pancreatic tumor cells known to express the rat sst_2 [75]. However, there was distinct tendency of increasing internalization rates from OC through TOC to TATE, which was also clearly reflected in the tumor uptake in AR42J tumor-bearing rats. Uptake in the experimental tumor and in the

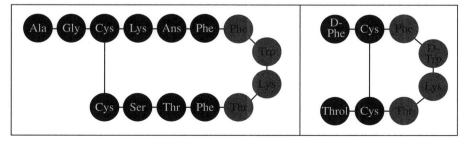

FIGURE 4.1.5   The amino acid sequence of native somatostatin-(14) and octreotide.

pancreas correlated well with the rate of internalization determined *in vitro*. Based on these results, the authors concluded that radioligands based on TATE are superior for *in vivo* sst_2-tumor targeting to those based on TOC and octreotide, respectively. Consequently, it was assumed that the higher hydrophilicity and the higher tumor-to-liver ratio of ^{99m}Tc-EDDA/HYNIC-TATE compared to ^{99m}Tc-EDDA/HYNIC-TOC may consequently lead to improved sensitivity in the detection of liver metastases. This assumption, however, was not confirmed in human study comparing ^{111}In-DOTA-TOC and ^{111}In-DOTA-TATE. Both tracers showed the expected high specific uptake in somatostatin receptor-positive tissue although better visualization of some liver metastases was found with ^{111}In-DOTA-TOC and a significantly higher mean absorbed dose to the liver was found for ^{111}In-DOTA-TATE [76].

The final verification of the diagnostic efficacy of ^{99m}Tc-EDDA/HYNIC-TOC and ^{99m}Tc-EDDA/HYNIC-TATE was performed by direct comparison of SRS using both tracers in the uniform group of 12 patients with confirmed GEP-NET [77]. Both ^{99m}Tc-EDDA/HYNIC-TOC and ^{99m}Tc-EDDA/HYNIC-TATE were found to be useful radiopharmaceuticals for SRS-SPECT, in neuroendocrine tumors, especially those expressing sst_2. Similar number of metastatic lesions was detected using either agent; 85% correlation was found when analyzing each of metastases individually. The uptake of ^{99m}Tc-HYNIC-TOC in the liver was higher than in the case of ^{99m}Tc-HYNIC-TATE, but the ratio of uptake in the lesion to background was comparable. No significant differences were observed in the uptake of these agents in the tumors and in the kidneys.

Promising preclinical results were obtained also with the ^{99m}Tc-Demotate series (e.g., [^{99m}Tc-N_4^0, Tyr^3]octreotate, ^{99m}Tc-Demotate 1) [78, 79] or with [^{99m}Tc-N_4^{0-1},Asp^0,Tyr^3]octreotate, ^{99m}Tc-Demotate 2, during a preclinical comparison with [^{111}In]DOTA-TATE in the detection of sst_2-positive tumors [80]. ^{99m}Tc-labeled octreotide analogues have been developed and clinically evaluated for SRS-SPECT imaging, such as HYNIC-TOC [81], HYNIC-TATE [82–84], and ^{99m}Tc-Demotate 1 [85, 86].

DEVELOPMENTS IN SRS-SPECT TRACERS

All of the so far mentioned analogues have high affinity for sst_2. The search for other somatostatin-based peptides having affinity for a broader range of somatostatin receptor subtypes, which might target a broader spectrum of tumors but also have a higher net tumor uptake, has been continued [87]. Several new compounds have been developed that show high affinity to sst_2, sst_3, and sst_5. These new compounds were modified at position 3 of octapeptide, the best of them containing the unnatural amino acids 1-naphthyl-alanine ^{111}In-DOTA-NOC (1-NaI^3-octreotide) [88, 89], ^{111}In-DOTANOC-ATE (1-NaI^3-Thr^8-octreotide), benzothienyl-alanine (DOTA-BOC) and ^{111}In-DOTABOC-ATE (Bz-Thi^3-Thr^8-octreotide); however, their applications were limited [90]. Modification of the physical properties of somatostatin analogues has led to the development of bicyclic somatostatin analogues with affinity to sst_2, sst_3, and sst_5, such as AM3 (DOTA-)Tyr-cyclo(DAB-Arg-cyclo(Cys-Phe-D-Trp-Lys-Thr-Cys)). These molecules show fast background clearance and high tumor-to-nontumor ratios are soon obtained; therefore, they look ideal for imaging with short-lived radionuclides such as ^{68}Ga [91]. Pansomatostatin radiopeptides with high-affinity binding for all five receptor subtypes have also been developed. The first such peptide, KE108 (Tyr-cyclo(DAB-Arg-Phe-Phe-D-Trp-Lys-Thr-Phe)), was modified by replacing Tyr as a prosthetic group for iodination NH_2-terminally [92] by DOTA, resulting in the analogue KE88 [93]. ^{111}In-KE88 was able to bind with high affinity to all five receptor subtypes (sst$_1$–sst$_5$) but was efficiently internalized only in sst$_3$-expressing cells. The sst$_3$-expressing tumors had a high and persistent tumor uptake, whereas the sst$_2$-expressing tumors showed low uptake and fast washout *in vivo*. It did not appear to offer multisubtype imaging properties, since the *in vitro* internalization and *in vivo* uptake in sst$_2$ tumors were very low, compared to sst3 tumors.

Presented studies have been based on the development of radiolabeled somatostatin agonists, assuming that the internalization of the receptor after radioligand binding is critical for efficient retention of the tracer in tumor cells, allowing for efficient imaging and therapy. The molecular–pharmacologic investigations showed that efficient internalization is usually provided by agonists [94]. Recent developments have indicated that receptor antagonists may be as good or even better than agonists for such purposes. A recent study showed that high-affinity somatostatin receptor antagonists that poorly internalize into tumor cells can, in terms of *in vivo* uptake in animal tumors, perform equally good or better than corresponding agonists, which highly internalized into tumor cells. The same tendency was seen for both sst$_2$ and sst$_3$ selective analogues, suggesting that this observation may be valid for more than just one particular G-protein-coupled receptor. The study demonstrated that the sst antagonists are preferable for *in vivo* tumor targeting [95]. The first clinical evaluation of SRS with an antagonist confirmed the preclinical data, as it showed higher tumor uptake of the antagonist ^{111}In-DOTA-sst$_2$-ANT (p-NO$_2$-Phe-cyclo(D-Cys-Tyr-D-Trp-Lys-Thr-Cys)D-Tyr-NH$_2$) compared to the agonist ^{111}In-DTPA^0-octreotide and improved tumor-to-background ratios, in particular tumor-to-kidney [96].

CLINICAL UTILITY OF SRS-SPECT IN NEUROENDOCRINE TUMORS

In vivo imaging of various tumors that overexpress somatostatin receptors, especially of neuroendocrine origin, using radiolabeled somatostatin analogs has become an accepted clinical tool in oncology [97]. Since its introduction, many applications of ^{111}In-pentetreotide have been reported. It has been successfully used in patients to visualize somatostatin receptor-positive tumors, sometimes identifying metastases that were not detected by conventional methods [16, 98].

The sensitivity and specificity of the ^{111}In-pentetreotide SRS-SPECT in the diagnosis of intestinal neuroendocrine tumors, including primary foci and metastatic lesions, were assessed as 71–96% and 76–95%, respectively [99–101]. On the basis of the 162 SRS studies with gastric carcinoid in a course of MEN-1 syndrome, the sensitivity of 75%, specificity of 95%, and negative predictive value of 97% were calculated in positive gastric SRS localization using ^{111}In-OctreoScan [102]. The potential value of SRS in identifying patients with gastric carcinoids in this study was supported by the analysis of value of clinical and laboratory characteristics in such patients. Except for the presence of MEN-1, no characteristics were helpful in identifying which patients might have gastric carcinoid. Therefore, the availability of SRS was assessed as very important in the treatment of patients with hypergastrinemic states and other diseases with an increased incidence of gastric carcinoids.

SRS allows the localization of distant metastases that are not visualized by other structural diagnostic methods revealing the superiority of SRS over CT or MRI in NET staging. SRS-SPECT with the use of ^{99m}Tc-labeled somatostatin analogues is very helpful in GEP-NET for staging, for primary focus and recurrent disease detection, and for therapy follow-up (Fig. 4.1.6).

Superimposed CT/MRI scans and scintigraphy images with fusion image techniques or the use of SPECT/CT hybrid devices provide optimal imaging. These techniques enable a precise anatomical localization of the lesion visible in SRS and improve sensitivity in differentiating lesions with physiological and pathological accumulation of the tracer. The combined anatomical–functional imaging with SPECT/CT and additional visual correlation to high-end CT has a high diagnostic accuracy and significantly improves tumor localization and characterization in patients with NET [103].

In comparison to ^{111}In-labeled agents, ^{99m}Tc-labeled radiopharmaceuticals show many advantages. In the group of 41 patients with somatostatin receptor-positive tumors, also higher sensitivity of ^{99m}Tc-EDDA/HYNIC-TOC in neuroendocrine tumor detection than that of the ^{111}In-pentetreotide (65.9 vs. 51.2%) was found [104]. More focal lesions were detected, especially small liver and lymph node metastases (105 vs. 91 lesions). The authors concluded that ^{99m}Tc-EDDA/HYNIC-TOC combines the advantages of favorable pharmacokinetics, higher spatial resolution, lower radiation dose, and improved availability of ^{99m}Tc with a simplified imaging procedure

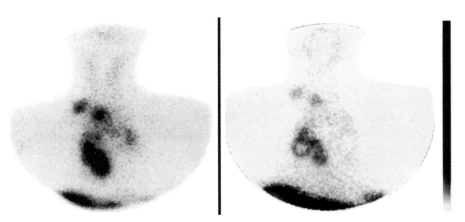

FIGURE 4.1.6 Anterior chest images obtained 3 h after administration of 370 MBq of ^{99m}Tc-HYNIC-TOC in a patient with metastatic medullary thyroid cancer. Before therapy (left panel) and after one cycle of therapy with ^{90}Y-DOTA-TOC (right panel) showing partial necrosis of some metastases and complete disappearance of others.

and could replace ^{111}In-OctreoScan for routine SRS. ^{99m}Tc-labeled somatostatin analogues are particularly helpful in detecting small tumors or tumors bearing a low density of somatostatin receptors.

Parisella and Signore [105] evaluated the use of ^{99m}Tc-EDDA/HYNIC-TOC in neuroendocrine tumors and concluded that SRS with ^{99m}Tc-EDDA/HYNIC-TOC is highly indicated for *in vivo* histological characterization of known NET lesions, previously identified by other imaging modalities or biopsy, to plan appropriate therapy especially for patients with inoperable disease.

Scintigraphy using ^{99m}Tc-EDDA/HYNIC-TOC was proved to be useful in the diagnostic of metastatic medullary thyroid cancer (MTC); however, it's important to note that sst2 expression is downregulated with the advancement of MTC. The method appeared to be more efficient than other radioisotope or radiologic techniques taking into account the number of detected metastatic foci of MTC [106]. In a significantly larger group of MTC patients, the sensitivity was 74.1%, therefore better than in the case of other diagnostic modalities [107–110].

CONCLUSIONS

SRS using SPECT radiopharmaceuticals opened a new era in the field of molecular imaging. When these new tracers were developed, they changed the approach to diagnosis and treatment of neuroendocrine tumors. In these days, however, the availability of ^{68}Ga-labeled sst-affine analogues for PET imaging improved significantly diagnostic sensitivity, and when available, they now represent the best option.

Not all nuclear medicine centers, however, have access to PET facilities and ^{68}Ga-labeled peptides, and therefore, SRS-SPECT remains a suitable and economical alternative, particularly when used for the molecular characterization of known lesions (larger than 1 cm), for differential diagnosis, and for therapy decision making but not for the detection of unknown sites of disease.

The use of SRS-SPECT for the study of inflammatory diseases is still an option, especially for the study of the state of activity of the disease and to early evaluation of therapy efficacy.

REFERENCES

[1] Behr, T. M.; Gotthardt, M.; Barth, A.; Behe, M. Quarterly Journal of Nuclear Medicine 2001, 45, 189–200.

[2] Fani, M.; Maecke, H. R. European Journal of Nuclear Medicine and Molecular Imaging 2012, 39 (Suppl 1), S11–S30.

[3] Fani, M.; Maecke, H. R.; Okarvi, S. M. Theranostics 2012, 2, 481–501.

[4] Lamberts, S. W. J.; Krenning, E. P.; Reubi, J. C. Endocrine Reviews 1991, 12, 450–482.

[5] Bauer, W.; Briner, U.; Doepfner, W.; et al. Life Sciences 1982, 31, 1133–1140.

[6] Pless, J.; Bauer, W.; Briner, U.; et al. Scandinavian Journal of Gastroenterology Supplement 1986, 119 (suppl. 119), 54–64.

[7] Murphy, W.; Lance, V. A.; Moreau, S. Life Sciences 1987, 40, 2515–2522.

[8] Veber, D. F.; Saperstein, R.; Nutt, R. F. Life Sciences 1984, 34, 1371–1378.

[9] Pless, J.; Bauer, W.; Briner, U.; et al. Scandinavian Journal of Gastroenterology 1986, 21(suppl. 119), 54–64.

[10] Breeman, W. A. P.; de Jong, M.; Kwekkeboom, D. J.; et al. European Journal of Nuclear Medicine 2001, 28, 1421–1429.

[11] Reubi, J. C.; Schar, J. C.; Waser, B.; et al. European Journal of Nuclear Medicine 2000, 27, 273–282.

[12] Bakker, W. H.; Krenning, E. P.; Breeman, W. A.; et al. Journal of Nuclear Medicine 1990, 31, 1501–1509.

[13] Bakker, W. H.; Krenning, E. P.; Breeman, W. A. P.; et al. Journal of Nuclear Medicine 1991, 32, 1184–1189.

[14] Kvols, L. K.; Brown, M. L.; O'Connor, M. K.; et al. Radiology 1993, 187, 129–133.

[15] Krenning, E. P.; Bakker, W. H.; Kooij, P. P. M.; et al. Journal of Nuclear Medicine 1992, 33, 652–658.

[16] Krenning, E. P.; Kewkkeboom, D. W.; Bakker, W. H.; et al. European Journal of Nuclear Medicine 1993, 20, 716–731.

[17] Hökfelt, T. Neuron 1991, 7, 867–879.

[18] Vallabhajosula, S.; Moyer, B. R.; Lister-James, J.; et al. Journal of Nuclear Medicine 1996, 37, 1016–1022.

[19] Menda, Y.; Kahn, D. Seminars in Nuclear Medicine 2002, 32, 92–96.

[20] Lebtahi, R.; Le Cloirec, J.; Houzard, C.; et al. Journal of Nuclear Medicine 2002, 43, 889–895.

[21] Bangard, M.; Béhé, M.; Guhlke, S.; et al. European Journal of Nuclear Medicine 2000, 27, 628–637.

[22] Decristoforo, C.; Mather, S. J.; Cholewinski, W.; Donnemiller, E.; Riccabona, G.; Moncayo, R. European Journal of Nuclear Medicine 2000, 27, 1318–1325.

[23] Płachcińska, A.; Mikołajczak, R.; Maecke, H. R.; et al. Cancer Biotherapy and Radiopharmaceuticals 2004, 19, 261–270.

[24] Płachcińska, A.; Mikołajczak, R.; Maecke, H. R.; et al. European Journal of Nuclear Medicine and Molecular Imaging 2004, 31, 1005–1010.

[25] Płachcińska, A.; Mikołajczak, R.; Maecke, H.; et al. Cancer Biotherapy and Radiopharmaceuticals 2004, 19, 613–620.

[26] Płachcińska, A.; Mikołajczak, R.; Kozak, J.; Rzeszutek, K.; Kuśmierek, J. Cancer Biotherapy and Radiopharmaceuticals 2006, 21, 61–67.

[27] Płachcińska, A.; Mikołajczak, R.; Kozak, J.; Rzeszutek, K.; Kuśmierek, J. European Journal of Nuclear Medicine and Molecular Imaging 2006, 33, 1041–1047.

[28] Płachcińska, A.; Mikołajczak, R.; Kozak, J.; Rzeszutek, K.; Kuśmierek, J. Nuclear Medicine Review 2004, 7, 143–150.

[29] Płachcińska, A.; Mikołajczak, R.; Kozak, J.; Rzeszutek, K.; Kuśmierek, J. Nuclear Medicine Review 2006, 9, 24–29.

[30] Stimmel, J. B.; Stockstill, M. E.; Kull, F. C. Bioconjugate Chemistry 1995, 6, 219–225.

[31] Liu, S.; Edwards, D. Bioconjugate Chemistry 2001, 12, 7–34.

[32] Viola-Villegas, N.; Doyle, R. P. Coordination Chemistry Reviews 2009, 253, 1906–1925.

[33] Paganelli, G.; Zoboli, S.; Cremonesi, M.; et al. European Journal of Nuclear Medicine 2001, 28, 426–434.

[34] Bodei, L.; Cremonesi, M.; Grana, C.; et al. European Journal of Nuclear Medicine and Molecular Imaging 2004, 31, 1038–1046.

[35] Mäcke, H. R.; Smith-Jones, P.; Maina, T.; et al. Hormone and Metabolic Research Supplement Series 1993, 27, 12–17.

[36] Stolz, B.; Smith-Jones, P. M.; Albert, R.; Reist, H.; Maecke, H.; Bruns, C. Hormone and Metabolic Research 1994, 26, 453–459.

[37] Fischman, A. J.; Babich, J. W.; Strauss, H. W. Journal of Nuclear Medicine 1993, 34, 2253–2263.

[38] Breeman, W. A.; Kwekkeboom, D. J.; Kooij, P. P.; et al. Journal of Nuclear Medicine 1995, 36, 623–627.

[39] Liu, S.; Edwards, D. S.; Barret, J. A. Bioconjugates Chemistry 1997, 8, 621–636.

[40] Liu, S.; Edwards, D. S. Chemical Reviews 1999, 99, 2235–2268.

[41] Hnatowich, D. J. Seminars in Nuclear Medicine 1990, 20, 80–91.

[42] Srivastava, S. C.; Mease, R. C. International Journal of Radiation Applications & Instrumentation B 1991, 18, 589–603.

[43] Eckelman, W. C. Cancer Research 1990, 3, 686–692.

[44] Thakur, M. L.; Eshbach, J.; Wilder, S.; John, E.; MsDevit, M. R. Journal of Labelled Compounds & Radiopharmaceuticals 1993, 32, 365–367.

[45] Wester, J. H., ed., *Pharmaceutical Radiochemistry*, SCINTOMICS, Munich, Vol. 1, 2010

[46] Liu, S. Advances in Drug Delivery Reviews 2008, 60, 1347–1370.

[47] Pearson, D. A.; Lister-James, J.; McBride, W. J.; et al. Journal of Medicinal Chemistry 1996, 39, 1361–1371.

[48] Blum, J.; Handmaker, H.; Lister-James, J.; Rinne, B. Chest 2000, 117, 1232–1238.

[49] Decristoforo, C.; Mather, S. J. Bioconjugate Chemistry 1999, 10, 431–438.

[50] Maina, T.; Stolz, B.; Albert, R.; Nock, B.; Bruns, C.; Maecke, H. European Journal of Nuclear Medicine 1994, 21, 437–444.

[51] Nicolini, M., Bandoli, G., Mazzi, U., ed., *Technetium and Rhenium Chemistry and Nuclear Medicine*, SGE Editoriali, Padova, 1995, p. 395–400.

[52] Thakur, M. L.; Kolan, H.; Li, J.; et al. Nuclear Medicine and Biology 1997, 24, 105–113.

[53] Alberto, R. European Journal of Inorganic Chemistry 2009, 1, 21–31.

[54] Egli, A.; Alberto, R.; Tannahill, L.; et al. Journal of Nuclear Medicine 1999, 40, 1913–1917.

[55] Krois, D.; Riedel, Ch.; Angelberger, P.; Kalchauser, H.; Virgolini, I.; Lehner, H. Liebigs Annalen 1996, 9, 1463–1469.

[56] Abrams, M. J.; Juwied, M.; tenKate, C. I.; et al. Journal of Nuclear Medicine 1990, 31, 2022–2028.

[57] Schwartz, D. A.; Abrams, M. J.; Hauser, M. M.; et al. Bioconjugate Chemistry 1991, 2, 333–336.

[58] Babich, J. W.; Solomon, H.; Pike, M. C.; et al. Journal of Nuclear Medicine 1993, 34, 1964–1974.

[59] Rennen, H. J.; Boerman, O. C.; Koenders, E. B.; Oyen, W. J.; Corstens, F. H. Nuclear Medicine and Biology 2000, 27, 599–604.

[60] Liu, S.; Edwards, D. S.; Looby, R. J.; et al. Bioconjugate Chemistry 1996, 7, 63–71.

[61] Von Guggenberg, E.; Sarg, B.; Lindner, H.; et al. Journal of Labelled Compounds & Radiopharmaceuticals 2003, 46, 307–318.

[62] Biechlin, M. L.; Bonmartin, A.; Gilly, F. N.; Fraysse, M.; du Moulinet d'Hardemare, A. Nuclear Medicine and Biology 2008, 35, 679–687.

[63] King, R. C.; Bashir-Uddin Surfraz, M.; Biagini, S. C.; Blower, P. J.; Mather, S. J. Dalton Transactions 2007, 43, 4998–5007.

[64] Meszaros, L. K.; Dose, A.; Biagini, S. C. G.; Blower, P. J. Inorganica Chimica Acta 2010, 363, 1059–1069.

[65] Decristoforo, C.; Melendez-Alafort, L.; Sosabowski, J. K.; Mather, S. J. Journal of Nuclear Medicine 2000, 41, 1114–1119.

[66] Decristoforo, C.; Mather, S. J. Nuclear Medicine and Biology 1999, 26, 869–876.

[67] Maecke, H. R.; Béhé, M. Journal of Nuclear Medicine 1996, 37, 1144.

[68] Béhé, M.; Maecke, H. R. European Journal of Nuclear Medicine 1995, 22, 791

[69] Decristoforo, C.; Mather, S. J. European Journal of Nuclear Medicine 1999, 26, 869–876.

[70] Pawlak, D.; Korsak, A.; Mikolajczak, R.; Janota, B.; Karczmarczyk, U.; Jakubowska, E. In *Technical Reports Series no 458. Comparative Evaluation of Therapeutic Radiopharmaceuticals*. International Atomic Energy Agency, Vienna, 2007, p. 217–232.

[71] Storch, D.; Behe, M.; Walter, M. A.; et al. Journal of Nuclear Medicine 2005, 46, 1561–1569.

[72] Von Guggenberg, E.; Mikolajczak, R.; Janota, B.; Riccabona, G.; Decristoforo, C. Journal of Pharmaceutical Sciences 2004, 93, 2497–2506.

[73] Gonzales-Vazquez, A.; Ferro-Flores, G.; de Murphy, A. C.; Gutierrez-Garcia, Z. Applied Radiation & Isotopes 2006, 64, 792–797.

[74] Huang, Z.; He, Y. B.; Raynor, K.; Tallent, M.; Reisine, T.; Goodman, M. Journal of American Chemical Society 1992, 114, 9390–9401.

[75] Froidevaux, S.; Heppeler, A.; Eberle, A. N.; et al. Endocrinology 2000, 141, 3304–3312.

[76] Forrer, F.; Uusijarvi, H.; Waldherr, C.; et al. European Journal of Nuclear Medicine and Molecular Imaging 2004, 31, 1257–1262.

[77] Ćwikła, J. B.; Mikołajczak, R.; Pawlak, D.; et al. Journal of Nuclear Medicine 2008, 49, 1060–1065.

[78] Nikolopoulou, A.; Maina, T.; Sotiriou, P.; Cordopatis, P.; Nock, B. A. Journal of Peptide Science 2006, 12, 124–131.

[79] Maina, T.; Nock, B.; Nikolopoulu, A.; et al. European Journal of Nuclear Medicine 2002, 29, 742–753.

[80] Maina, T.; Nock, B. A.; Cordopatis, P.; et al. European Journal of Nuclear Medicine and Molecular Imaging 2006, 33, 831–840.

[81] Płachcińska, A.; Mikołajczak, R.; Maecke, H. R.; et al. European Journal of Nuclear Medicine and Molecular Imaging 2003, 30, 1402–1406.

[82] Hubalewska-Dydejczyk, A.; Fröss-Baron, K.; Mikołajczak, R.; et al. European Journal of Nuclear Medicine and Molecular Imaging 2006, 33, 1123–1133.

[83] Hubalewska-Dydejczyk, A.; Fröss-Baron, K.; Gołkowski, F.; Sowa-Staszczak, A.; Mikołajczak, R.; Huszno, B. Experimental and Clinical Endocrinology and Diabetes 2007, 115, 47–49.

[84] Hubalewska-Dydejczyk, A.; Szybiński, P.; Fröss-Baron, K.; Mikołajczak, R.; Huszno, B.; Sowa-Staszczak, A. Nuclear Medicine Review 2005, 8, 155–156.

[85] Gabriel, M.; Decristoforo, C.; Maina, T.; Nock, B.; von Guggenberg, E.; Cordopatis, P.; Moncayo, R. Cancer Biotherapy and Radiopharmaceuticals 2004, 19, 73–79.

[86] Decristoforo, C.; Maina, T.; Nock, B.; Gabriel, M.; Cordopatis, P.; Moncayo, R. European Journal of Nuclear Medicine and Molecular Imaging 2003, 30, 1211–1219.

[87] Reubi, J. C.; Maecke, H. R. Journal of Nuclear Medicine 2006, 49, 1735–1738.

[88] Wild, D.; Schmitt, J. S.; Ginj, M.; et al. European Journal of Nuclear Medicine and Molecular Imaging 2003, 30, 1338–1347.

[89] Wild, D.; Maecke, H. R.; Waser, B.; et al. European Journal of Nuclear Medicine and Molecular Imaging 2005, 32, 724.

[90] Ginj, M.; Chem, J.; Walter, M. A.; Eltschinger, V.; Reubi, J. C.; Maecke, H. R. Clinical Cancer Research 2005, 11, 1136–1145.

[91] Fani, M.; Mueller, A.; Tamma, M. L.; et al. Journal of Nuclear Medicine 2010, 51, 1771–1779.

[92] Reubi, J. C.; Eisenwiener, K. P.; Rink, H.; Waser, B.; Maecke, H. R. European Journal of Pharmacology 2002, 456, 45–49.

[93] Ginj, M.; Zhang, H.; Eisenwiener, K. P.; et al. Clinical Cancer Research 2008, 14, 2019–2027.

[94] Cescato, R.; Schulz, S.; Waser, B.; et al. Journal of Nuclear Medicine 2008, 51, 4030–4037.

[95] Ginj, M.; Zhang, H.; Waser, B.; et al. Proceedings of the National Academy of Sciences of the USA 2006, 103, 16436–16441.

[96] Wild, D.; Fani, M.; Behe, M.; et al. Journal of Nuclear Medicine 2011, 52, 1412–1417.

[97] Eising, E. G.; Bier, D.; Kunst, E. J.; Reiners, C. Radiologie 1996, 36, 81–88.

[98] Dörr, U.; Räth, U.; Sautter-Bihl, M. L.; et al. European Journal of Nuclear Medicine 1993, 20, 431–433.

[99] Kwekkeboom, D. J.; Krenning, E. P.; Bakker, W. H.; Oei, H. Y.; Kooij, P.P.; Lamberts, S. W. European Journal of Nuclear Medicine 1993, 20, 283.

[100] Kälkner, K. M.; Jansson, E. T.; Nilsson, S. Cancer Research 1995, 55, 5801.

[101] Westlin, J.; Janson, E. T.; Arnberg, H.; Ahlstrom, H.; Oberg, K.; Nilsson, S. Acta Oncologica 1993, 32, 783.

[102] Gabriel, M.; Muehllechner, P.; Decristoforo, C.; et al. Quarterly Journal of Nuclear Medicine and Molecular Imaging 2005, 49, 237–244.

[103] Pfannenberg, A. C.; Eschmann, S. M.; Horger, M.; et al. European Journal of Nuclear Medicine and Molecular Imaging 2003, 30, 835–843.

[104] Gabriel, M.; Decristoforo, C.; Donnemiller, E.; et al. Journal of Nuclear Medicine 2003, 44, 708–716.

[105] Parisella, M. G.; Chianelli, M.; D'Alessandria, C.; et al. Quarterly Journal of Nuclear Medicine 2012, 56, 90–98.

[106] Parisella, M. G.; D'Alessandria, C.; Van de Bossche, B.; et al. Cancer Biotherapy and Radiopharmaceuticals 2004, 19, 211–217.

[107] Czepczyński, R.; Parisella, M. G.; Kosowicz, J.; et al. European Journal of Nuclear Medicine and Molecular Imaging 2007, 34, 1635–1645.

[108] Czepczyński, R.; Kosowicz, J.; Ziemnicka, K.; Mikołajczak, R.; Gryczyńska, M.; Sowiński, J. Endokrynologia Polska 2006, 4, 431–435.

[109] Czepczyński, R.; Kosowicz, J.; Mikołajczak, R.; Ziemnicka, K.; Gryczyńska, M.; Sowiński, J. Polskie Archiwum Medycyny Wewnętrznej 2006, 116, 853–860.

[110] Kosowicz, J.; Mikołajczak, R.; Czepczyński, R.; Ziemnicka, K.; Gryczyńska, M.; Sowiński, J. Cancer Biotherapy and Radiopharmaceuticals 2007, 5, 613–628.

4.2

MOLECULAR IMAGING OF SOMATOSTATIN RECEPTOR-POSITIVE TUMORS USING PET/CT

RICHARD P. BAUM AND HARSHAD R. KULKARNI

THERANOSTICS Center for Molecular Radiotherapy and Molecular Imaging, ENETS Center of Excellence, Zentralklinik Bad Berka, Germany

ABBREVIATIONS

APUD	amine precursor uptake and decarboxylation
BBQ-MIT	Bad Berka Molecular Imaging Tool
CNL	cognition network language
CT	computed tomography
CUP	cancer of unknown provenience
DOTA-BOC-ATE	(DOTA-BzThi^3, Thr^8)-octreotide
DOTANOC	DOTA-1-NaI^3-octreotide
DOTA-NOC-ATE	(DOTA-1-NaI^3, Thr^8)-octreotide
DOTATATE	DOTA-D-Phe^1-Tyr^3-Thr^8-octreotide
DOTATOC	DOTA-D-Phe^1-Tyr^3-octreotide
GEP	gastroenteropancreatic system
MORE	molecular response
MRI	magnetic resonance imaging
MTD	molecular tumor diameter
MTV	molecular tumor volume
NET	neuroendocrine tumor

Somatostatin Analogues: From Research to Clinical Practice, First Edition. Edited by Alicja Hubalewska-Dydejczyk, Alberto Signore, Marion de Jong, Rudi A. Dierckx, John Buscombe, and Christophe Van de Wiele.

PET   positron emission tomography
pNETs   pancreatic NETs
PRRT   peptide receptor radionuclide therapy
SI-NETs   small intestinal (ileum/jejunum/duodenum) NETs
SMS   somatostatin
SSTR   somatostatin receptor
SUV   standardized uptake value
SUV T/S   SUVmax tumor-to-spleen ratio
TACE   transarterial chemoembolization
USG   ultrasonography
WHO   World Health Organization

INTRODUCTION

Neuroendocrine tumors (NETs) are heterogeneous group of tumors originating from pluripotent stem cells or differentiated neuroendocrine cells. The hallmark of NETs is the expression of somatostatin receptors (SSTRs), which forms the rationale for the use of radiolabeled somatostatin (SMS) analogues for imaging and therapy. Another unique feature is their endocrine metabolism, that is, decarboxylation of amine precursors, hence previously referred to as amine precursor uptake and decarboxylation (APUD)-omas. NETs occur predominantly in the lungs and the gastroenteropancreatic system (GEP). Taking into account the complex and diverse histology of NETs and to allow optimal prognostic stratification, a new system of classification was devised in 2010 by the WHO (Table 4.2.1) [1].

The heterogeneous nature, the indolent course, and the possibility of multiple and variable anatomic site of primary make it difficult to evaluate patients with NETs. The clinical manifestations due to the secretion of a wide range of biogenic amines typify NETs. However, they do not provide adequate information to allow the clinician to decide upon a treatment regime, which therefore requires imaging.

TABLE 4.2.1   WHO 2010 classification of NET with corresponding values for mitoses per 10 high-power fields (HPF) and proliferation rate (as Ki-67/MIB-1 index in %), determining the grade (G) of NET

Categories	Mitoses (/10HPF)	Ki-67/MIB-1 index (%)
Neuroendocrine tumor		
G1	<2	≤2
G2	2–20	3–20
Neuroendocrine carcinoma		
G3	>20	>20
Mixed adenoneuroendocrine carcinoma (MANEC)		
Hyperplastic and preneoplastic lesion		

Computed tomography (CT) scans, magnetic resonance imaging (MRI), and endoscopic ultrasonography (USG) (EUS) are the morphological imaging modalities in the diagnostics of NETs. However, these do not give the functional status of the tumor, which is often essential for defining the prognosis, as molecular changes precede the morphologic changes. One of the biggest disadvantages of USG is its operator dependency. It also fails to differentiate liver metastases of NETs from other type of liver metastases. Hypervascularity, one of the most common features of NET liver metastases, can be very well documented using Doppler techniques and contrast-enhanced USG. However, these hypervascular solitary lesions may be misdiagnosed as hemangiomas. A combination of functional and morphological imaging often helps in initial diagnosis and staging, deciding upon the treatment regime, and monitoring therapy response [2–4].

The discovery of overexpression of SSTRs for peptide hormones in NETs, more than two decades ago, has revolutionized the role of nuclear medicine in the diagnosis and therapy of NETs [5, 6]. This was indeed paralleled by the development of various radiopharmaceuticals targeting these tumor-related receptors, as well as various metabolic pathways peculiar to the NET cells [7].

Over the past two decades, ^{111}In-DTPA-octreotide scintigraphy has been the foremost functional imaging in the diagnostics of NETs [3, 8].

More recently, molecular imaging using diverse positron emission tomography (PET) radiopharmaceuticals has gained popularity in the diagnostic workup of patients with NET [9–11]. In addition, the amalgamation of PET with CT (PET/CT) enables fast and high-resolution functional imaging with more accurate anatomical localization.

With the ever-growing list of PET tracers currently being employed for the imaging of NETs, it is essential to first describe their respective molecular targets so as to understand and compare the results of this wide range of PET radiopharmaceuticals (Table 4.2.2). The potential molecular events/targets that can be currently targeted by PET-based radiopharmaceuticals are:

a) SSTR expression
b) Serotonin production pathway
c) Biogenic amine storage
d) Catecholamine transport
e) Glucose metabolism
f) Miscellaneous peptide receptor expression

SMS ANALOGUES FOR PET/CT IMAGING

SMS is a cyclic peptide hormone with primary action of inhibition of hormone secretion and modulation of neurotransmission and cell proliferation through specific membrane-bound G-protein-coupled receptors. These SSTRs, which are also normally expressed in different organs such as the pituitary, thyroid, adrenals,

TABLE 4.2.2   Diagnostic PET radiopharmaceuticals for NET

Radiopharmaceutical	Receptor/metabolic target	Indication and comments
^{18}F-FDG	Glycolytic pathway	All NETs. Observation of flip-flop mechanism with SMS-R PET
^{68}Ga-DOTANOC	Somatostatin receptor (pansomatostatin, high affinity for SSTR2, SSTR3, and SSTR5)	All SSTR +ve NETs
^{68}Ga-DOTATOC	Somatostatin receptor (high affinity for SSTR2)	
^{68}Ga-DOTATATE	Somatostatin receptor (highest affinity for SSTR2)	
^{11}C-5-HTP	Serotonin production pathway	All serotonin-producing NETs
^{11}C-DOPA	Dopamine production pathway	Pheochromocytoma, paraganglioma, neuroblastoma; short half-life, cost of production, and difficulty in obtaining ^{11}C major obstacle
^{18}F-DOPA	Dopamine production pathway	Pheochromocytoma, paraganglioma, neuroblastoma, glomus tumor
^{18}F-FDA	Catecholamine precursor	Pheochromocytoma, paraganglioma, neuroblastoma
^{64}Cu-TETA-octretoide	Somatostatin receptor	All SSTR +ve NETs
^{18}F-FP-Gluc-TOCA	Somatostatin receptor	All SSTR +ve NETs
^{11}C-Ephidrine	Catecholamine transporter	Pheochromocytoma, neuroblastoma, study of the sympathetic nervous system
^{11}C-Hydroxyephidrine	Catecholamine transporter	Pheochromocytoma, neuroblastoma, study of the sympathetic nervous system

spleen (activated lymphocytes), kidneys, and gastrointestinal tract in different quantities, have generated immense clinical interest due to their expression on various tumor types. This offers the potential of labeling SMS and its analogues with different radionuclides for imaging and also for therapy. The advantages of small peptides are better pharmacokinetic characteristics and no (or very low) antigenicity as compared to antibodies, making them nearly ideal ligands for receptor-based radionuclide imaging.

The basis of peptide receptor imaging using radiolabeled SMS is the overexpression in NETs of SSTRs [5]. There are five different types of SSTR proteins, which have been cloned (SSTRs 1 through 5); SSTR2 consists of two subtypes, SSTR2A and SSTR2B. Though most of the tumors predominantly express SSTR2, it has been demonstrated that SSTR1, SSTR3, SSTR4, and SSTR5 are also expressed on many tumors with varying percentage of expression [12]. The prevalence of expression of SSTRs by NET of midgut origin was found to be maximum for SSTR2 (95%), followed by SSTR1 (80%) and SSTR5 (75%).

^{111}In-DTPA-D-Phe1-octreotide (^{111}In-pentetreotide; OctreoScan, Mallinckrodt, Inc., St. Louis, Missouri) was the first radiolabeled SMS analogue to be approved for scintigraphy of NETs and has been shown to be well suited for the scintigraphic localization of primary and metastatic NET [13, 14]. [99m]Tc (technetium) has also been labeled with SMS analogues to enable conventional nuclear medicine imaging [15–18]. ^{99m}Tc-EDDA/HYNIC-TOC has been demonstrated to be promising for the detection of SSTR-positive tumors and metastases [19].

DOTA-D-Phe[1]-Tyr[3]-octreotide (DOTATOC) and DOTA-D-Phe1-Tyr[3]-Thr[8]-octreotide (DOTATATE) have a very high affinity for the SSTR2, low or negligible affinity for SSTR3 and SSTR5, and no significant affinity for SSTR1 and SSTR4 [20, 21]. The development of this next generation of SMS analogues opened the prospect for convenient radiolabeling with ^{68}Ga for PET imaging [22]. DOTA-1-NaI[3]-octreotide (DOTANOC) was developed by amino acid exchange at position 3 of octreotide as a pansomatostatin analogue, covering a broader spectrum of SSTRs. This compound has not only a higher affinity to SSTR3 and SSTR5 but also binds more avidly to SSTR2 [23].

Fourth-generation analogues to be have been studied preclinically are (DOTA-1-NaI[3], Thr[8])-octreotide (DOTA-NOC-ATE) and (DOTA-BzThi[3], Thr[8])-octreotide (DOTA-BOC-ATE) [24]. They have been shown to have very high affinity for SSTR2, SSTR3, and SSTR5 and intermediate high affinity to SSTR4. SSTR antagonists [NH(2)-CO-c(DCys-Phe-Tyr-DAgl(8)(Me,2-naphthoyl)-Lys-Thr-Phe-Cys)-OH (SST(3)-ODN-8) and (SST(2)-ANT)] have also been labeled with ^{111}In, and their superiority over SSTR agonists (in murine models) for *in vivo* targeting of SSTR2- and SSTR3-rich tumors has resulted in the shift in paradigm, and they are now being contemplated for use in tumor diagnosis [25]. Recently, the SSTR antagonist ^{111}In-DOTA-BASS has been demonstrated to have a favorable human biodistribution in five patients with metastatic thyroid carcinoma or neuroendocrine neoplasms [26].

SSTR PET/CT USING [68]Ga

[68]Ga is a diagnostic trivalent radiometal and is feasible for labeling with SMS analogues (DOTATOC, DOTATATE, or DOTANOC) with the help of chelator DOTA [27]. [68]Ga is prepared from a TiO_2-based [68]Ge/^{68}Ga generator system, which has a half-life of 288 days [28]. A GMP-compliant, fully automated click-and-start cassette-based synthesis system with easy handling is now available (EZAG, Berlin, Germany) for the daily routine production of ^{68}Ga-labeled radiopharmaceuticals. Postprocessing of [68]Ge/^{68}Ga radionuclide generators using cation exchange resin provides chemically and radiochemically pure [68]Ga ($97 \pm 2\%$) within a few minutes, ready for on-site labeling with high overall product yields.

The most consequential feature of PET/CT is its ability to quantify the disease at a molecular level. In addition, superior resolution gives it a distinct edge over SPECT/CT using gamma-emitting radionuclides like Tc-99m. The recent tremendous increase in the number of diagnostic imaging studies with ^{68}Ga has indeed demonstrated its potential to become the Tc-99m for PET/CT. Apart from detection of primary and metastatic disease (staging), assessment of molecular response to therapy, and long-term follow-up, PET/CT using ^{68}Ga-labeled SMS analogues like DOTATOC, DOTATATE, and DOTANOC (SSTR PET/CT) also helps to select patients who are likely to benefit from peptide receptor radionuclide therapy (PRRT) using the same analogue labeled with a beta emitter like ^{177}Lu or [90]Y. The successful use of ^{68}Ga and ^{177}Lu/^{90}Y, respectively, for diagnosis and radionuclide therapy using the same peptide targeting SSTRs, has demonstrated that THERANOSTICS of neuroendocrine neoplasms is already a fact today and not a fiction.

IMAGING PROTOCOL

The guidelines for ^{68}Ga-SSTR PET/CT have been outlined [29]. Sandostatin LAR injections must be stopped 3–4 weeks prior to the scan, and subcutaneous (s.c.) treatment with octreotide should be stopped at least 1 day before. However, there are some centers that do not recommend stopping Sandostatin injection before imaging. Care is taken for proper hydration of the patient. Just prior to the acquisition, the patient should be requested to void. Use of oral contrast media is recommended. The maximum tumor activity is reached within 70 ± 20 min after injection. PET/CT acquisition should start at 45–90 min (depending upon the radiolabeled analogues) after intravenous injection of approximately 120 MBq of the radiolabeled peptides. In order to increase renal elimination and to reduce radiation exposure to the urinary bladder, furosemide may be given at the time of injection of ^{68}Ga-DOTANOC. In order to use the full potential of the modern PET/CT cameras equipped with multislice CT, a three-phase CT should be performed.

DIAGNOSIS, STAGING, AND RESTAGING

In a recent study in normal human tissues, expression of SSTR2 at the level of mRNA was found to correlate with the SUVmax obtained from ^{68}Ga-DOTATOC PET/CT [30]. Another recent study provided for the first time the proof of concept of the utility of SSTR PET/CT for quantification of the SSTR density on tumor cells: a close correlation between maximum SUV and immunohistochemical scores used for the quantitative assessment of the density of subtypes of SSTR in NET tissue [31]. This underlines the crucial role of molecular imaging of the SSTR expression by PET/CT using [68]Ga.

^{68}Ga-DOTATOC has been demonstrated to be superior to ^{111}In-octreotide SPECT (CT was taken as the reference for comparison) in detecting upper abdominal metastases more than 10 years ago [22]. In a recent study, ^{68}Ga-DOTATOC PET/CT was proven to be superior to ^{111}In-octreotide in the detection of NET metastases in the lung and skeletal system and similar for the detection of NET metastases in the liver

and brain [32]. On a patient basis, the accuracy of ^{68}Ga-DOTATOC PET (96%) was found to be significantly higher than that of CT (75%) and ^{111}In-DOTATOC SPECT (58%) [33]. In 32/88 patients, ^{68}Ga-DOTATOC PET was not only able to detect more lesions than SPECT and CT but also was true positive where SPECT results were false negative. It was observed that for the staging of patients, PET was better than CT or SPECT as it could pick up more lesions in the lymph node (LN), in the liver, and in the bone. In addition, in comparison to the ^{111}In-octreotide scan, ^{68}Ga-DOTATOC PET has been established to be superior especially in detecting small tumors or tumors bearing only a low density of SSTRs [34]. In patients with equivocal or negative OctreoScan, ^{68}Ga-DOTATATE PET/CT detected additional lesions and changed the management [35]. Of the 51 patients included in the study, 47 showed evidence of disease on cross-sectional imaging or biochemically. ^{68}Ga-DOTATATE PET was found to be positive in 41 of these 47 patients (87.2%), detecting 168 of the 226 lesions (74.3%) that were identified with cross-sectional imaging. ^{68}Ga-DOTATATE PET also identified significantly more lesions than ^{111}In-DTPA-octreotide scintigraphy ($P < 0.001$) and changed management in 36 patients (70.6%), who were subsequently deemed suitable for peptide receptor-targeted therapy. ^{68}Ga-DOTATOC was also shown to perform better than CT or SRS for the early detection of skeletal metastases of NETs [36]. In a larger subgroup of patients ($n = 90$) with pathologically confirmed NET, a comparison of ^{68}Ga-DOTANOC PET/CT with conventional imaging (CI) CT and EUS showed the superiority of PET/CT over CI [37]. Considering PET/CT and CI concordant cases (47/90 [52.2%]), PET findings affected the therapeutic management in 17 of 47 (36.2%) patients. Although PET did not result in modification of disease stage, ^{68}Ga-DOTANOC detected a higher lesion number in most patients. PET resulted in a modification of stage in 12 patients (28.6%) and affected the treatment plan in 32 patients (76.2%). ^{68}Ga-DOTANOC PET/CT thus affected either stage or therapy in 50 of 90 (55.5%) patients. The most frequent impact was the initiation or continuance of PPRT, followed by the initiation or continuance of SMS analogue medical treatment and referral to surgery. Of importance is that PET could avoid unnecessary surgery in 6 patients and excluded from treatment with SMS analogues two patients with NET lesions that did not express SSTRs. Due to the broader range of SSTR expression, ^{68}Ga-DOTANOC is an excellent tracer for imaging SSTR-positive tumors, which in addition, due to the high target to nontarget ratios, allows the detection of very small lesions, especially of LN and bone metastases [38]. PET using ^{68}Ga-DOTATOC has been found to be superior to ^{18}F-FDG PET in the detection of NETs, imaging 57/63 lesions in 15 patients, as compared with only 43/63 on FDG PET [39]. In malignant neural crest tumors (pheochromocytoma, paraganglioma, and medullary thyroid cancer), a direct comparison with ^{123}I-MIBG study showed the superiority of the ^{68}Ga-DOTATATE PET/CT in terms of sensitivity [40]. In a study in pulmonary endocrine tumors, ^{68}Ga-DOTATATE was shown to have a definite incremental value over ^{18}F-FDG for typical bronchial carcinoids than in atypical carcinoids and higher grades of tumors [41]. Also, due to the probability of development of concomitant NETs, SSTR PET/CT with ^{68}Ga could be useful in the detection and follow-up of pulmonary NETs [42]. Indeed, ^{68}Ga-SSTR PET/CT provides a whole-body

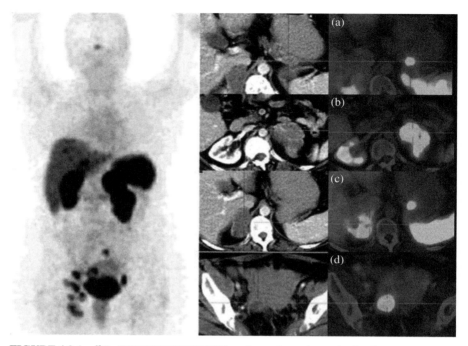

FIGURE 4.2.1   ^{68}Ga-DOTATATE PET/CT (maximum intensity projection image on extreme left) shows multiple primary neuroendocrine tumors in pancreatic tail with an SUVmax of 15.4 (a); in both adrenals, massive enlargement and partial necrosis in the left adrenal with an SUVmax of 29.7 (b) and comparatively smaller and less dense SSTR expression in the right adrenal with an SUVmax of 10.3 (c); and right ovary with an SUVmax of 11.6 (d). There was also widespread metastasis in the liver, multiple abdominal/extra-abdominal lymph nodes, and possibly the pituitary gland (SUVmax of 5.4) (a, b, c, d—images in transverse view; CT on the left and fused PET/CT on the right). (*See insert for color representation of the figure.*)

one-stop-shop approach to the identification and localization of NETs and their metastases (Figs. 4.2.1 and 4.2.2).

^{68}Ga-DOTATOC PET/CT has been also found to be considerably cheaper than ^{111}In-DTPA-octreotide with respect to both material and personnel costs [43]. In clinical practice, apart from higher resolution and excellent quality of the images, the other advantages of PET imaging with the ^{68}Ga-labeled SMS analogues over ^{111}In-DTPA0-octreotide scintigraphy are easy availability of the ^{68}Ga generator, relative short scanning time, and low radiation exposure to the patient [44]. Two recent studies have taken into account the comparison between the ^{68}Ga-labeled SMS analogues. A preliminary intraindividual study comparing ^{68}Ga-DOTANOC and ^{68}Ga-DOTATATE demonstrated that ^{68}Ga-DOTANOC localized more lesions in, especially, the liver and pancreas, due to its broader SSTR affinity profile [45]. Another study demonstrated comparable diagnostic accuracy of ^{68}Ga-DOTATATE and ^{68}Ga-DOTATOC for detection of NET lesions [46].

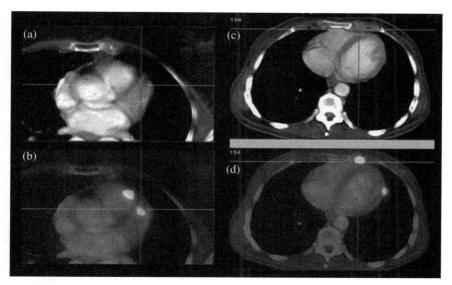

FIGURE 4.2.2 Identification of rare metastases of neuroendocrine tumors with high-sensitivity on ^{68}Ga-SSTR PET/CT: not only were myocardial metastases, which were otherwise difficult to appreciate on CT (a), localized on ^{68}Ga-DOTA-SSTR PET/CT (b) but also pericardial metastases (c, CT; d, fused PET/CT). (*See insert for color representation of the figure.*)

Gluc-Lys ^{18}F FP-TOCA is an ^{18}F-based radiopharmaceutical that targets SSTRs. In a preliminary comparative study, Gluc-Lys ^{18}F FP-TOCA PET was found to be superior to ^{111}In-DTPA-octreotide scan in the diagnosis of NETs. The results also suggested that the sensitivity and specificity of Gluc-Lys ^{18}F FP-TOCA is comparable to the reported sensitivity and specificity of ^{68}Ga-DOTATOC PET findings in NETs [47]. [64]Cu, with a half-life of 12.7 hours, is another potential positron-emitting radio-nuclide for PET imaging [48]. The possibility of performing dosimetry for PRRT based upon ^{64}Cu is one other possible advantage. In a preliminary study, ^{64}Cu-TETA-octreotide PET was found to have high sensitivity and favorable dosimetry and pharmacokinetics [49].

DETECTION OF UNKNOWN PRIMARY TUMOR

In a bicentric study, the role of ^{68}Ga-DOTANOC PET/CT in the detection of unknown primary NETs has been demonstrated (Fig. 4.2.3) [50].

Overall, 59 patients (33 men and 26 women, aged 65 ± 9 years) with documented NETs and unknown primary were enrolled. PET/CT was performed after injection of approximately 100 MBq (46–260 MBq) of ^{68}Ga-DOTANOC. The SUVmax were calculated and compared with SUVmax in known pancreatic NETs (pNETs) and ileum/jejunum/duodenum NETs (SI-NETs). The results of PET/CT were also corre-lated with CT alone. In 35 of 59 patients (59%), ^{68}Ga-DOTANOC PET/CT localized

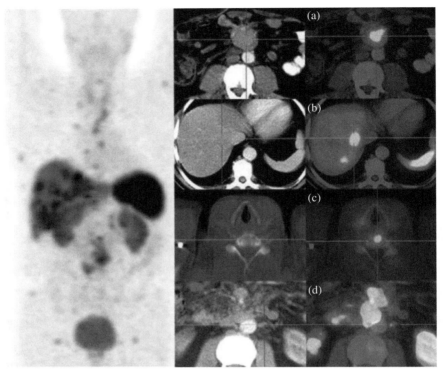

FIGURE 4.2.3   In this case of CUP syndrome, ^{68}Ga-DOTATOC PET/CT (MIP image on extreme left) revealed somatostatin receptor-positive primary tumor in the jejunum with SUVmax of 13.8 (a), along with multiple metastases in the liver (b), bone (c), and lymph nodes (d), and a metastasis in the left adrenal. (a, b, c, d—images in transverse view; CT on the left and fused PET/CT on the right.) (*See insert for color representation of the figure.*)

the site of the primary: ileum/jejunum (14), pancreas (16), rectum/colon (2), lungs (2), and paraganglioma (1). CT alone (on retrospective analyses) confirmed the findings in 12 of 59 patients (20%). The mean SUVmax of previously unknown (cancer of unknown provenience (CUP)) pNETs and SI-NETs were 18.6 ± 9.8 (range: 7.8–34.8) and 9.1 ± 6.0 (range: 4.2—27.8), respectively. SUVmax in patients with previously known pNET and SI-NET were 26.1 ± 14.5 (range: 8.7–42.4) and 11.3 ± 3.7 (range: 5.6–17.9). The SUVmax of the unknown pNETs and SI-NETs were significantly lower ($P < 0.05$) as compared to the ones with known primary tumor sites; 19% of the patients had high-grade NET and 81% low-grade NET. Based on ^{68}Ga-DOTANOC receptor PET/CT, 6 of 59 patients were operated, and the primary was removed (4 pancreatic, 1 ileal, and 1 rectal tumor) resulting in a management change in approximately 10% of the patients. In the remaining 29 patients, because of the far advanced stage of the disease (due to distant metastases), the primary tumors were not operated. Additional histopathological sampling was available from one patient with bronchial carcinoid (through bronchoscopy). In this study, ^{68}Ga-DOTANOC

PET/CT was found to be highly superior to ^{111}In-OctreoScan (39% detection rate for CUP according to the literature). It therefore has a major role to play in the management of patients with CUP-NET.

THERAPY PLANNING

Curative treatment of NETs usually requires the possibility of complete surgical resection of the primary tumor and perhaps regional LN metastases. However, effective palliative therapies are also available at all stages of the disease and can be applied even to advanced stage. Depending upon tumor stage, size, localization, and degree of differentiation, treatment protocols for NET are currently based upon the following therapeutic options:

1. Surgery
2. Immunological therapy (interferon)
3. Trans-arterial chemoembolization
4. Chemotherapy
5. Therapy with SMS analogues
6. PRRT
7. Intra-arterial PRRT

Among these, surgical resection and cold SMS analogues (intramuscular or s.c. octreotide) are most commonly used as the first line of treatment; chemotherapy is used as the last option. The recent years have seen the development of PRRT as a highly effective treatment option for metastasized progressive NET, and PRRT is now regarded as the third leg of treatment protocol. For effective management of patients with NET, receptor and metabolic PET/CT play a very important role. As indicators of prognosis and by directing the surgeons toward the site of primary tumor, PET/CT with ^{68}Ga-DOTA-peptides and ^{18}F-FDG play an important role in the management of NETs. For the institution of "cold" SMS therapy as well as SSTR-based radionuclide therapy, it is important to document the expression of SSTRs on the tumor cells. The therapy schedule (amount of administered radioactivity and timing) of PRRT using [177]Lu- or ^{90}Y-DOTATATE/DOTATOC is highly dependent on the semiquantitative/visual interpretation of ^{68}Ga-DOTA-peptide PET/CT. Although dosimetry still is the best way to "individualize" PRRT, in our own experience, semiquantitative (SUVmax) evaluation is a good measure of the degree of receptor expression and hence prediction of response. Intra-arterial PRRT is also an exciting option specifically for localized and bulky liver metastases. Although partial hepatectomy and hepatic transplantation remains a possibility, at Zentralklinik Bad Berka, intra-arterial PRRT has been found to be highly successful. The dosing of such treatment is also dependent on receptor expression, since size on CT and MRI is not a reliable parameter because of the possibility of presence of cystic degeneration.

EVALUATION OF THERAPY RESPONSE AND PROGNOSIS

PET/CT is increasingly being used for therapy monitoring of various tumors. Most of these response parameters are based on morphologic imaging. However off-late, there has been upsurge in the use of molecular response criteria for the early and accurate detection of response to therapy. SUVmax has been correlated with the prognosis in several cancers. Despite the fact that several studies have been published regarding the prognostic factors of NETs, there are some cases in which available data are not sufficient to predict disease progression and to define a correct therapeutic approach. In a study involving 47 patients, the SUVmax of ^{68}Ga-DOTANOC was validated for its potential to predict prognoses [51]. SUVmax was significantly higher in patients with pNET and in those with well-differentiated NET, which indicated an elevated expression of SSTR2A. During the follow-up, whereas disease was stable or presented a partial response in 25 patients, it progressed in 19 cases. The patients with stable disease or a partial response had SUVmax significantly higher than did those in the progressive disease group, with the best cutoff ranging from 17.9 to 19.3. Univariate and multivariate analyses demonstrated significant positive prognostic factors to be well-differentiated NET, a SUVmax of 19.3 or more, and a combined treatment with long-acting SMS analogues and radiolabeled SMS analogues. This study thus showed that SUVmax correlated with the clinical and pathologic features of NETs and is also an accurate prognostic index.

The RECIST and World Health Organization (WHO) criteria for classifying tumor response work best with fast-growing cancers and when the therapy for which the response is being assessed is cytotoxic and not cytostatic [52]. Treatment evaluation of slow-growing NETs based on size changes alone is far more difficult. Also, a high percentage of NETs are nonfunctional, and clinical response parameters are often insufficient. Biochemical markers, such as CgA and 5-HIAA, may be misleading too, owing to their poor sensitivity. Rather than using RECIST criteria alone when monitoring NET, combined imaging approach that takes into consideration both molecular response parameter (MORE) and morphological information provides a much better measure of early treatment response since molecular response precedes morphology [53].

The role of metabolic PET/CT in the assessment of response to therapy is limited, primarily because NETs are slow-growing tumors and glucose metabolism does not necessarily increase in slow-growing and well-differentiated tumors. In addition, no definitive therapy that directly influences the glucose metabolism so as to be assessed by ^{18}F-FDG exists for these tumors. It had already been postulated that ^{18}F-FDG PET should be performed only if SSTR imaging is negative [54]. The main use of ^{18}F-FDG PET in the diagnosis of NETs depends on the grade of differentiation and/or aggressiveness of NETs and has been proposed for comprehensive tumor assessment in intermediate- and high-grade tumors [39, 55]. Intense metabolic activity, reflected on ^{18}F-FDG PET scans, can still be an important prognostic indicator, being related to an outgrowth of aggressive tumor clones, suggesting a poor prognosis. However, functional imaging with both ^{68}Ga-DOTATATE and ^{18}F-FDG has been shown to address different biological properties of the NET lesions in patients planned for PRRT

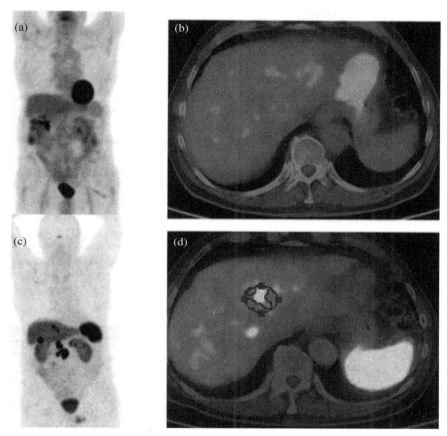

FIGURE 4.2.4 Nonfunctional, well-differentiated neuroendocrine neoplasm of the pancreas head with hepatic and abdominal lymph node metastases and proliferation rate (Ki-67) of 7%. ^{18}F-FDG PET/CT (a, b) demonstrated no increased glucose metabolism in the metastases. ^{68}Ga-DOTATOC PET/CT (c, d), on the other hand, demonstrated intense expression of somatostatin receptors. This mismatch is a sign of good prognosis and an indication for PRRT (a, c, MIP images; b, d, fused PET/CT images in transverse view, demonstrating hepatic metastases). (*See insert for color representation of the figure.*)

and has been proposed as for comprehensive tumor assessment in intermediate- and high-grade tumors (Fig. 4.2.4) [39, 55].

The finding of increased L-DOPA decarboxylase activity in 80% of NETs has resulted in the use of this parameter as a marker of tumor activity [56]. ^{18}F-DOPA PET/CT has been shown to have a promising role in GEP-NET patients with negative or inconclusive findings at conventional radiological imaging and ^{111}In-pentetreotide scintigraphy [57]. ^{18}F-DOPA may have a role in the evaluation of functionally active NETs, especially pancreatic tumors, and also pheochromocytoma, paraganglioma, medullary thyroid cancer, and neuroblastoma [9]. In comparison with ^{18}F-DOPA,

[68]Ga-labeled SMS analogues have been demonstrated to have a higher sensitivity for the identification of metastases as well as occult primary [58, 59].

In a preliminary study at the Zentralklinik Bad Berka, investigators selected 25 subjects at random from a group of 505 patients with metastasized neuroendocrine cancer who were scheduled for treatment with PRRT (138 lesions) and compared pre- and posttreatment images acquired using ^{68}Ga-DOTANOC PET/CT (molecular response), ^{18}F-FDG PET/CT (metabolic response), and contrast-enhanced CT (morphological response) [60]. A response index was calculated for each lesion from PET images based on the pre- and posttreatment SUVmax. RECIST criteria were applied to the contrast-enhanced CT data. All lesions were categorized as partial responders, stable disease, or progressive disease. No correlation was observed between any of the three modalities. For example, ^{68}Ga-DOTANOC PET classified 70.6% of the lesions as partial responders, while ^{18}F-FDG PET put 43.8% into this category and CT just 17.6%. The sensitivity and specificity of ^{68}Ga-DOTANOC PET to predict response to radiopeptide therapy were calculated as 89 and 71%, respectively. ^{68}Ga-DOTANOC PET/CT was found to be superior to ^{18}F-FDG PET/CT and morphological imaging in early and better prediction of response to PRRT. Furthermore, a matching pattern between receptor expression and glucose metabolism was observed to increase with the grade of NETs, and therefore in high-grade NETs, a concurrence between the changes in glucose metabolism and SSTR expression, that is, on ^{18}F-FDG PET/CT and ^{68}Ga-DOTANOC PET/CT, respectively, after PRRT was noticed. Also, higher tumor remission rate was correlated with a high-baseline SUVmax on SSTR PET/CT. This finding is consistent with previous studies, and PRRT seems to be quite an effective therapy option for NET patients expressing adequate densities of SSTRs on the tumors (Figs. 4.2.5 and 4.2.6) [61].

In another recent study, ^{68}Ga-DOTATATE PET/CT was validated for its potential to predict response to PRRT at an early stage [62]. Thirty-three consecutive patients (22 men and 11 women; mean age ±SD, 57.8 ± 12.1 years) were investigated at baseline and again 3 months after initiation of the first cycle of PRRT. ^{68}Ga-DOTATATE receptor expression was assessed using 2 measures of standardized uptake value (SUV): SUVmax and tumor-to-spleen SUV ratio (SUV T/S). Percentage change in SUV scores after PRRT relative to baseline (SUV) was calculated. After completing 1–3 cycles of PRRT, patients entered the follow-up study for estimation of time to progression. According to the RECIST criteria, progression was defined on the basis of contrast-enhanced CT. Clinical symptoms as well as the tumor markers chromogranin A and neuron-specific enolase were also recorded during regular follow-up visits. The 23 of 31 patients with decreased SUV T/S after the first PRRT cycle had longer progression-free survival than did the 8 of 31 patients with stable or increased scores (median survival not reached vs. 6 months, $P = 0.002$). For the 18 of 33 patients showing a reduction in SUVmax, there was no significant difference in progression-free survival (median survival not reached vs. 14 months, $P = 0.22$). Multivariate regression analysis identified SUV T/S as the only independent predictor for tumor progression during follow-up. In the 17 of 33 patients with clinical symptoms before PRRT, SUV T/S correlated with clinical improvement ($r = 0.52$, $P < 0.05$), whereas

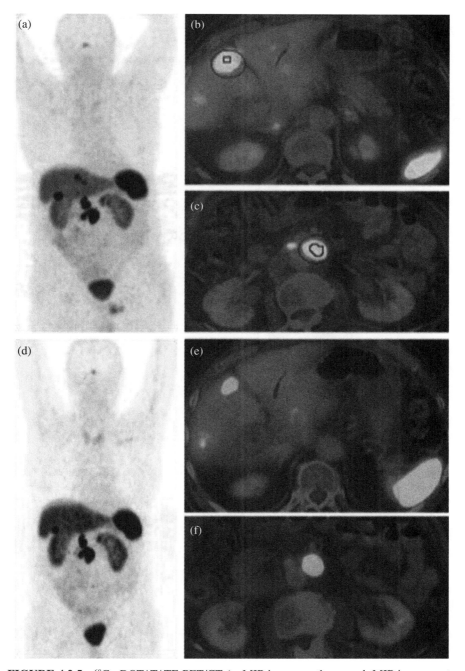

FIGURE 4.2.5 ^{68}Ga-DOTATATE PET/CT (a, MIP image pretherapy; d, MIP image post-therapy) demonstrated significant therapeutic response in the same patient (described in Fig. 4.2.4) (partial response according to molecular imaging criteria) after PRRT with 2700 MBq ^{90}Y-DOTATOC. Hepatic metastasis in segment S5 inferoventral segment of the liver (b, fused PET/CT image in transverse view pretherapy; e, fused PET/CT image in transverse view posttherapy) showed an SUV fall of 68% from 61.1 to 19.4 posttherapy. Preaortic lymph node metastasis (c, fused PET/CT image in transverse view pretherapy; f, fused PET/CT image in transverse view posttherapy) showed an SUV fall of 29% from 55.5 to 39.5 posttherapy. (*See insert for color representation of the figure.*)

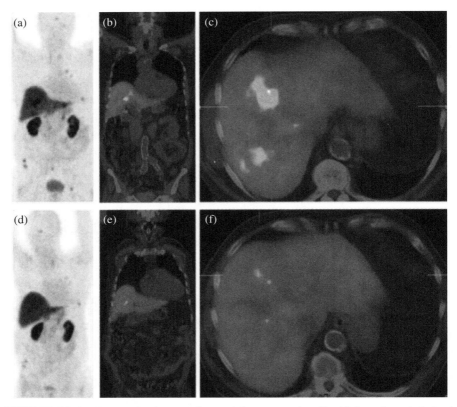

FIGURE 4.2.6 Neuroendocrine neoplasm of the pancreatic tail with hepatic and skeletal metastases: ^{68}Ga-DOTATATE PET/CT (pretherapy images—a, MIP; b, coronal PET/CT fused; c, transverse PET/CT fused) demonstrated excellent response in the liver metastases after two cycles of PRRT with a cumulative administered activity of 8000 MBq of [177]Lu (posttherapy images—d, MIP; e, coronal PET/CT fused; f, transverse PET/CT fused). (*See insert for color representation of the figure.*).

SUVmax did not ($r = 0.42$, $P = 0.10$). Changes in the tumor markers (chromogranin A and neuron-specific enolase) did not predict SUV scores, clinical improvement, or time to progression. This study showed that decreased ^{68}Ga-DOTATATE uptake in tumors after the first cycle of PRRT predicted time to progression and correlated with an improvement in clinical symptoms among patients with well-differentiated NETs; SUV T/S was superior to SUVmax for prediction of outcome. However, more data is needed to substantiate this observation. The other potential application of receptor and metabolic PET/CT would be in the assessment of response to transarterial chemoembolization (TACE), chemotherapy, and Sandostatin therapy (to predict relapse). Biochemical markers are not very good indicators for early and accurate response of therapy necessitating the need to finalize the role of PET/CT in NET therapy response.

SUMMARY

Receptor PET/CT using ^{68}Ga-labeled SMS analogues enables molecular imaging of NETs and their metastases with very high diagnostic sensitivity and specificity. It provides quantitative, reproducible data (SUV) that can be used for selecting patients for PRRT and evaluation of therapy response. Among other advantages are fast protocol (60–90 min), low radiation burden (10–12 mSv), flexibility in daily use, and lower cost than octreotide scintigraphy. As we move toward personalized medicine, the diagnostic information obtained from PET/CT must be improved, that is, by fast and routine quantification of lesions. The Bad Berka Molecular Imaging Tool (BBQ-MIT) has been developed based on the cognition network language (CNL), provided by Definiens AG (Munich, Germany). The BBQ-MIT is an automatic, user-independent routine for segregation and quantification of neoplastic lesions in molecular PET/CT DICOM sets. This prototype routine built on CNL for PET/CT images enables the automatic analysis of lesions, for example, by calculating SUV, molecular tumor volume (MTV), molecular tumor diameter (MTD), molecular tumor index (SUV x MTD), whole-body and organ tumor burden, and many other parameters. It seems especially promising for shortening the analysis time for reading a PET scan with many tumor lesions, improving reproducibility, as well as increasing the sensitivity in lesion detection. The BBQ-MIT is a definite step forward and should set a trend toward the fast and accurate analysis of serial PET/CT, allowing monitoring of tumor response and assessment of therapy effect early in the course of therapy, thus enabling effective personalized patient management.

REFERENCES

[1] Anlauf, M.; Gerlach, P.; Raffel, A.; et al. Der Onkologe 2011, 17, 572–582.

[2] Bombardieri, E.; Maccauro, M.; de Deckere, E.; et al. Annals of Oncology 2001, 12, 51–61.

[3] Rufini, V.; Calcagni, ML.; Baum, R. P. Seminars in Nuclear Medicine 2006, 36, 228–247.

[4] Baum, R. P.; Kulkarni, H. R.; Carreras, C. Seminars in Nuclear Medicine 2012, 42, 190–207.

[5] Reubi, J. C. Journal of Nuclear Medicine 1995, 36, 1825–1835.

[6] Koopmans, K. P.; Neels, O.N.; Kema, I. P.; et al. Critical Reviews in Oncology/ Hematology 2009, 71, 199–213.

[7] Carrasquillo, J. A.; Chen, C. C. Seminars in Oncology 2010, 37, 662–679.

[8] Kwekkeboom, D. J.; Krenning, E. P. Seminars in Nuclear Medicine 2002, 32, 84–91.

[9] Wahl, R. L., ed., *Principles and Practice of PET and PET/CT*. Wolters Kluwer/Lippincott Williams & Wilkins, Philadelphia; 2008.

[10] Ambrosini, V.; Tomassetti, P.; Franchi, R.; et al. The Quarterly Journal of Nuclear Medicine and Molecular Imaging 2010, 54, 16–23.

[11] Miederer, M.; Weber, M. M.; Fottner, C. Gastroenterology Clinics of North America 2010, 39, 923–935.

[12] Reubi, J.C.; Waser, B.; Schaer, J. C.; et al. European Journal of Nuclear Medicine and Molecular Imaging 2001, 287, 836–846.

[13] Krenning, E. P.; Kwekkeboom, D. J.; Bakker, W. H.; et al. European Journal of Nuclear Medicine and Molecular Imaging 1993, 20, 716–731.

[14] Bombardieri, E.; Ambrosini, V.; Aktolun, C.; et al. European Journal of Nuclear Medicine and Molecular Imaging 2010, 37, 1441–1448.

[15] Decristoforo, C.; Melendez-Alafort, L.; Sosabowski, J. K.; et al. Journal of Nuclear Medicine 2000, 41, 1114–1119.

[16] Lebtahi, R.; Le Cloirec, J.; Houzard, C.; et al. Journal of Nuclear Medicine 2002, 43, 889–895.

[17] Maina, T.; Nock, B.; Nikolopoulou, A.; et al. European Journal of Nuclear Medicine and Molecular Imaging 2002, 29, 742–753.

[18] Storch, D.; Béhé, M.; Walter, M. A.; et al. Journal of Nuclear Medicine 2005, 46, 1561–1569.

[19] Gabriel, M.; Muehllechner, P.; Decristoforo, C.; et al. The Quarterly Journal of Nuclear Medicine and Molecular Imaging 2005, 49, 237–244.

[20] De Jong, M.; Bakker, W. H.; Krenning, E. P.; et al. European Journal of Nuclear Medicine and Molecular Imaging 1997, 24, 368–371.

[21] Forrer, F.; Uusijärvi, H.; Waldherr, C.; et al. European Journal of Nuclear Medicine and Molecular Imaging 2004, 31, 1257–1262.

[22] Hofmann, M.; Maecke, H.; Borner, R.; et al. European Journal of Nuclear Medicine and Molecular Imaging 2001, 28, 1751–1757.

[23] Wild, D.; Schmitt, J.S.; Ginj, M.; et al. European Journal of Nuclear Medicine and Molecular Imaging 2003, 30, 1338–1347.

[24] Ginj, M.; Chen, J.; Walter, M. A.; et al. Clinical Cancer Research 2005, 11, 1136–1145.

[25] Ginj, M.; Zhang, H.; Waser, B.; et al. Proceedings of the National Academy of Sciences 2006, 103, 16436–16441.

[26] Wild, D.; Fani, M.; Behe, M.; et al. Journal of Nuclear Medicine 2011, 52, 1412–1417.

[27] Roesch, F.; Baum, R. P. Generator-based PET radiopharmaceuticals for molecular imaging of tumors: on the way to THERANOSTICS. Dalton Transactions 2011, 40, 6104–6111.

[28] Zhernosekov, K. P.; Filosofov, D. V.; Baum, R. P.; et al. Journal of Nuclear Medicine 2007, 48, 1741–1748.

[29] Virgolini, I.; Ambrosini, V.; Bomanji, J.B.; et al. European Journal of Nuclear Medicine and Molecular Imaging 2010, 37, 2004–2010.

[30] Boy, C.; Heusner, T. A.; Poeppel, T. D.; et al. European Journal of Nuclear Medicine and Molecular Imaging 2011, 38, 1224–1236.

[31] Kaemmerer, D.; Peter, L.; Lupp, A.; et al. European Journal of Nuclear Medicine and Molecular Imaging 2011, 8, 1659–1668.

[32] Buchmann, I.; Henze, M.; Engelbrecht, S.; et al. European Journal of Nuclear Medicine and Molecular Imaging 2007, 34, 1617–1626.

[33] Gabriel, M.; Decristoforo, C.; Kendler, D.; et al. Journal of Nuclear Medicine 2007, 48, 508–518.

[34] Kowalski, J.; Henze, M.; Schuhmacher, J.; et al. Molecular Imaging and Biology 2003, 5, 42–48.

[35] Srirajaskanthan, R.; Kayani, I.; Quigley, A. M.; et al. Journal of Nuclear Medicine 2010, 51, 875–882.

[36] Putzer, D.; Gabriel, M.; Henninger, B.; et al. Journal of Nuclear Medicine 2009, 50, 1214–1221.

[37] Ambrosini, V.; Campana, D.; Bodei, L.; et al. Journal of Nuclear Medicine 2010, 51, 669–673.

[38] Prasad, V.; Baum, R. P. The Quarterly Journal of Nuclear Medicine and Molecular Imaging 2010, 54, 61–67.

[39] Koukouraki, S.; Strauss, L. G.; Georgoulias, V.; et al. European Journal of Nuclear Medicine and Molecular Imaging 2006, 33, 1115–1122.

[40] Naji, M.; Zhao, C.; Welsh, S. J.; et al. Molecular Imaging and Biology 2011, 13, 769–775.

[41] Kayani, I.; Conry, B. G.; Groves, A. M.; et al. Journal of Nuclear Medicine 2009, 50, 1927–1932.

[42] Kaemmerer, D.; Khatib-Chahidi, K.; Baum, R. P.; et al. Cancer Imaging 2011, 11, 179–183.

[43] Schreiter, N. F.; Brenner, W.; Nogami, M.; et al. European Journal of Nuclear Medicine and Molecular Imaging 2012, 39, 72–82.

[44] Krausz, Y.; Freedman, N.; Rubinstein, R.; et al. Molecular Imaging and Biology 2011, 13, 583–593.

[45] Wild, D.; Bomanji, B. J.; Reubi, J. C.; et al. European Journal of Nuclear Medicine and Molecular Imaging 2009, 36, S201.

[46] Poeppel, T. D.; Binse, I.; Petersenn, S.; et al. Journal of Nuclear Medicine 2011, 52, 1864–1870.

[47] Meisetschläger, G.; Poethko, T.; Stahl, A.; et al. Journal of Nuclear Medicine 2006, 47, 566–573.

[48] Hanaoka, H.; Tominaga, H.; Yamada, K.; et al. Annals of Nuclear Medicine 2009, 23, 559–567.

[49] Anderson, C. J.; Dehdashti, F.; Cutler, P. D.; et al. Journal of Nuclear Medicine 2001, 42, 213–221.

[50] Prasad, V.; Ambrosini, V.; Hommann, M.; et al. European Journal of Nuclear Medicine and Molecular Imaging 2010, 37, 67–77.

[51] Campana, D.; Ambrosini, V.; Pezzilli, R.; et al. Journal of Nuclear Medicine 2010, 51, 353–359.

[52] Chalian, H.; Töre, H. G.; Horowitz, J. M.; et al. Radiographics 2011, 31, 2093–2105.

[53] Baum, R. P.; Prasad, V.; Hommann, M.; et al. Recent Results in Cancer Research 2008, 170, 225–242.

[54] Adams, S.; Baum, R.; Rink, T.; et al. European Journal of Nuclear Medicine 1998, 25, 79–83.

[55] Kayani, I.; Bomanji, J. B.; Groves, A.; et al. Cancer 2008, 112, 2447–2455.

[56] Eldrup, E.; Clausen, N.; Scherling, B.; et al. Scandinavian Journal of Clinical & Laboratory Investigation 2001, 61, 479–490.

[57] Ambrosini V, Tomassetti, P.; Rubello, D.; et al. Nuclear Medicine Communications 2007, 28, 473–477.

[58] Ambrosini, V.; Tomassetti, P.; Castellucci, P.; et al. European Journal of Nuclear Medicine and Molecular Imaging 2008, 35, 1431–1438.

[59] Haug, A.; Auernhammer, C. J.; Wängler, B.; et al. European Journal of Nuclear Medicine and Molecular Imaging 2009, 36, 765–770.

[60] Oh, S.; Prasad, V.; Lee, D. S.; et al. International Journal of Molecular Imaging 2011, 2011, 524130.

[61] Kwekkeboom, D. J.; Bakker, W. H.; Kooij, P. P.; et al. European Journal of Nuclear Medicine 2001, 28, 1319–1325.

[62] Haug, A. R.; Auernhammer, C. J.; Wängler, B.; et al. Journal of Nuclear Medicine 2010, 51, 1349–1356.

4.3

OTHER RADIOPHARMACEUTICALS FOR IMAGING GEP-NET

Klaas Pieter Koopmans[1], Rudi A. Dierckx[2], Philip H. Elsinga[2], Thera P. Links[3], Ido P. Kema[4], Helle-Brit Fiebrich[5], Annemiek M.E. Walekamp[5], Elisabeth G.E. de Vries[5], and Adrienne H. Brouwers[2]

[1]Department of Radiology and Nuclear Medicine, Martini Hospital Groningen, Groningen, The Netherlands
[2]Department of Nuclear Medicine and Molecular Imaging, University of Groningen, and University Medical Center of Groningen, Groningen, The Netherlands
[3]Department of Endocrinology, University of Groningen, and University Medical Center of Groningen, Groningen, The Netherlands
[4]Department of Laboratory Center, University of Groningen, and University Medical Center of Groningen, Groningen, The Netherlands
[5]Department of Medical Oncology, University of Groningen, and University Medical Center of Groningen, Groningen, The Netherlands

ABBREVIATIONS

^{11}C-5-HTP	β-[^{11}C]-5-hydroxy-L-tryptophan
^{18}F-DOPA	6-^{18}F -L-3,4 -dihydroxyphenylalanine
^{18}F-FDG	^{18}F-fluorodeoxyglucose
5-HIAA	5-hydroxyindoleacetic acid
5-HT	serotonin
5-HTP	5-hydroxytryptophan
AADC	aromatic amino acid decarboxylase

Somatostatin Analogues: From Research to Clinical Practice, First Edition. Edited by Alicja Hubalewska-Dydejczyk, Alberto Signore, Marion de Jong, Rudi A. Dierckx, John Buscombe, and Christophe Van de Wiele.
© 2015 John Wiley & Sons, Inc. Published 2015 by John Wiley & Sons, Inc.

GEP   gastroenteropancreatic
LAT   system L large amino acid transporters
L-DOPA   L-3,4 -dihydroxyphenylalanine
MIBG   metaiodobenzylguanidine
MIP   maximum intensity projection
NET   neuroendocrine tumor
PET   positron emission tomography
SRS   somatostatin receptor scintigraphy with ^{111}In-octreotide
VMAT   vesicular monoamine transporters

INTRODUCTION

Gastroenteropancreatic neuroendocrine tumors (GEP-NETs) are tumors that arise in the pancreas and gastrointestinal system and are derived from neuroendocrine cells.

In contrast to many other malignancies, well-differentiated GEP-NETs generally have a low glucose metabolism [1, 2]. ^{18}F-fluorodexyglucose (^{18}F-FDG) PET scanning has limited value for the primary staging of patients with well-differentiated GEP-NET, showing only moderately increased glucose uptake in primary tumors and also often missing metastases. Besides, these primary tumors can be very small in size, thus failing the detection limit of PET camera systems. Therefore, as a staging tool, ^{18}F-FDG does not seem to have a place in staging the general GEP-NET patient population [1, 3]. A few studies with small heterogeneous GEP-NET tumor groups have shown that in patients with rapidly progressive disease, dedifferentiation of GEP-NET tumors can lead to a higher cellular glucose metabolism in tumor cells. In these patients, ^{18}F-FDG PET can be of benefit for tumor staging. The ^{18}F-FDG uptake in tumor lesions in these patients could possibly play a role in predicting outcome and in assessing therapy response [4]. ^{18}F-FDG PET can be of value when other malignancies are suspected in patients with GEP-NETs, since these patients experience a higher incidence of these malignancies compared to the general population [5]. However, data for this indication are scarce. Due to the limited applicability of ^{18}F-FDG PET in GEP-NETs, this technique will not be discussed in depth.

In contrast to most tissues in the human body, GEP-NETs often show increased synthesis and secretion of hormones and neurotransmitters. Nontumorous neuroendocrine cells regulate a variety of body functions through paracrine stimulation with a large variety of hormones and neurotransmitters. Serotonin, catecholamine, and histamine are examples of compounds that share specific steps in their biosynthesis and storage, such as decarboxylation prior to storage in granules [6]. In GEP-NETs, especially the catecholamine and serotonin biosynthetic pathways are upregulated. Therefore, increased biosynthesis of these specific amines in GEP-NETs enables imaging with specific amine precursors. In this chapter, imaging of cells that show increased production of catecholamine and/or serotonin will be discussed for GEP-NETs encompassing neuroendocrine tumors (NETs) from foregut origin (bronchus, lung, thymus, stomach, pancreas, and proximal duodenum), midgut NETs (from distal half of second part of the duodenum to the proximal two-thirds of the transverse colon), and hindgut NETs (descending colon, sigmoid, and rectum).

CATECHOLAMINE PATHWAY

Catecholamines act as neurotransmitters especially in the brain, or as hormones, for example, adrenaline when it is released from the adrenals, via α- and β-adrenergic receptors located on vessels and internal organs.

In the catecholamine pathway, phenylalanine and intermediate products such as L-3,4 -dihydroxyphenylalanine (L-DOPA) are taken up via system L large amino acid transporters (LAT) (Fig. 4.3.1). After entering the cell, decarboxylation to dopamine takes place intracytoplasmatically via the enzyme aromatic amino acid decarboxylase (AADC). Vesicular monoamine transporters (VMAT) then transport dopamine into intracellular storage vesicles. In these vesicles, dopamine can, dependent on the activity of specific enzymes, be further metabolized to noradrenaline and adrenaline. The resulting end products dopamine, noradrenaline, and adrenaline can be released in the extracellular environment from these vesicles. Selective

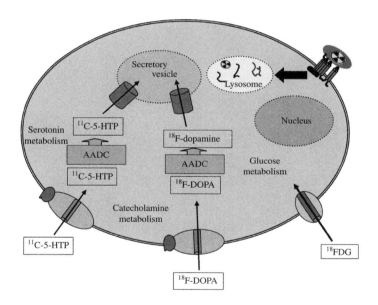

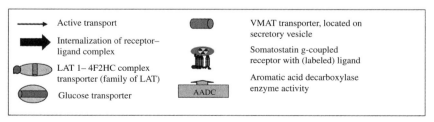

FIGURE 4.3.1 Overview. In this figure, a schematic overview is presented on the uptake mechanisms of the main tracers used in GEP-NET imaging. GEP-NET, gastroenteropancreatic neuroendocrine tumor.

reuptake transporter systems, for example, dopamine and noradrenaline transporters, can thereafter transport these hormones back into the cell.

Catecholamines can be metabolized mainly intracellularly but also extracellularly into various metabolites, which can be measured in plasma and urine as acid and basic catecholamine metabolites, such as homovanillic acid (HVA), vanillylmandelic acid (VMA), and metanephrines. Increased LAT activity seems to play a role in NETs, where due to a high precursor turnover and high AADC activity, a high precursor need has to be satisfied. The exact mechanism for precursor uptake regulation is not yet clear though [6, 7].

For the catecholamine pathway, 6-^{18}F-L-3,4-dihydroxyphenylalanine (^{18}F-DOPA) and 6-^{18}F-dopamine are available as tracers. Since 6-^{18}F-dopamine will be described extensively in Chapter 4.4.2, we will limit this chapter to ^{18}F-DOPA.

^{18}F-DOPA is most commonly produced via regioselective fluorodestannylation (electrophilic fluorination). Another possibility is nucleophilic fluorination, a multistep procedure that, although more time consuming, has the advantage of having readily available large quantities of no-carrier-added ^{18}F-fluoride [6, 8].

^{18}F-DOPA tracer has been developed and first studied in men at McMaster University, Hamilton, Canada, during the 1970s and 1980s [9]. It has since then first been used for the imaging of the dopaminergic system in the striatum in patients with Parkinson's disease and related disorders. Several years later, this tracer has been used as a whole-body imaging technique in patients with NETs (Fig. 4.3.2) [10, 11].

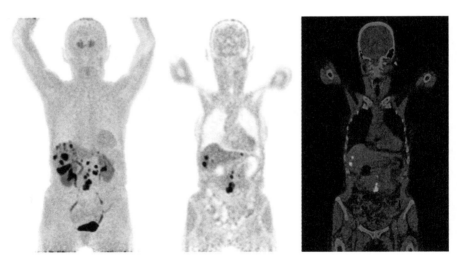

FIGURE 4.3.2 ^{18}F-DOPA PET/CT. Images of a 73-year-old female with a serotonin-producing metastasized well-differentiated NET of the small bowel (carcinoid). On the left, a maximum intensity projection (MIP) image is depicted; in the middle, a coronal slice; and on the right, a coronal fusion slice of PET and low-dose CT. Physiological uptake of ^{18}F-DOPA is visible in the striata and excretion via kidneys with physiological activity in the ureters and urinary bladder. Metastases are visible in the liver and mesenteric tissue/lymph nodes. ^{18}F-DOPA, 6-^{18}F-L-3,4-dihydroxyphenylalanine; PET/CT, positron emission tomography/computed tomography. (*See insert for color representation of the figure.*)

In most centers, patients are required to fast 2–6 h with amino acid-free fluid intake before the ^{18}F-DOPA PET scan is performed [12, 13]. ^{18}F-DOPA doses used are either a fixed dose or a body weight-dependent dose, up to 5 MBq/kg. Scanning after injection of ^{18}F-DOPA PET scan is usually performed 60 min after tracer injection with or without SUV quantification. Scanning at other time points has not been proven to be advantageous [14].

The use of carbidopa as pretreatment is still in debate. When used, it is given as 2 mg/kg body weight or a fixed dose (i.e., 100–200 mg) 1 h before tracer injection [15–17]. Carbidopa is a peripheral inhibitor of the enzyme AADC. It decreases the peripheral decarboxylation of ^{18}F-DOPA and also β-[^{11}C]-5-hydroxy-L-tryptophan (^{11}C-5-HTP), a tracer of the serotonin pathway, thus reducing renal excretion and subsequently improving tracer uptake in metastatic carcinoid tumors most likely due to increased tracer availability in the circulation. This resulted in better image quality due to decreased streaky image reconstruction artifacts caused by high physiological excretion of the radiotracer via kidneys and urinary bladder and increasing SUV values of tumor lesions [17]. Why carbidopa pretreatment results in a generally decreased tracer uptake in pancreatic tissue is not quite understood, but may be related to differences in AADC activity in NETs and differences in metabolic handling of these PET tracers by exocrine and endocrine pancreatic tissue [16]. Because of the likely strong similarities between ^{11}C-5-HTP and ^{18}F-DOPA tracer handling in patients with gastrointestinal NETs, carbidopa pretreatment is also advocated for ^{18}F-DOPA PET imaging in this patient group to improve image quality especially in the region of the kidneys and bladder and to improve lesion detectability via further increased SUV values of lesions. Indeed, high accuracy rates for the detection of (metastatic) NETs have been reported by the institutes that do use carbidopa pretreatment [18]. However, pancreatic islet cell tumors may be an exception to the rule [15]. This should be further investigated [16]. With carbidopa pretreatment, the estimated mean radiation dose is 1.9 mSv per 100 MBq ^{18}F-DOPA [19].

A possible complication of administering catecholamines to patients with a large tumor load is the development of a carcinoid crisis. Thus far, the development of a carcinoid crisis after injection of ^{18}F-DOPA has only been described in one case and is possibly related to a lower specific activity of ^{18}F-DOPA tracer [20].

Results of ^{18}F-DOPA PET(/CT) imaging in patients with midgut well-differentiated NETs (carcinoids) and other GEP-NETs, such as (non)functioning pancreatic NETs (islet cell tumors), have been described in the literature [6, 18, 21–23]. In these studies, most patients had proven (recurrent) disease, and ^{18}F-DOPA PET(/CT) has been compared with current anatomical imaging techniques, mostly CT and/or MRI, and other functional imaging techniques, such as somatostatin receptor scintigraphy (SRS) with ^{111}In-octreotide, [123/131]I-labeled metaiodobenzylguanidine (MIBG), and ^{18}F-FDG PET, and more recently also with ^{68}Ga-labeled PET-based variants of octreotide. Morphological imaging techniques seem to be complementary to the functional imaging techniques under study in these patient series. In carcinoids, patient-based reported sensitivities for ^{18}F-DOPA PET(/CT) are (very) high, ranging from 65 to 100% [6, 18, 22, 23], and therefore, it seems to be an excellent staging method for this patient group [6, 23]. ^{11}C-5-HTP PET, however, was better than

^{18}F-DOPA PET on a patient-based analysis (sensitivity of 100% vs. 96%, respectively), whereas per-lesion analysis showed the opposite (sensitivity of 78% for ^{11}C-5-HTP and 87% for ^{18}F-DOPA, respectively) in patients with carcinoids and islet cell tumors [22]. Another study of Fiebrich et al. showed that in patients with carcinoid tumors, ^{18}F-DOPA uptake, defined as total body tumor load, corresponds well to the tumor markers of the serotonin and catecholamine pathway in the urine and plasma in carcinoid patients, thereby reflecting metabolic tumor activity [24].

The ^{68}Ga-labeled analogues of somatostatin performed better than ^{18}F-DOPA PET in two studies with mixed NET types. A partial explanation for this result can be that in patients with (non)functioning islet cell tumors, ^{18}F-DOPA PET does not seem to have a good detection rate for these tumors [13, 21, 22].

Since the published results for ^{18}F-DOPA PET were mainly achieved in patients with proven NET, it can be expected that performance of ^{18}F-DOPA PET will be lower in clinically more difficult cases where diagnosis of a NET is suspected but has yet to be confirmed [6, 25]. This was addressed in a recent published study in which patients ($n = 119$) with only biochemical proof of the disease and various types of NETs were investigated [25]. In this study, the diagnostic accuracy of ^{18}F-DOPA PET/CT in patients who were entered for primary staging of abdominal NETs was 88%, whereas the diagnostic accuracy for patients entered for restaging of abdominal NETs ($n = 61$) was 92%. Specificity of ^{18}F-DOPA PET in this tumor type has not yet been studied, since most research has been performed in patients with proven NETs where it is unlikely that histological proof will be obtained [6]. Thus far, no studies have been published on the influence of ^{18}F-DOPA PET on therapeutic management or the prognostic value for early response monitoring with ^{18}F-DOPA PET. These are all areas in which more research is warranted.

There is a clear need for in-depth comparison of the other recently emerging functional PET imaging techniques, especially the ^{68}Ga-labeled somatostatin analogues in (subsets of) NETs with ^{18}F-DOPA PET. Advantages of ^{68}Ga-labeled somatostatin analogues are the relatively easy generator-based synthesis and the possibility to evaluate whether peptide receptor radionuclide therapy (PRRT) for NETs can be considered. However, ^{18}F-DOPA PET may have a broader clinical applicability, for example, for studying the dopaminergic system of the human brain, malignant tumors that do not (usually) express somatostatin receptors [26], and nononcologic settings, such as infants with hyperinsulinism. Since ^{18}F-DOPA is already commercially available in many countries, the increased availability may lead to a better understanding of the place of ^{18}F-DOPA PET in clinical decision making for NETs.

SEROTONIN PATHWAY

Serotonin is a monoamine neurotransmitter that is present in blood platelets and in the intestinal wall where it regulates contractions. In the brain, it acts as a neurotransmitter and plays a role in feelings of well-being.

Both the serotonin and the catecholamine pathways share common transporter systems, the LAT and the VMAT systems. Also, in both pathways, the enzyme AADC plays a key role in the final step for the synthesis of a functional hormone, serotonin and dopamine, respectively.

In the serotonin pathway, the amino acid tryptophan and the intermediate product 5-HTP are precursors for serotonin after uptake via the LAT system. How uptake is regulated is still unclear. After decarboxylation via the enzyme AADC, serotonin is taken up via VMAT into storage vesicles from which it can be released extracellularly (Fig. 4.3.1). Serotonin is eventually metabolized to 5-hydroxyindoleacetic acid (5-HIAA), which is secreted in the urine. The serotonin pathway is overactive in many NETs, which makes this pathway a good candidate for imaging with PET tracers [6].

Thus far, only a carbon-11-labeled tracer has been developed for the serotonin pathway, β-[^{11}C]-5-HTP. This tracer was developed in Uppsala, Sweden, during the 1980s. It is a complex tracer to produce, demanding an on-site cyclotron for the production of the ^{11}C isotope, which has a half-life of only 20 min. The synthesis of ^{11}C-5-HTP itself is rather complex since it requires two multienzyme steps [27, 28]. Only a few centers worldwide produce ^{11}C-5-HTP. However, in experienced hands, quantities up to 1000 MBq can be reliably synthesized for (clinical) use. Image interpretation is mostly easy due to high tumor tracer uptake and low background uptake in normal tissues. Clinical results justify the use of this tracer [7].

In 1993, the first results with this tracer were published [29]. Nowadays, ^{11}C-5-HTP PET scanning is typically performed 10–20 min after injection of ^{11}C-5-HTP with carbidopa pretreatment, 2 mg/kg body weight or with a fixed dose of 200 mg, 1 h prior to injection (Fig. 4.3.3). In contrary to ^{18}F-DOPA, no adverse reactions have thus far been reported after intravenous administration of ^{11}C-5-HTP.

For ^{11}C-5-HTP PET scans, carbidopa proved to be essential in order to improve image quality. Without carbidopa pretreatment, scans showed streaky image reconstruction artifacts due to high urinary tract ^{11}C-5-HTP uptake and excretion. Theoretically, carbidopa decreases the peripheral decarboxylation of ^{11}C-5-HTP to ^{11}C-5-HT (serotonin) and subsequent urinary excretion of serotonin metabolites. The effect of carbidopa was tested in patients with midgut carcinoids, who were scanned with and without carbidopa pretreatment 1 h prior to injection of ^{11}C-5-HTP. It was found that oral pretreatment significantly reduced urinary radioactivity concentration from a mean SUV of the pelvis of 155 ± 195 to 39 ± 14 SD. This led to an improved image quality in that area. Tumor uptake of ^{11}C-5-HTP significantly increased, from 11 ± 3 to 14 ± 3 SD. Interestingly, pancreatic uptake decreased slightly to an SUV of 4.4 ± 0.8 SD. In liver tissue, a small increase of ^{11}C-5-HTP uptake was found, now reaching an SUV of 3.6 ± 0.8 [17].

Due to the low number of centers using ^{11}C-5-HTP PET, only few studies with a reasonable number of patients with GEP-NETs have thus far been published. In the first published study with ^{11}C-5-HTP PET, 18 patients with histopathologically verified NETs were included: midgut ($n=14$), foregut ($n=1$), and hindgut carcinoid ($n=1$) and endocrine pancreatic tumors ($n=2$) [30]. ^{11}C-5-HTP PET without oral

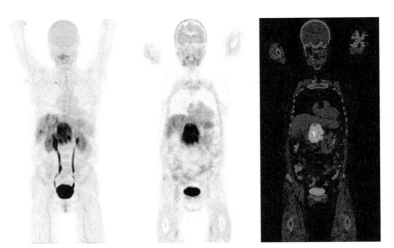

FIGURE 4.3.3 ^{11}C-5-HTP PET/CT. Images of a 62-year-old female with a gastrinoma. In contrary to ^{18}F-DOPA PET, the striata are not visualized. On the left, an MIP is depicted; in the middle, a coronal slice; and on the right, a coronal fusion slice of PET and low-dose CT. Physiological excretion of ^{11}C-5-HTP metabolites is via kidneys with physiological activity in the ureters and urinary bladder. Tumor is visible in the head of the pancreas and metastases in the liver. ^{11}C-5-HTP, β-[^{11}C]-5-hydroxy-L-tryptophan; ^{18}F-DOPA, 6-^{18}F-L-3,4-dihydroxyphenylalanine; PET/CT, positron emission tomography/computed tomography. (*See insert for color representation of the figure.*)

carbidopa pretreatment was compared to CT. All 18 patients showed increased ^{11}C-5-HTP uptake in tumor tissue; interestingly, a patient with a hindgut carcinoid and a patient with nonfunctioning endocrine pancreatic tumor with normal urinary 5-HIAA levels also showed increased tumor uptake. ^{11}C-5-HTP PET detected more tumor lesions than CT in 10 patients and was equal in five patients (four midgut, one foregut), with missing data in three patients with midgut NET. In the 10 patients that were on treatment (interferon-α ± octreotide, or somatostatin analogue only), a close correlation between the changes in ^{11}C-5-HTP transport rate constant in the tumors and urinary 5-HIAA was noted. It was therefore suggested that ^{11}C-5-HTP PET may serve as a means to monitor therapy, although it is still unknown whether these PET findings under medication relate to changes in tumor metabolism or in changes in amine processing [30].

In another patient series, 38 patients were evaluated, again with a variety of NETs, consisting of midgut carcinoids ($n=13$), lung carcinoids ($n=7$), nonfunctioning endocrine pancreatic tumors ($n=5$), and other NETs [30]. Whole-body ^{11}C-5-HTP PET imaging with oral carbidopa pretreatment was compared to both CT and the functional imaging technique SRS, which is often used in standard clinical care in NETs and more widely available than ^{11}C-5-HTP PET. Whole-body ^{11}C-5-HTP PET scanning detected tumor lesions in 36 of 38 (95%) patients. In 84% of patients, SRS was positive, whereas CT was positive in 79%. More lesions were detected with

[11]C-5-HTP PET in 22 of 38 (58%) patients compared to SRS and CT, whereas the imaging modalities showed equal numbers of lesions in 13 of 38 patients (34%). In three patients, SRS or CT showed more lesions than ^{11}C-5-HTP PET. Patients with a nonfunctioning endocrine pancreatic tumor and a pancreatic carcinoma with some endocrine differentiation on immunohistochemistry were PET negative. A patient with metastasized thymus carcinoma only showed the primary tumor on ^{11}C-5-HTP PET and SRS, while CT scan also detected metastases. From these patients, it was speculated that in the case of high proliferation rate and dedifferentiation of NETs, ^{18}F-FDG PET is probably the imaging modality of choice. In the 17 patients who had their primary tumor still *in situ*, ^{11}C-5-HTP PET was positive in 16, compared to SRS in 9 and CT in 8 patients. PET could detect surgically removed lesions as small as six mm. The main conclusion of this study was that PET imaging with ^{11}C-5-HTP can be universally applied in NETs, also in patients without elevated 5-HIAA excretion in urine, as long as the tumor is not highly proliferating and/or dedifferentiating [31].

In another study with 24 patients with carcinoid tumors and 23 patients with pancreatic islet cell tumors, ^{11}C-5-HTP PET was compared to conventional imaging with CT and SRS and a ^{18}F-DOPA PET scan [22]. Whole-body PET images with carbidopa pretreatment were recorded 10 min after intravenous injection of ^{11}C-5-HTP or 60 min after intravenous administration of ^{18}F-DOPA. The PET findings were compared, per patient and per lesion, with a composite reference standard derived from all available imaging data along with clinical and cytological/histological information. Results indicated that indeed ^{11}C-5-HTP PET can be seen as a universal imaging agent for carcinoid and pancreatic islet cell tumor patients. It was the only imaging modality that was positive in all patients (100% sensitivity). Especially in islet cell tumor patients, more tumor-positive patients and lesions were found with ^{11}C-5-HTP (100 and 67%) compared to ^{18}F-DOPA (89 and 41%), SRS (78 and 46%), and CT (87 and 68%). In carcinoid patients, the per-lesion analysis showed that ^{18}F-DOPA PET outperformed all other imaging techniques. Adding CT to both imaging techniques resulted in a further improvement in sensitivity in a per-lesion analysis, since both types of imaging techniques were complementary to each other. Therefore, for pancreatic islet cell tumor patients, ^{11}C-5-HTP PET/CT was considered the optimal imaging technique, whereas for carcinoid patients, this was ^{18}F-DOPA PET/CT. Furthermore, it was stated that in carcinoid patients SRS scanning can be omitted without missing any lesions. However, in islet cell tumors in a minority of patients (8%), SRS performed better than both PET techniques and therefore remains of additional value.

A direct comparison between ^{11}C-5-HTP, ^{18}F-DOPA, and the recently developed ^{68}Ga-labeled somatostatin analogues for PET imaging in various subtypes of NETs would be much of interest.

Since tracer availability is scarce due to the difficult tracer production, it will be even more important for this tracer to assess its place in staging and monitoring disease progression in GEP-NETs. An easier to produce serotonin tracer analogue, preferably labeled with a PET isotope with a longer half-life such as ^{18}F, would be beneficial for increasing the understanding of the role of serotonin PET analogues in the imaging of GEP-NETs.

CONCLUSION

In conclusion, ^{11}C-5-HTP PET and ^{18}F-DOPA PET are excellent functional imaging techniques for evaluating patients with (suspected) NETs. There is however only limited evidence for the use of ^{18}F-FDG PET in patients with GEP-NETs. Thus far, the indications for ^{18}F-FDG PET seem limited to specific patients in which dedifferentiation of GEP-NET tumors is suspected, the evaluation of secondary tumors, and possibly therapy monitoring.

Convincing data are available that show the benefits of ^{11}C-5-HTP PET and ^{18}F-DOPA PET for staging patients with proven pancreatic islet cell tumors and carcinoids. More data is needed to determine the place of these tracers in therapy evaluation. For both tracers, the place in diagnostic workup in patients where only a suspicion of NET has risen has yet to be determined. For both tracers, the combination with CT further improves the detection rate of NET, which shows that performing PET scans with these tracers in PET/CT scanners is beneficial for patients. Also, a direct comparison in large homogeneous patient groups of these tracers with ^{68}Ga-labeled somatostatin analogues for PET imaging would be very useful to help the clinician determine when to use which technique.

The major drawbacks of both ^{11}C-5-HTP PET and ^{18}F-DOPA PET are the difficult syntheses and therefore limited availability. However, the commercial availability of ^{18}F-DOPA PET is increasing, which might lead to a better understanding of the place of this tracer in the clinic for patients with (suspected) NET. Since the availability of ^{11}C -5-HTP PET is very limited, the development of a ^{18}F-labeled serotonin precursor that performs at least equal to the ^{11}C-5-HTP tracer would be very helpful.

REFERENCES

[1] Adams, S.; Baum, R.; Rink, T.; et al. European Journal Nuclear Medicine 1998, 25, 79–83.

[2] Belhocine, T.; Foidart, J.; Rigo, P.; et al. Nuclear Medicine Communications 2002, 23, 727–734.

[3] Pasquali, C.; Rubello, D.; Sperti, C.; et al. World Journal of Surgery 1998, 22, 588–592.

[4] Garin, E.; Le Jeune, F.; Devillers, A.; et al. Journal of Nuclear Medicine 2009, 50, 858–864.

[5] Fiebrich, H. B.; Brouwers, A. H.; Koopmans, K. P.; et al. European Journal of Cancer 2009, 45, 2312–2315.

[6] Jager, P. L.; Chirakal, R.; Marriott, C. J.; et al. Journal of Nuclear Medicine 2008, 49, 573–586.

[7] Koopmans, K. P.; Neels, O. N.; Kema, I. P.; et al. Critical Reviews in Oncology/Hematology 2009, 71, 199–213.

[8] Lemaire, C.; Damhaut, P.; Plenevaux, A.; et al. Journal of Nuclear Medicine 1994, 35, 1996–2002.

[9] Garnett, E. S.; Firnau, G.; Hahmias, C. Nature 1983, 305, 137–138.

[10] Ahlström, H.; Eriksson, B.; Bergström, M.; et al. Radiology 1995, 195, 333–337.

[11] Hoegerle, S.; Schneider, B.; Fraft, A.; et al. Nuklearmedizin 1999, 38, 127–130.

[12] Hoegerle, S.; Ghanem, N.; Althehoefer, C.; et al. European Journal of Nuclear Medicine and Molecular Imaging 2003, 30, 689–694.

[13] Haug, A.; Auernhammer, C. J.; Wängler, B.; et al. European Journal of Nuclear Medicine and Molecular Imaging 2009, 36, 765–770.

[14] Becherer, A.; Szabó, M.; Karanikas, G.; et al. Journal of Nuclear Medicine 2004, 45, 1161–1167.

[15] Kauhanen, S.; Seppänen, M.; Nuutila, P.; et al. Journal of Clinical Oncology 2008, 26, 5307–5308.

[16] Kema, IP.; Koopmans, KP.; Elsinga, PH.; et al. Journal of Clinical Oncology 2008, 26, 5308–5309.

[17] Örlefors, H.; Sundin, A.; Lu, L.; et al. European Journal of Nuclear Medicine and Molecular Imaging 2006, 33, 60–65.

[18] Koopmans, K. P.; De Vries, E. G. E.; Kema, I. P.; et al. Lancet Oncology 2006, 7, 728–734.

[19] Brown, W. D.; Oakes, T. R.; DeJesus, O. T.; et al. Journal of Nuclear Medicine 1998, 39, 1884–1891.

[20] Koopmans, K. P.; Brouwers, A. H.; De Hooge, M. N.; et al. Journal of Nuclear Medicine 2005, 46, 1240–1243.

[21] Ambrosini, V.; Tomassett, P.; Castellucci, P.; et al. European Journal of Nuclear Medicine and Molecular Imaging 2008, 35, 1431–1438.

[22] Koopmans, K. P.; Neels, O. C.; Kema, I. P.; et al. Journal of Clinical Oncology 2008, 26, 1489–1495.

[23] Yakemchuk, V. N.; Jager, P. L.; Chirakal, R.; et al. Nuclear Medicine Communications 2012, 33, 322–330.

[24] Fiebrich, H. B.; de Jong, J. R.; Kema, I. P.; et al. European Journal of Nuclear Medicine and Molecular Imaging 2011, 38, 1854–1861.

[25] Kauhanen, S.; Seppänen, M.; Ovaska, J.; et al. Endocrine Related Cancer 2009, 16, 255–265.

[26] Ambrosini, V.; Tomassetti, P.; Rubello, D.; et al. Nuclear Medicine Communications 2007, 28, 1599–1606.

[27] Bjurling, P.; Watanabe, Y.; Tokushige, M.; et al. Journal Chemical Society Perkin Transactions 1989, 1331–1334.

[28] Neels, O. C.; Jager, P. L.; Koopmans, K. P.; et al. Journal of Labelled Compounds and Radiopharmaceuticals 2006, 49, 889–895.

[29] Eriksson, B.; Bergström, M.; Anders, L.; et al. Acta Oncologica 1993, 32, 189–196.

[30] Örlefors, H.; Sundin, A.; Ahlström, H.; et al. Journal of Clinical Oncology 1998, 16, 2534–2541.

[31] Orlefors, H.; Sundin, A.; Garske, U.; et al. Journal of Clinical Endocrinology & Metabolism 2005, 90, 3392–3400.

4.4

THE PLACE OF SOMATOSTATIN RECEPTOR SCINTIGRAPHY IN CLINICAL SETTING: INTRODUCTION

ALICJA HUBALEWSKA-DYDEJCZYK

Department of Endocrinology with Nuclear Medicine Unit, Medical College, Jagiellonian University, Krakow, Poland

ABBREVIATIONS

CT	computed tomography
EUS	endoscopic ultrasound
MANEC	mixed adenoneuroendocrine cancer
MRI	magnetic resonance imaging
NEN	neuroendocrine neoplasm
NET	neuroendocrine tumor
SRS	somatostatin receptor scintigraphy
Sstr	somatostatin receptor
VIP	vasoactive intestinal peptide
WHO	World Health Organization

Overexpression of somatostatin receptors on the cells of neuroendocrine tumors (NETs) gives the possibility of the specific diagnostic and therapeutic options. Somatostatin receptors are present in 80–100% of functioning and nonfunctioning neuroendocrine neoplasms (NENs) [1–3]. Though the distribution of somatostatin receptors in NETs is generally homogenous, there are differences in the incidence and

Somatostatin Analogues: From Research to Clinical Practice, First Edition. Edited by
Alicja Hubalewska-Dydejczyk, Alberto Signore, Marion de Jong, Rudi A. Dierckx,
John Buscombe, and Christophe Van de Wiele.
© 2015 John Wiley & Sons, Inc. Published 2015 by John Wiley & Sons, Inc.

density of these receptors depending on the type of the tumor and the degree of tumor's differentiation [1–3]. According to Reubi et al., pancreatic NENs such as gastrinomas express sstr2 in about 100%, sstr5 in 35%, sstr3 in 20%, and sstr1 in 10%; insulinomas express especially sstr2 but only in 70%, sstr1 in 60%, sstr3 in 35%, sstr5 in 15%, and sstr4 in about 3%; and jejunoileal NENs express sstr2 in about 95%, sstr1 in 50%, sstr5 in 48%, sstr3 in 15%, and sstr4 in 3% [1]. Moreover, somatostatin receptor subtypes differ in their affinity for the radioligand, which also influences tumor detectability.

The current WHO 2010 classification system divides NENs by their mitotic index or Ki67 into NET G1 (with Ki67 ≤2%), NET G2 (Ki67 3–20%), and neuroendocrine cancer (NEC) with Ki67 over 20% [4, 5]. There are also mixed adenoneuroendocrine cancers (MANEC) distinguished [4, 5]. The sensitivity of standard somatostatin receptor scintigraphy (SRS) depending on the type of NET and the degree of its differentiation is presented in Table 4.4.1.

Sensitivity of SRS differs also between primary lesion and metastases. In some cases of NENs, liver metastases may appear isointense due to a similar degree of tracer accumulation by the normal hepatic tissue. Therefore, correlation with anatomic imaging and SPECT imaging may be helpful [3, 6]. Hybrid imaging results in more accurate characterization of foci showing elevated radiopharmaceutical uptake and a more precise anatomical localization. CT can be also used to generate attenuation maps to correct the SPECT imaging. The usefulness of this technology is especially visible in the abdominal lesions with sensitivity of 95% and specificity of 92% [3, 7].

TABLE 4.4.1   Sensitivity of standard SRS depending on the type of neuroendocrine tumor and the degree of its differentiation

High sensitivity > 75%
1. GEP-NEN
   (a) NET G1 (according to WHO 2010) (excluding insulinoma)
       – Functioning endocrine tumors, that is, gastrinoma
       – Nonfunctioning endocrine pancreatic tumors and jejunoileal NENs
   (b) NET G2 (according to WHO 2010)
       – Functioning tumors of pancreatic and extrapancreatic origin (gastrinoma, VIPoma, glucagonoma, jejunoileal NETs)
       – Nonfunctioning pancreatic tumors
2. Other endocrine tumors
       – Paraganglioma
       – Malignant pheochromocytoma
       – Small cell lung cancer

Intermediate sensitivity 40–75%
1. GEP-NEN
   (a) NET G1 (according to WHO 2010)
       – Insulinoma
   (b) NEC G3 (according to WHO 2010)
   (c) MANEC (mixed adenoneuroendocrine cancer)
2. Other endocrine tumors
       – Medullary thyroid cancer
       – Differentiated thyroid cancer including Hurthle cell cancer
       – Pheochromocytoma

The advantage of SRS is that it can examine all body regions, whereas conventional imaging can only examine suspected areas. In normal scintigraphic imaging, the thyroid, spleen, liver, kidney, pituitary, and adrenal glands are visualized due to the sstr expression in those glands. The urinary bladder and bowel are usually visualized to variable degrees. Uptake in the kidneys is mainly the consequence of the reabsorption of the radiolabeled peptide in the renal tubular cells after glomerular filtration. While interpreting results of SRS, it has to be taken into consideration that somatostatin receptors exist also on white body cells. This may lead to false-positive results in cases of inflammation or infection and existing healing processes after surgical treatment. False-positive uptake in SRS may be visible in radiation pneumonia, bacterial pneumonia, respiratory infections, accessory spleen, surgical scar tissue, nodular goiter, focal collection of stool, gallbladder, ventral hernia, cerebrovascular accident, concomitant granulomatous disease, urine contamination, and concomitant second primary tumor. The negative results may be connected with the lack of sstr on tumor cell membrane, sstr type, and sstr functional status. The receptor-negative lesions may be poorly differentiated and characterized by aggressive growth and poor prognosis.

Indications to SRS includes detection and localization of different types of NENs and their metastases, staging in patients with NEN, selection of patients with metastatic tumors and/or inoperable primary lesion for treatment with "cold" and radiolabeled somatostatin analogues, prediction of the effect of PRRT, and follow-up of patients to evaluate potential recurrence [1, 8–10].

Sensitivity of different imaging modalities varies for locating specific NETs [10]. SRS specificity for detection of primary gastrointestinal NETs is 86–95% and is higher than for location of pancreatic gastrin-/VIP-/somatostatin-secreting NETs (75%) and insulinomas (50–60%) [10]. For pancreatic NETs, the imaging modality with higher specificity is endoscopic ultrasound (EUS) (82–93%) [10]. Specificities of dual-phase multidetector computed tomography (CT) and magnetic resonance imaging (MRI) in the case of pancreatic NENs are 57–94% and 74–94%, respectively [10]. In the case of gastrointestinal NENs, specificities of CT enteroclysis and MRI enteroclysis are 85 and 86%, respectively [10]. For detection of neuroendocrine liver metastases, specificities of CT and MRI are 44–82% and 82–95%, respectively [10].

It is worth mentioning here that the assessment of somatostatin receptor expression is possible also with the use of positron emission tomography/computed tomography (PET/CT) with different somatostatin analogues (DOTA-NOC, DOTA-TOC, DOTA-TATE) labeled with ^{68}Ga [10, 11], and this imaging method is being increasingly introduced into clinical practice.

REFERENCES

[1] Reubi, J.C. Neuroendocrinology 2004, 80, 51–56.

[2] Appetecchia, M.; Baldelli, R. Journal of Experimental and Clinical Cancer Research 2010, 29, 19.

[3] Pepe, G.; Moncayo, R.; Bombardieri, E.; et al. European Journal of Nuclear Medicine and Molecular Imaging 2012, 39, 41–51.

[4] Bosman, F.; Carneiro, F.; Hruban R. H.; et al. *WHO Classification of Tumours of the Digestive System; IARC Press*, Lyon: 2010.

[5] Salazar, R.; Wiedenmann, B.; Rindi, G.; et al. Neuroendocrinology 2012, 95, 71–73.

[6] Kwekkeboom, D. J.; Kam, B. L.; van Essen M.; et al. Endocrine-Related Cancer 2010, 17, 53–73.

[7] Bombardieri, E.; Coliva, A.; Maccauro, M.; et al. Quarterly Journal of Nuclear Medicine and Molecular Imaging 2010, 54, 3–15.

[8] Kwekkeboom, D. J.; Krenning, E. P.; Scheidhauer, K.; et al. Neuroendocrinology 2009, 90, 184–189.

[9] Őberg, K.; Reubi, J. C.; Kwekkeboom, D. J.; et al. Gastroenterology 2010; 139, 742–753.

[10] Ramage, J. K.; Ahmed, A.; Ardill, J.; et al. Gut 2010, 61, 6–32.

[11] Ambrosini, V.; Campana, D.; Tomasetti, P. et al. European Journal of Nuclear Medicine and Molecular Imaging 2012, 39, 52–60.

4.4.1

SOMATOSTATIN RECEPTOR SCINTIGRAPHY IN MANAGEMENT OF PATIENTS WITH NEUROENDOCRINE NEOPLASMS

ANNA SOWA-STASZCZAK[1], AGNIESZKA STEFAŃSKA[1], AGATA JABROCKA-HYBEL[1,2], AND ALICJA HUBALEWSKA-DYDEJCZYK[3]

[1] *Department of Endocrinology, University Hospital in Krakow, Krakow, Poland*
[2] *Department of Endocrinology, Medical College, Jagiellonian University, Krakow, Poland*
[3] *Department of Endocrinology with Nuclear Medicine Unit, Medical College, Jagiellonian University, Krakow, Poland*

ABBREVIATIONS

^{18}F-DOPA	6-[fluoride-18]fluoro-levodopa
5-HIAA	5-hydroxyindoleacetic acid
5-HTP	5-hydroxytryptophan
ACTH	adrenocorticotropic hormone
CEUS	contrast-enhanced ultrasonography
CgA	chromogranin A
CT	computed tomography
d-NENs	duodenal neuroendocrine neoplasms
EUS	endoscopic ultrasonography
FDG PET	fluorodeoxyglucose positron emission tomography
FNAC/B	fine-needle aspiration cytology/biopsy
GCC	goblet cell carcinoma

Somatostatin Analogues: From Research to Clinical Practice, First Edition. Edited by Alicja Hubalewska-Dydejczyk, Alberto Signore, Marion de Jong, Rudi A. Dierckx, John Buscombe, and Christophe Van de Wiele.

GLP-1R   glucagon-like peptide type 1 receptors
g-NENs   gastric neuroendocrine neoplasms
IOUS   intraoperative ultrasound
LCNEC   large cell neuroendocrine cancers
MANEC   mixed adenoneuroendocrine carcinoma
MEN-1   multiple endocrine neoplasia type 1
MRI   magnetic resonance imaging
MRT   magnetic resonance tomography
NEC   neuroendocrine cancer
NEN   neuroendocrine neoplasm
NET   neuroendocrine tumor
NF-NENs   nonfunctioning neuroendocrine neoplasms
PET   positron emission tomography
p-NEN   pancreatic neuroendocrine neoplasm
PRRT   peptide receptor radionuclide therapy
RGS   radio-guided surgery
SCLC   small cell lung cancer
SPECT   single-photon emission computed tomography
SRI   somatostatin receptor imaging
SRS   somatostatin receptor scintigraphy
sstr   somatostatin receptors
THPVS   transhepatic portal venous sampling
UGI   upper gastrointestinal
USG   ultrasound examination
VIP   vasoactive intestinal peptide
ZES   Zollinger–Ellison syndrome

GASTRIC NEUROENDOCRINE NEOPLASMS

Gastric neuroendocrine neoplasms (g-NENs) are nowadays revealed more often due to expanding indications to upper gastrointestinal (UGI) endoscopy [1, 2]. Usually silent and benign, gastric NEN may, however, be aggressive and may sometimes mimic the course of gastric adenocarcinoma [1, 2]. The main cause of type 1 gastric NEN is achlorhydria secondary to autoimmune atrophic fundic gastritis [1–5]. These neoplasms present usually as multiple (2–10) polyps, <1 cm in diameter in the gastric fundus with little risk of deep invasion of the gastric wall [5]. Type 2 gastric NENs are related to hypergastrinemia resulting from tumoral secretion from gastrinomas (Zollinger–Ellison syndrome), mostly in patients presenting with multiple endocrine neoplasia type 1 (MEN-1) [1–3]. These neoplasms present usually as small polyps (<1–2 cm), may involve the entire fundic mucosa, and are generally asymptomatic [1–3]. Type 3 neoplasms are usually solitary neuroendocrine cancers (NEC) G3 tumors, above 2 cm in diameter with infiltrative growth [1–4]. In type 3 pain, weight loss and iron-deficiency anemia occur very often; also, distal metastases are observed in more than 50% of cases [1–4].

The minimal biochemical tests in patients with type 1 and 2 g-NENs include serum gastrin and chromogranin A (CgA), and in patients with type 3 tumors (especially in case of rare in this group well-differentiated tumors) assessment of CgA level may be useful [1–3]. Moreover, in type 1 tumors, antiparietal cell and anti-intrinsic factor autoantibodies as well as thyroid functional tests and thyroperoxidase antibodies should be assessed [2, 3].

g-NENs are revealed by UGI endoscopy. Endoscopy and biopsy are usually sufficient for type 1 and small type 2 tumors [1]. Biopsy samples should be taken from antrum (2 biopsies) and fundus (4 biopsies) in addition to biopsies of the largest polyps [1–5]. When the tumor size is above 1 cm, endoscopic ultrasonography (EUS) should be performed to assess regional lymph node involvement and for histological confirmation by fine-needle aspiration [1–5]. Computed tomography (CT) and magnetic resonance imaging (MRI) have limited value for small type 1 and 2 tumors [1–5]. However, CT, MRI, and transabdominal ultrasonography have high sensitivity/specificity to detect liver metastases. Therefore, these imaging procedures should be considered in case of larger tumors or tumors invasive in EUS [2]. While cells of the g-NENs, as in the other cases of neuroendocrine tumors (NET), express all subtypes of somatostatin receptors (sstr), except sstr4, somatostatin receptor scintigraphy (SRS) is also useful diagnostic tool [1]. SRS is particularly recommended in patients with well-differentiated tumors to search for liver, lymph node, and bone metastases [2, 4, 5]. In case of type 3 g-NENs, which are usually more aggressive and poorly differentiated tumors (NEC G3), UGI endoscopy is not sufficient diagnostic tool [2, 4, 5]. In these tumors, use of EUS, CT, MRI, SRS, and/ or positron emission tomography (PET) should be considered to assess advancement of the disease; to detect lymph nodes, liver, or bone metastases; and also in case of disseminated tumors to qualify patients to peptide receptor radionuclide therapy (PRRT) [2, 4, 5].

Figures 4.4.1.1 and 4.4.1.2 present the use of different techniques, including SRS, in diagnosis (Fig. 4.4.1.1) and follow-up (Fig. 4.4.1.2) in case of g-NENs.

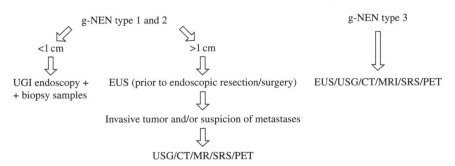

FIGURE 4.4.1.1 Gastric neuroendocrine neoplasms (g-NENs): diagnostic procedures and imaging (according to [2]). CT, computed tomography; EUS, endoscopic ultrasonography; MRI, magnetic resonance imaging; PET, positron emission tomography; SRS, somatostatin receptor scintigraphy; UGI, upper gastrointestinal; USG, ultrasound examination.

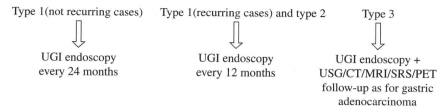

Type 1(not recurring cases)	Type 1(recurring cases) and type 2	Type 3
UGI endoscopy every 24 months	UGI endoscopy every 12 months	UGI endoscopy + USG/CT/MRI/SRS/PET follow-up as for gastric adenocarcinoma

FIGURE 4.4.1.2   Gastric neuroendocrine neoplasms (g-NENs): follow-up (according to [2]). CT, computed tomography; MRI, magnetic resonance imaging; PET, positron emission tomography; SRS, somatostatin receptor scintigraphy; UGI, upper gastrointestinal; USG, ultrasound examination.

DUODENAL NEUROENDOCRINE NEOPLASMS

Duodenal neuroendocrine neoplasms (d-NENs) are generally small, >75%, and have <2 cm in diameter [6]. Tumors are usually limited to the submucosa or mucosa, in which 40–60% are associated with regional lymph node metastases. Liver metastases occur in <10% of all patients with d-NENs [7]. These tumors may produce hormones such as gastrin and somatostatin, which is associated with specific clinical syndromes. In some asymptomatic cases, hormone's production (serotonin and calcitonin) is revealed by the immunohistochemical examination [7]. More than 90% of d-NENs do not cause any clinical syndrome. Therefore, usually, tumor-related symptoms such as pain, jaundice, nausea/vomiting, bleeding, anemia, diarrhea, duodenal obstruction, or the incidental discovery of the tumor (usually at UGI endoscopy) lead to diagnosis [6, 7]. Lesions are usually single; multiple tumors are only found in about 9% of cases. Multiple lesions should lead to a suspicion of MEN-1, which occurs in 6±2.5% of all patients with d-NENs [6].

Similarly as in case of g-NENs to assess the primary d-NEN, UGI endoscopy with biopsy is the most sensitive modality [2, 7]. Endoscopic ultrasonography (EUS) examination is used for confirmation of the diagnosis and for staging of the disease [2, 7]. Gastrinomas may be primarily submucosal and may be therefore not revealed by both UGI endoscopy and EUS, which results in low detection rate (30–60%) for duodenal tumors causing Zollinger–Ellison syndrome (ZES), which was diagnosed by hormone assays [7]. In case of d-NENs, SRS might play an important role not only in searching for lymph node, liver, and bone metastases but, different than in case of g-NENs, also in detection of the primary tumor [2, 7–9]. The imaging methods (CT, MRI, ultrasound) are generally not useful in conventional diagnostic of the usually small primary duodenal tumors [2, 7, 8]. However, helical CT and MRI are used to fully assess disease extent and to reveal distant metastases [2, 7] although studies with gastrinomas suggest SRS may be more sensitive [7]. In patients with advanced disease, especially in suspicion of bone metastases (bone metastases are often present in patients with liver metastases), whole-body SRS and MRI of the spinal column should be performed [7]. In diagnostic approach of d-NENs, there is also a place for PET/CT imaging, which is one of the most sensitive examinations to assess small lesions (<1 cm) and metastases to the lymph nodes.

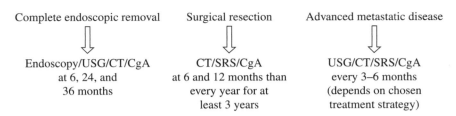

Complete endoscopic removal	Surgical resection	Advanced metastatic disease
⇓	⇓	⇓
Endoscopy/USG/CT/CgA at 6, 24, and 36 months	CT/SRS/CgA at 6 and 12 months than every year for at least 3 years	USG/CT/SRS/CgA every 3–6 months (depends on chosen treatment strategy)

FIGURE 4.4.1.3 Figure 1.3 Duodenal neuroendocrine neoplasms (d-NENs): follow-up (according to [2]). CgA, chromogranin A; CT, computed tomography; EUS, endoscopic ultrasonography; SRS, somatostatin receptor scintigraphy; USG, ultrasound examination.

SRS, among other imaging techniques, might be useful also in the follow-up of d-NENs (Fig. 4.4.1.3).

DIAGNOSTIC PROCEDURES IN ZES

In case of patients with ZES, sensitive and specific diagnostic tools are needed to choose the proper treatment option. Tumor localization studies are necessary to localize the primary tumor, to determine whether the surgical resection is indicated, and to determine the extent of the disease and the presence of metastases [10, 11]. Tumor localization studies should be performed in all patients with biochemically established ZES. UGI endoscopy with inspection of the duodenum followed by a helical CT and SRS is recommended as initial study. If these studies are negative, EUS should be performed. EUS is particularly sensitive for pancreatic lesions, and its use in detection of small duodenal focuses is controversial [10–12]. If results of aforementioned studies are negative, selective angiography with secretin stimulation and hepatic venous sampling should be considered [10, 11]. This is worth to emphasize that SRS plays an important role in diagnosis of primary tumor and metastatic lesions in case of gastrinomas. Prospective studies for primary gastrinomas show that conventional imaging studies localize 10–40%, angiography 20–50%, and SRS 60–70% of the tumors [10–13]. The use of SRS changes management in 15–45% of patients [10, 13]. Prospective studies in metastatic gastrinoma show that CT and ultrasonography detect 30–50% of patients with metastases, MRI and angiography 60–75%, and SRS 92% [13]. Promising are results of the use of PET/CT with ^{68}Ga-labeled somatostatin analogues (SSA), ^{11}C-5-hydroxytryptophan (5-HTP) or 6-[fluoride-18]fluoro-levodopa (^{18}F-DOPA), but the availability of this techniques is still limited [11].

PANCREATIC NEUROENDOCRINE NEOPLASMS

Pancreatic neuroendocrine neoplasms (p-NENs) account for about 2–10% of all pancreatic tumors [14, 15]. p-NENs can be divided into nonfunctioning and functioning tumors. Nonfunctioning neoplasms (NF-NENs) account for about 50% of all pancreatic NET, and most of them (60–100%) are classified as NET G1 and

G2 [14–16]. Nonfunctioning tumors are usually located in the head of the pancreas. Clinical presentation relates to the anatomic site of the lesion. Predominant symptoms can be abdominal pain, weight loss, and jaundice. These symptoms are similar as those found in other pancreatic tumors such as adenocarcinoma. The term "nonfunctioning" refers to the absence of clinical symptoms of hormonal hypersecretion, but these tumors may show immunohistochemical positivity for hormones, neuropeptides, or neurotransmitters [14, 15]. About 8% of patients with nonfunctioning p-NENs have MEN-1 syndrome, while the prevalence of NF-NEN in MEN-1 patients is about 55%. Functioning tumors will be discussed in the following text.

In diagnosis of p-NEN, abdominal ultrasonography is usually the first-step examination. But the sensitivity of ultrasonography in diagnosis of small lesions such as gastrinoma or insulinoma varies between 19 and 70% [14, 15]. However, the recent implementation of contrast-enhanced ultrasonography (CEUS) has led to improvement in the diagnostic capabilities of B-mode sonography of the liver and pancreas [14, 17, 18]. Moreover, there is a correlation observed between CEUS enhancement pattern and the Ki-67 index [14, 17, 18]. However, the standard imaging procedures for p-NENs are contrast-enhanced helical CT or MRI and EUS in combination with SRS. These methods are used to detect the primary tumor and metastases. EUS provides high-resolution images of structures within or just beyond the wall of the gastrointestinal wall and is very effective method for detection of NENs [14, 15, 19]. The sensitivity of CT and MRI in localization of the primary tumor is about 75–79% [14, 15, 20]. In difficult situation, magnetic resonance tomography (MRT) can be used. The aforementioned techniques enable differentiation of the hypervascular pancreatic NET from hypovascular pancreatic adenocarcinoma (multidetector CT or MRT). CT and MRT are helpful in determination of the mean larger volume of the tumor and assessment of the presence of the cystic component and the lack of infiltration of peripancreatic fat and vessels in case of NET in comparison to the more aggressively growing adenocarcinoma [14, 15]. In patients with a high degree of clinical suspicion and negative noninvasive imaging like USG, CT, and/or MRI/MRT, further diagnostic investigations may include contrast-enhanced USG, where sensitivity and specificity are 94 and 96%, respectively, or EUS with EUS-guided fine-needle aspiration cytology/biopsy (FNAC/B) with sensitivity of 82–86% [14, 15].

Pancreatic NET show high expression of receptor subtypes 2, 3, and 5, while the expression of subtype 1 is usually intermediate. A lot of gastrinoma and glucagonoma tumors express sstr subtype 2, while all somatostatinomas express subtype 5 receptor. These facts enable use of SRS as important diagnostic tool to detect both primary tumor and metastases and should be performed prior to the treatment. SRS has a high sensitivity and specificity for p-NENs, 90 and 80%, respectively [14, 21]. SRS can be used for the localization of the primary disease and for assessment of the extent of the disease [14, 15]. It is the most sensitive method for assessment of the presence of extrahepatic disease. SRS is an important diagnostic procedure when the demonstration of extrahepatic metastases is necessary for making therapeutic decision [14, 15]. Image-fusion data combining CT and SRS (single-photon emission computed tomography (SPECT)) appears promising in helping to accurately locate residues and plan surgery and to detect lesions especially in the tail of the pancreas, in staging, and in qualification to PRRT. Following standard SRS procedure is recommended: a double- or triple-head gamma camera and a medium-energy, parallel-hole collimator, peaks at 172 and 245 keV with the window of 20% [14, 15, 22]. At an acquisition time of 15 min and 4 hours (h) after injection, anterior and posterior abdominal views should be performed [14, 15, 22]. At 24 h after injection, anterior and posterior views of the upper abdomen, head, chest, and pelvis as well as left and right lateral, anterior, and posterior oblique views of the upper abdomen should be performed [14, 15, 22]. Whole-body imaging should be performed with a scanning speed of 3 cm/min [14, 15, 22]. SPECT images should be acquired at 24 h after injection with a 6-step rotation for 360°/40–60 s [14, 15, 22]. Optional delayed images at 30–48 h after injection are recommended [14, 15, 22].

PET/CT with ^{68}Ga-labeled SSA is another sensitive diagnostic method, but its use is still limited (Fig. 4.4.1.4). In comparison to scintigraphy, PET has a two- to threefold higher spatial resolution and facilities quantification of tracer uptake [14, 23]. PET with the use of 5-hydroxytryptophan (5-HTP) or ^{18}F-DOPA has also shown promising results and may be an option for detection of small well-differentiated tumors [14, 24]. Standard PET with ^{18}F-glucose is not efficient in detecting well-differentiated tumors, but can be helpful in the detection of aggressive poorly differentiated p-NENs (NEC G3) [14]. NEC G3 of the pancreas is very rare tumors. Patients present with jaundice, weight loss, abdominal pain, and hepatomegaly. Overproduction of the hormones is rare, but Cushing's syndrome and carcinoid syndrome were reported in few cases. Histopathological features of pancreatic NEC G3 include small- to intermediate-sized tumor cells growing diffusely or in irregular nests, often with extensive necrosis and high mitotic rate [25, 26]. In diagnosis of these neoplasms, similarly as for well-differentiated tumors, CT, MRI, or EUS with biopsy was used. Fluorodeoxyglucose positron emission tomography (FDG PET) may be useful in the diagnosis of the primary tumor and for staging [25, 26]. SRS is not recommended, but should be evaluated in the clinical setting.

Somatostatin receptor imaging (SRI), including SRS and/or PET/CT with the use of labeled SSA, is useful in the follow-up of p-NENs (Fig. 4.4.1.5).

FUNCTIONING p-NENS

In contrast to NF-NENs, where due to lack of specific symptoms, diagnosis is made usually in the advanced stage of the disease; in case of hormonally active pancreatic tumors, revealing of the small primary lesion might be a diagnostic challenge.

For both nonfunctioning and functioning p-NENs, CgA is a tumor marker, and its serum concentration should be measured for diagnostic and follow-up purposes [14, 15].

(a)   (b)

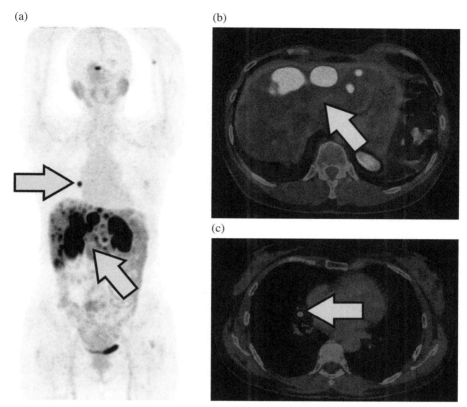

(c)

FIGURE 4.4.1.4 PET/CT with radiolabeled somatostatin analogue (^{68}Ga-DOTATOC) in patient with disseminated pancreatic neuroendocrine tumor. Arrows indicate liver (a, b) and lymph node metastases (a, c). (*See insert for color representation of the figure.*)

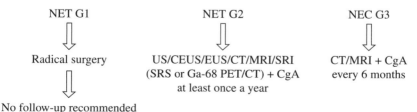

FIGURE 4.4.1.5 Pancreatic neuroendocrine tumors: follow-up (according to [14]). NET, neuroendocrine tumor; NEC, neuroendocrine cancer; CEUS, contrast-enhanced ultrasound; EUS, endoscopic ultrasonography; CT, computed tomography; MRI, magnetic resonance imaging; SRI, somatostatin receptor imaging; SRS, somatostatin receptor scintigraphy; PET, positron emission tomography; CgA, chromogranin A.

Insulinomas are the most common functioning NET of the pancreas [11]. They are usually small, solitary tumors, and the majority of the insulinoma tumors have pancreatic localization [11, 27]. These small tumors might be difficult to localize radiologically. The best diagnostic methods to localize insulinoma are MRI, 3-phase CT, and EUS. But usually more than one diagnostic tool has to be used.

In CT, insulinomas are usually hypervascular, and therefore, these tumors and their metastases are better visible in arterial phase. The sensitivity of CT for the detection of insulinomas ranges from 30 to 85%, depending on tumor size [28]. MRI shows sensitivity from 85 to 95% [28]. Compared to CT, MRI is superior in the detection of small lesions [28]. The enhancement pattern of these tumors on MRI depends primarily on their hypervascularity [28]. Small metastases and the primary tumor show homogeneous enhancement [28]. EUS is effective but invasive preoperative procedure to localize insulinomas with sensitivity of 94% [11, 27]. The high spatial resolution of this technique allows the detection of very small lesions and their precise anatomical localization. It is easier to localize lesions in the head and body of the pancreas than in the tail, for lesions in the pancreatic tail, sensitivity is about 60% [27]. Other invasive investigations such as angiography combined with calcium stimulation and transhepatic portal venous sampling (THPVS) might be useful for insulinoma localization, when the noninvasive techniques have failed [11]. Angiography combines anatomic localization with functional information provided by THPVS, which can confirm angiographic abnormality as insulinomas [27]. The rate of false-positive results is low; sensitivity ranges from 63 to 94%. Intraoperative localization techniques, which include careful palpation of the pancreas and the use of intraoperative ultrasound (IOUS), remain the most reliable way to localize insulinomas and to determine the correct surgical procedure [27]. Combination of palpation and IOUS enables detection of about 92% of tumors [28]. SRS might be less useful in localization of insulinomas, because these tumors have a low density of sstr and generally do not express the somatostatin subtype 2 and 5 cell-surface receptor [11, 29]. In malignant insulinomas, the relative distribution of sstr subtypes is different than in benign tumors, and a higher rate of scan positivity with this technique can be expected [27, 30].

To detect tumors such as insulinoma, gastrinoma, or medullary thyroid cancer, in which cells express not only sstr but also glucagon-like peptide type 1 receptors (GLP-1R), a new radiopharmaceutical—labeled exendin-4—might be used [31, 32]. Christ et al. showed the usefulness of ^{111}In-labeled GLP-1R agonist ^{111}In-DOTA-exendin-4 in localizing insulinomas using SPECT with combination of CT images [33].

In our center, the first clinical study with the use of ^{99m}Tc-labeled analogue of GLP-1 to diagnose primary insulinoma was performed (Fig. 4.4.1.6). This method enabled detection of the primary insulinoma tumor, local recurrence, and metastases in cases in which other diagnostic methods have failed. We also detected medullary thyroid cancer and glucagonoma using this method.

PET/CT with the use of ^{68}Ga-labeled SSA or ^{11}C-5-HTP might be considered in experienced centers in case of doubtful results of aforementioned methods [11].

Diagnostics of **gastrinomas** was discussed in the preceding text.

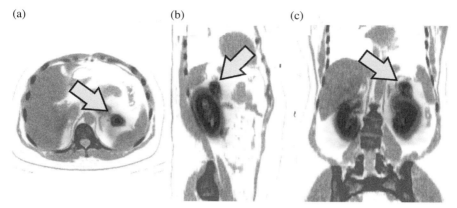

FIGURE 4.4.1.6 A 65-year-old woman with benign insulinoma: scintigraphy with ^{99m}Tc-labeled glucagon-like peptide 1 (GLP-1) analogue. A study showing a small lesion of insulinoma in tail of the pancreas. Arrows indicate lesion in the tail of the pancreas (a - transverse plane, b - sagittal plane, c - coronal plane).

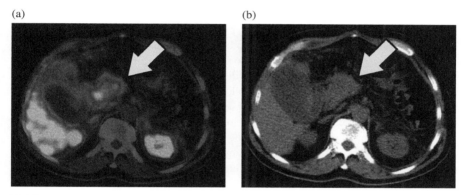

FIGURE 4.4.1.7 A 60-year-old man with recurrence of pancreatic glucagonoma. Somatostatin receptor scintigraphy with ^{99m}Tc-EDDA/HYNIC-TOC showing a recurrence of the disease in the head of the pancreas. Arrows indicate the lesion in the head of the pancreas (a - SPECT/CT image; b - CT image). (*See insert for color representation of the figure.*)

Similarly as in case of NF-NENs to diagnose rare functioning tumors of the pancreas, such as glucagonoma, VIP-oma, ACTH-oma, and somatostatinoma, combined use of multidetector CT (or MRI) and SRS-SPECT is always recommended [11, 34]. EUS and EUS-guided fine-needle aspiration may be useful in cases in which previously mentioned techniques are inconclusive [11, 34]. PET/CT methods are not recommended on a routine basis; however, particularly ^{68}Ga-labeled SSA PET, if available, might be helpful in doubtful cases [11, 34]. **Glucagonomas** are usually large tumors, and CT scanning is the imaging modality of choice for the detection of these neoplasms. SRS is helpful in confirming the diagnosis and in detecting metastatic disease, which is present in up to 50% of patients [35, 36]. SRS enables localization of

95–100% of primary and metastatic sites (Fig. 4.4.1.7) [35, 36]. Similarly, **VIP-omas** are localized in CT scanning in most cases; SRS is useful to confirm the diagnosis and to identify metastatic disease [29]. Nikou et al. showed that SRS was superior to conventional anatomic imaging modalities and detected 91% of primary tumors and 75% of cases with metastases [37]. **Somatostatinomas** are usually solitary tumors localized in the head of the pancreas. Standard imaging such as CT, MRI, and EUS are useful for localization and staging. SRS is used to confirm the diagnosis and also for staging [29].

BRONCHIAL NENS

Bronchial NENs consist of enterochromaffin or Kulchitsky cells and are located in the bronchial mucosa. Bronchial NENs account for 2–5% of all bronchial tumors. Bronchial NENs are usually well or intermediate differentiated, rarely metastatic [38, 39]. This is the description of typical and atypical bronchial carcinoids, which are NET G1 and NET G2 tumors, respectively [38, 39]. But in some percentage there are poorly differentiated large cell neuroendocrine cancer (LCNEC) and small cell lung cancers (SCLC) [38, 39]. The most common symptoms are recurrent pneumonia, cough, hemoptysis, and chest pain. Peripheral tumors are usually asymptomatic. In some cases, hormonal secretion leads to paraneoplastic syndromes or atypical carcinoid syndrome. Most bronchial NENs express sstr, and therefore, SRS can be used as diagnostic tool. CT detects 94% of primary and 89% of recurrent/metastatic sites [40]. SRS is a useful method to detect primary tumor, metastases, and recurrent disease. The results of this examination are helpful in planning the treatment and for follow-up. In doubtful cases, SRS may identify neuroendocrine origin of the tumor. In some cases, SRS might be more sensitive than conventional anatomic imaging, allowing proper treatment planning or enabling the assessment of local recurrence [40, 41]. The sensitivity of SRS does not depend exclusively on tumor size, but rather on a high target-to-background ratio associated with the density of the sstr on the cells' surface compared with adjacent tissue. Yellin et al. presented for SRS the sensitivity of 90% and specificity of 83% in diagnosis of primary and metastatic disease, respectively [41]. The sstr exist on white body cells, which should be taken into consideration in the assessment of the postoperative SRS. In some cases, the sites of inflammation or infection may lead to false-positive results of SRS. Similarly as for NENs in other localization imaging, another modality of SRI, with still limited use but promising results, is PET/CT with ^{68}Ga-labeled SSA [40]. FDG PET might be a useful diagnostic tool in case of aggressive, atypical LCNEC or SCLC [40], while a majority of typical benign and well-differentiated bronchial NENs are FDG PET negative, which was reported for the first time by Erasmus et al. [40].

JEJUNOILEAL NENS

Jejunoileal NENs are usually slowly growing, but even the tumor smaller than 1 cm in diameter may be metastatic at presentation. The size of the primary lesion correlates with the presence of lymph nodes and distant metastases, and an early diagnosis is clearly imperative [42]. Jejunoileal NENs are often detected while searching for a primary tumor in asymptomatic but metastatic patients [43]. The most frequent initial symptom is nonspecific abdominal pain [43–49]. Other nonspecific symptoms include weight loss, fatigue, fever of unknown origin, and tumor mass-related symptoms such as nausea, vomiting, jaundice, or gastrointestinal bleeding [43–48]. The rate of functionality and the presence of the carcinoid syndrome in patients with jejunoileal NENs are about 20–30% [43, 45, 46, 48]. The carcinoid syndrome is associated with presence of liver metastases (in at least 95% of patients) [43, 47, 48]. The minimally required biochemical tests in case of suspicion of jejunoileal NEN are plasma CgA and urinary 5-hydroxyindoleacetic acid (5-HIAA) [43, 50, 51]. CgA has been shown to be a prognostic factor with higher levels of this marker indicating worse prognosis [43, 50, 51].

In diagnosis of jejunoileal NENs, SRI plays very important role. Similarly as for NENs in other localization, imaging examinations used to search for the primary tumor are abdominal USG, CT, and/or MRI, but these imaging modalities should be followed by SRS, especially in combination with SPECT/CT [43]. Jejunoileal NENs are usually small tumors; therefore, abdominal USG is usually not sufficient method to reveal the primary lesion [43, 52]. However, it may be used for screening of hepatic metastases [43, 52]. CT (3-phase, contrast-enhanced, multislice-detector CT) and/or MRI (also with contrast media) enable not only imaging of the primary tumor but also detection of the lymph node and/or distant metastases [22, 43, 52–54]. CT or MR enteroclysis is more sensitive and specific method for detection of the primary tumor in the small intestine, but it is not widely available [43, 55, 56]. Colonoscopy and gastroscopy are not useful diagnostic tools in case of tumors of the small intestine [43]. However, lesions localized close to the ileocecal valve may be revealed with the use of colonoscopy [43]. Video capsule endoscopy and double-balloon enteroscopy are more effective techniques to search for primary tumors of the ileum and jejunum, but the limitation is also their availability [43, 57, 58].

As it was mentioned earlier, SRI is very useful and important diagnostic procedure in case of jejunoileal NENs for detection of primary tumors, for staging, for detection of bone metastases, for qualification to PRRT, and for follow-up [43]. The sensitivity of SRS ranges between 80 and 95% and specificity between 80 and 100% [59]. OctreoScan has the affinity for receptor types 2 and 5 expressed on the cell membranes of NET [60]. Namwongprom et al. have been searching for the correlation between CgA level and SRS findings in the evaluation of metastases in NET [61]. SRS proved to be more specific, more sensitive, and more accurate than CgA to assess metastatic disease in NET [61]. Positive SRS correlated with elevation of serum CgA levels, but serum CgA might be elevated also in patients with negative SRS studies. Nevertheless, both SRS and CgA should be taken into consideration in the evaluation of metastases in patients with NENs [61]. Except the SRS, another SRI is PET with ^{68}Ga with simultaneous CT, which has a high sensitivity ranging from 82 to 100% [43, 62]. Results of this examination may change management in 20–30% of patients [43]. Studies with the use of newer tracers used for PET imaging such as 11-carbon-5-HTP or 18-fluoro-dihydroxyphenylalanine (^{18}F-DOPA) are also promising, but availability of these methods is even lower than for PET/CT with

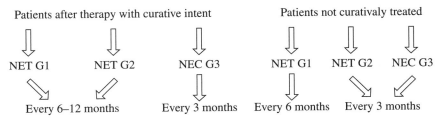

Minimal examination — CgA, 5-HIAA, triphasic CT and SRI (SRS or PET-CT with labeled somatostatin analogues)

FIGURE 4.4.1.8   Jejunoileal Neuroendocrine Neoplasms (NENs) - Lifelong follow-up (according to [43]). NET, neuroendocrine tumor; NEC, neuroendocrine cancer; CgA, chromogranin A; 5-HIAA, 5-hydroxyindoloeacetic acid; CT, computed tomography; SRI, somatostatin receptor imaging; SRS, somatostatin receptor scintigraphy; PET, positron emission tomography.

^{68}Ga-labeled SSA [43, 62, 63]. However, there are also studies showing that ^{18}F-DOPA detected 57–64% of NENs and had lower per-lesion sensitivity and less intense uptake than ^{68}Ga. ^{18}F-DOPA is reserved for somatostatin-negative and serotonin-positive tumors [64]. In two separate reports of patients with jejunoileal NENs, the sensitivity of ^{18}F-DOPA ranged from 93 to 97% with an impact on management in 50% of patients; in patients with noncarcinoid NENs, the sensitivity was 25%. In contrast, SRS sensitivity was similar in both groups, with values 81 and 75%, respectively. PET/CT with the use of FDG had low sensitivity for slow-growing tumors; therefore, it does not have an important role for the usually well- or moderately differentiated jejunoileal NENs [43]. FDG uptake increases with increasing proliferation rate and may be a useful marker of prognosis in patients with poorly differentiated tumors. It may be a prognostic factor for early progression of the disease. Early progression was seen in 93% of FDG-positive patients and 82% of SRS-negative patients [64]. In contrast, only 9% of FDG-negative patients and 26% of SRS-positive patients had early progression [64]. Belhocine et al. compared SRS and FDG in patients with NET. Those authors found out that 86% of primary tumors were localized with SRS, but only 57% with FDG. Patients with metastatic disease were positive with SRS in 69% of cases, with FDG in 47%, and with anatomic imaging in 56% [65].

Figure 4.4.1.8 presents the use of different imaging techniques, including SRI (both SRS and ^{68}Ga-PET/CT), in the follow-up of jejunoileal NENs.

APPENDICEAL NENS

Appendiceal NENs are usually diagnosed incidentally on histopathological examination after appendicectomy; therefore, there are no specific symptoms connected with these tumors [43]. The carcinoid syndrome is very rarely described in metastatic NENs, so cases with typical carcinoid syndrome should raise a suspicion of jejunoileal NEN [43, 66]. Usually, appendiceal NENs are well-differentiated tumors with

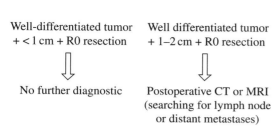

Well-differentiated tumor + < 1 cm + R0 resection

⇩

No further diagnostic

Well differentiated tumor + 1–2 cm + R0 resection

⇩

Postoperative CT or MRI (searching for lymph node or distant metastases)

Tumor > 2 cm and/or localization at the base of the appendix, mesoappendiceal invasion < 3 mm or angioinvasion

Long time follow up: contract-enhanced multiphase CT or MRI + SRI (SRS with SPECT-CT or somatostatin receptor PET with CT) 6 and 12 months postoperatively, and then once a year

FIGURE 4.4.1.9 Appendiceal NENs – Follow-up (according to [43]). CT, computed tomography; MRI, magnetic resonance imaging; SRI, somatostatin receptor imaging; SRS, somatostatin receptor scintigraphy; PET, positron emission tomography.

good prognosis, but there are some features connected with the higher risk of recurrence or dissemination, such as location at the base of the appendix, size >2 cm, deep mesoappendiceal invasion, or margin invasion [43, 66–68]. In case of appendiceal NENs, diagnostic procedures are used usually for postoperative staging and follow-up [43, 69]. CgA can be used as a tumor marker, particularly to differentiate NEN from goblet cell carcinoids/carcinoma (GCC) [43]. Colonoscopy or CT colonography might be considered to diagnose possible synchronous tumors [43, 69]. Follow-up has to be lifelong, as recurrence has been seen even 20 years after diagnosis [69].

SRI has also its place in the follow-up of patients with appendiceal NENs (Fig. 4.4.1.9).

GCC

GCC is a rare subtype of mixed adenoneuroendocrine carcinomas (MANEC) with malignant behavior, similar to the clinical course of adenocarcinoma [43, 67–69]. These tumors are also usually incidentally found after appendectomy; therefore, diagnostic procedures are rather used for postoperative staging and follow-up [43, 67–69]. Imaging examinations will involve similar procedures as for high-risk (>2 cm) appendiceal NENs [43, 70]. CT or MRI of the abdomen, pelvis, and also thorax should be performed [43, 70]. SRI with the use of SRS or PET scanning (with CT) should be considered; however, their sensitivity decreases in case of poorly differentiated tumors [43, 70]. In those cases, FDG PET might be useful [43]. Colonoscopy should be performed for screening of synchronous or metachronous tumors, due to the potentially increased risk of secondary neoplasms [43, 69, 70]. In case of GCC serum CgA has no value for detection and monitoring of these tumors [43, 70]. More useful are markers such as carcinoembryonic antigen (CEA), CA-19-9 and CA-125 [43, 70]. Follow-up should be lifelong and should mimic the guidelines for colorectal adenocarcinoma [43, 70].

RADIO-GUIDED SURGERY

Radio-guided surgery (RGS) is an intraoperative localizing technique using target-specific radiotracer that is accumulated in tissues. SRS followed by RGS enables detection of the occult tumors and improves the effectiveness of surgical treatment of patients suffering from gastrointestinal NENs expressing sstr. The sensitivity of SRS in localization of NENs ranges (according to different studies results) between 51 and 96% and depends on the type of analogue, which is used. The sensitivity of PET examination is about 95%, which is especially high with the use of ^{68}Ga-labeled SSA [71]. In a few studies based on examination of small group of patients, the sensitivity of scintigraphy followed by RGS to detect gastrointestinal NENs was assessed for 90% (Fig. 4.4.1.10) [71, 72]. Some authors concluded that intraoperative detection of NEN after ^{111}In-labeled SSA administration is more sensitive than SRS and intraoperative palpation [69]. Finding of primary tumors gives the possibility of the radical removal of the tumors' mass.

Some NENs, especially tumors localized in the small intestine, with a relative high malignancy rate and small tumors of the pancreas such as insulinoma cause the most difficulties in preoperative localization. In these cases, the use of a handheld

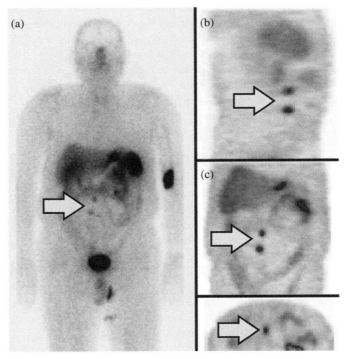

FIGURE 4.4.1.10   A 72-year-old man with primary unknown origin. ^{99m}Tc-EDDA/HYNIC-TOC revealed a primary neuroendocrine tumor of the small intestine confirmed later by radio-guided surgery and histopathological examination. Arrows indicate the primary tumor (a, b, c).

gamma probe enables the localization of tumors, especially in cases with positive SRS and negative results of other imaging examinations. SRS followed by RGS can localize primary tumors and their metastases, and the density and distribution of cell membrane-bound receptors in a lesion are more determinant than the tumor size [69, 73]. Adams et al. presented that RGS seems to be the most sensitive diagnostic tool for detecting microscopic and occult endocrine tumors [74]. The authors detected lesions of 6 mm in diameter [74]. The limitation of this method is false-positive results due to tracer accumulation in the activated lymphocytes within inflammatory infiltrates, which expresses sstr [73]. The major limitation of intraoperative gamma probe detection is high background activity from the liver, kidneys, and spleen. It is extremely important to avoid directing the probe toward such physiological tracer accumulation. During exploration of the pancreas, the probe should be placed toward the posterior surfaces of the pancreas, directing it up toward the abdominal wall. In the case of bowel tumors, the examination is easier. Moving the bowel loop beyond the peritoneal cavity, the examination can be performed while avoiding abdominal background activity. Intraoperative gamma probe scanning requires careful, slow searching of suspected areas, which is time consuming owing to subtle activity differences occurring due to background type [69]. In the literature, successful RGS tumor detection requires tumor-to-nontumor tissue count ratios of at least 1.5–2.0; however, a ratio above 2.4 makes detection more reliable [73–75]. In our experience in group of patients with midgut tumors, the target/nontarget ratio reached 50 for primary lesions and average 2.7 for lymph node metastases. The target/nontarget ratio was substantially lower for islet cell pancreatic tumors.

COLORECTAL NENS

Colonic NENs in about 30–40% present with metastases to the lymph nodes, liver, mesentery, and peritoneum at the time of diagnosis [76–78]. Patients with metastases have a 5-year survival of about 50% [76, 77]. The clinical symptoms in colonic tumors are not specific and include diarrhea, abdominal pain, gastrointestinal bleeding, or weight loss [75, 77].

Rectal tumors are usually small, polypoid lesions located between 4 and 20 cm above the dentate line on the anterior or lateral rectal wall, most often diagnosed incidentally during the routine sigmoidoscopy [74, 77]. At present, increasing incidence of rectal NENs is observed. Small tumors rarely metastasize, but larger tumors, above 2 cm in diameter, may present with metastases to the bones, lymph nodes, and liver [76, 77]. Overall distant metastases of rectal NENs occur in only 2.3% of cases [76, 77]. Rectal NENs may present as an incidental finding on sigmoidoscopy or colonoscopy [72, 73].

The base diagnostic tool for colorectal NENs is endoscopy [76, 77]. Full colonoscopic assessment is required to exclude concomitant colonic disease, as part of staging, and to exclude metastases [76, 77]. Another first-step examination may be CT colonography or barium enema (which has lower sensitivity for colorectal neoplasms) [76, 77]. However, tumors detected with the use of these methods require histopathological confirmation; therefore, endoscopic excision has to be performed [76, 77].

Standard diagnostic tool such as abdominal ultrasound has low sensitivity for the detection of primary lesion and assessment of the local advancement of the disease but is useful in searching for metastases and to perform the biopsy of suspected liver lesions [76, 77]. Multislice triple-phase CT is most useful for revealing metastases in the thorax, abdomen, and pelvis, but MRI is superior to detect liver metastases [76, 77, 79].

In diagnostic management of rectal NENs, endoanal/rectal ultrasound (EUS) is very useful in preoperative assessment, and results of the examination influence the choice of treatment option [76, 78]. It provides information about the tumor size, the depth of invasion, and the presence of pararectal lymph node metastases [76, 78].

SRI, with either SRS or ^{68}Ga-labaled SSA PET/CT, is usually used for staging if residual or metastatic disease is suspected [76, 78]. Detection of primary tumor localized in the rectum with background activity can be difficult with SRI [76, 78]. Poorly differentiated colorectal NENs, similarly as in other localization, are often negative in SRI. In those cases, PET with FDG might be used [76, 77].

The minimum biochemical marker is serum CgA [76, 77].

The use of SRS and PET/CT with labeled SSA in the follow-up of patients with colorectal NENs is presented in the following text (Fig. 4.4.1.11).

Methods of follow-up:
Rectal NENs — EUS, colonoscopy, MRI
Colonic NENs — CT, colonoscopy
Disseminated colorectal NENs — CT, MRI, SRI,
(SRS or Ga-68-labeled somatostatin analogues PET/CT), FDG-PET

NET G1, G2	NET G1, G2	NEC G3	NET G1, G2	NEC G3
+ <1 cm	+ 1–2 cm	+ < 1 cm	+ > 2 cm	+ > 2 cm
+ No lymph nodes metastases				
+ No invasion of muscularis				
⇓	⇘	⇗	⇓	⇓
No data to recommend regular follow-up	Annual follow-up then as per adenomatous polyp follow-up protocol	Endoscopy + CT/MRI + CgA within 1st year	Endoscopy + CT/MRI + CgA every 4–6 months then annually

FIGURE 4.4.1.11 Colorectal NENs—Follow-up (according to [76]). NEN, neuroendocrine neoplasm; EUS, endoscopic ultrasonography; CT, computed tomography; MRI, magnetic resonance imaging; SRI, somatostatin receptor imaging; SRS, somatostatin receptor scintigraphy; PET, positron emission tomography; CgA, chromogranin A.

NENS OF UNKNOWN ORIGIN

NENs, especially neoplasms of the pancreas and intestine (small intestine and right hemicolon), are frequently metastatic at the time of initial diagnosis [79, 80]. It is also not uncommon that the patient presents with metastases, usually liver lesions, of unknown origin [79, 80]. All available diagnostic methods such as endoscopy, CT, MRI, SRI (SRS and/or PET/CT with labeled SSA), or FDG PET might be used to search for the primary lesion in case of NEN of unknown origin [79, 80]. However, in about 5–10% of all such cases, it is not possible to detect the primary lesion [79, 80]. In patients with liver metastases of unknown origin, the initial diagnostic approach includes histological and immunohistochemical examination of the hepatic lesions with assessment of Ki-67 and/or count of mitoses per 10 high-power fields as the basis for grading of the tumors [79–82]. Analysis of hormones, monoamines, and transcription factors may also provide important information in searching for the primary site [79, 80]. To search for the primary tumor, the most common primary sites of metastatic NENs should be taken into consideration, and adequate diagnostic procedures should be performed, that is, colonoscopy for the investigation of the large bowel, double-balloon enteroscopy or video capsule endoscopy for the small intestine, and EUS for pancreatic tumors [79, 80]. For staging, imaging examinations such as CT and/or MRI should be performed [79, 80]. In case of well-differentiated NENs (NET G1 and G2), SRI—either SPECT/CT SRS or PET/CT using ^{68}Ga-labeled SSA such as DOTATOC, DOTATATE, or DOTANOC—is required [79, 80, 83–88]. These examinations might be helpful in searching for the primary tumor and are important in the assessment of the stage of the disease and very useful in planning the treatment [79, 80]. Other promising and useful diagnostic tools are ^{18}F-DOPA PET/CT or 5-HTP-PET/CT, and these examinations might be considered in case of negative results of SRS; however, their availability is limited [35, 79]. Cells of the low-differentiated NENs (NEC G3) do not usually express sstr; therefore, in those cases, FDG PET might be more useful [79, 80]. Use of FDG PET in case of NET G2 is not recommended as a routine imaging method; however, there are studies indicating its prognostic value [79, 89].

SUMMARY

SRS is a useful imaging modality in case of patients with NENs. SRS provides information about both the receptor status of the primary tumor and the extent of the disease. On the basis of SRS results (high tumor uptake and favorable target/nontarget ratios), patients with sstr-expressing tumors are qualified to PRRT or to the treatment with long-acting SSA [90]. The level of the uptake in SRS is also a prognostic factor. Usually, a high uptake in SRS indicates a well-differentiated histotype more likely to respond well to therapy, whereas a low uptake, or no uptake at all, is generally correlated with a poorly differentiated histotype and worse prognosis for the patients [90]. It is also known that the degree of differentiation may vary between primary tumor and metastases. Moreover, it can change also during the treatment [90]. This fact should be taken into consideration in the assessment of SRS after therapy.

SRS can visualize primary tumors and possible metastatic lesions by virtue of studying a variety of molecular processes with high sensitivity in the body. Mapping of the results with those of anatomic imaging may give individualized information about heterogeneity between metastases as well [91]. Due to the aforementioned characteristic features, interpretation of SRS images can be difficult due to physiological uptake in different organs.

SRS has an established role in diagnostic algorithm in different types of NENs. In 25–30%, SRS plays an important role in making a therapeutic decision especially such as qualification to PRRT. In some cases, SRS is the best method in postoperative follow-up, in assessment of the extent of the disease (staging), and in follow-up to evaluate potential recurrence. PET scanning with the use of ^{68}Ga-labeled SSA is more sensitive for detection and follow-up of patients with NENs; therefore, the time needed for investigation in comparison to SPECT is reduced. The PET examination is performed in a short-time window, in which it is only possible to assess the presence of the receptors. SRS can be performed after the longer interval time, according to the type of isotopes that is used for the study. It means that, in addition to ascertaining the expression of the receptor on membrane surface, internalization, which occurs after binding, can be assessed. And internalization of the radiocompound is a predictor of an efficacy of the radiometabolic therapy. The proof of internalization is a better guarantee of response to the therapy than early PET imaging, and therefore, a further discussion and clinical studies are necessary [91]. It is emphasized in the literature that PET imaging had significantly higher sensitivity than SRS in patients with NENs [90]. FDG PET might be considered in patients with negative SRS [90].

REFERENCES

[1] Ruszniewski, P.; Fave, G. D.; Cadiot, G.; et al. Neuroendocrinology 2006, 84, 158–116.

[2] Fave, G. D.; Kwekkeboom, D. J.; van Cutsem, E.; et al. Neuroendocrinology 2012, 95, 74–87.

[3] Merola, E.; Sbrozzi-Vanni, A.; Panzuto, F.; et al. Neuroendocrinology 2012, 95, 207–213.

[4] Borch, K.; Ahren, B.; Ahlman, H.; et al. Annals of Surgery 2005, 242, 64–73.

[5] Rindi, G.; Azzoni, C.; La Rosa, S.; et al. Gastroenterology 1999, 116, 532–542.

[6] Hoffmann, K. M.; Furukawa, M.; Jensen, R. T.; et al. Best Practice & Research Clinical Gastroenterology 2005, 19, 675–697.

[7] Jensen, R. T.; Rindi, G.; Arnold, R.; et al. Neuroendocrinology 2006, 84, 165–172.

[8] Frucht, H.; Doppman, J. L.; Norton, J. A.; et al. Radiology 1989, 171, 713–717.

[9] Kloppel, G.; Perren, A.; Heitz, P. U.; et al. Annals of the New York Academy of Science 2004, 1014, 13–27.

[10] Jensen, R. T.; Niederle, B.; Mitry, E.; et al. Neuroendocrinology 2006, 84, 173–182.

[11] Jensen, R. T.; Cadiot, G.; Brandi, M. L.; et al. Neuroendocrinology 2012, 95, 98–119.

[12] Norton, J. A.; Jensen, R. T. Annals of Surgery 2004, 240, 757–773.

[13] Gibril, F.; Reynolds, J. C. Digestive and Liver Disease 2004, 36, 106–120.

[14] Falconi, M.; Bartsch, D. K.; Eriksson, B.; et al. Neuroendocrinology 2012, 95, 120–134.

[15] Falconi, M.; Plöckinger, U.; Kwekkeboom, D. J.; et al. Neuroendocrinology 2006, 84, 196–211.

[16] Marion-Audibert, A. M.; Barel, C.; Gouysse, G.; et al. Gastroenterology 2003, 125, 1094–1104.

[17] Quaia, E.; Stacul, F.; Gaiani, S.; et al. Radiology Medical Journal 2004, 108, 71–81.

[18] Malago, R.; D'Onforio, M.; Zamboni, G. A.; et al. American Journal of Roentgenology 2009, 192, 424–430.

[19] McLean, A. M.; Fairclough, P. D. Best Practice & Research Clinical Endocrinology & Metabolism 2005, 19, 177–193.

[20] Rappeport, E. D.; Hansen, C. P.; Kjaer, A.; et al. Acta Radiologica 2006, 47, 248–256.

[21] Lebtahi, R.; Cadiot, G.; Sarda, L.; et al. Journal of Nuclear Medicine 1997, 38, 853–858.

[22] Ricke, J.; Klose, K. J.; Mignon, M.; et al. European Journal of Radiology 2001, 37, 8–17.

[23] Hofmann, M.; Maecke, H.; Borner, R.; et al. European Journal of Nuclear Medicine 2001, 28, 1751–1757.

[24] Hoegerle, S.; Altehoefer, C.; Ghanem, N.; et al. Radiology 2001, 220, 373–380.

[25] Nilsson, O.; Cutsem, E. V.; Fave, G. D.; et al. Neuroendocrinology 2006, 84, 212–215.

[26] DeLellis, R. A.; Lloyd, R. V.; Heitz, P. U.; et al. Pathology and genetics of tumors of endocrine organs. IARC Press, Lyon, 2004.

[27] de Herder, W.; Niederle, B.; Scoazec, J. Y.; et al. Neuroendocrinology 2006, 84, 183–188.

[28] Noone, T. C.; Hosey, J.; Firat, T.; et al. Best Practice & Research: Clinical Endocrinology & Metabolism 2005, 19, 195–211.

[29] Divies, K.; Conlon, K. Current Gastroenterology Reports 2009, 11, 119–127.

[30] Vezzosi, D.; Bennet, A.; Rochaix, P.; et al. European Journal of Endocrinology 2005, 152, 757–767.

[31] Laverman, P.; Sosabowski, J. K.; Boerman, O.; et al. European Journal of Nuclear Medicine and Molecular Imaging 2012, 39, 78–92.

[32] Reubi, J. C. Neuroendocrinology 2004, 80, 51–56.

[33] Christ, E.; Wild, D.; Forrer, F.; et al. Journal of Clinical Endocrinology & Metabolism 2009, 94, 4398–4405.

[34] O'Toole, D.; Salazar, R.; Falconi, M.; et al. Neuroendocrinology 2006, 84, 189–195.

[35] Orlefors, H.; Sundin, A.; Garske, U.; et al. Journal of Clinical Endocrinology & Metabolism 2005, 90, 3392–3400.

[36] Carrasquillo, J.; Chen, C. Seminars in Oncology 2010, 37, 662–679.

[37] Nikou, G. C.; Toubanakis, C.; Nikolaou, P.; et al. Hepatogastroenterology 2005, 52, 1259–1265.

[38] Travis, W. D.; Linnoila, R. I.; Tsokos, M. G.; et al. American Journal of Surgical Pathology 1991, 15, 529–553.

[39] Travis, W. D.; Rush, W.; Flieder, D. B.; et al. American Journal of Surgical Pathology 1998, 22, 934–944.

[40] Fanti, S.; Farsad, M.; Battista, G.; et al. Clinical Nuclear Medicine 2003, 28, 548–552.

[41] Yellin, A.; Zwas, S. T.; Rozenman, J.; et al. Israel Medical Association 2005, 7, 712–716.

[42] Scarpa, M.; Prando, D.; Pozza, A.; et al. Journal of Surgical Oncology 2010, 102, 877–888.

[43] Pape, U. F.; Perren, A.; Niederle, B.; et al. Neuroendocrinology 2012, 95, 135–156.

[44] Landerholm, K.; Falkmer, S.; Jarhult, J. World Journal of Surgery 2010, 34, 1500–1505.

[45] Pape, U. F.; Berndt, U.; Muller-Nordhorn, J.; et al. Endocrine-Related Cancer 2008, 15, 1083–1097.

[46] Helland, S. K.; Prosch, A. M.; Viste, A. Scandinavian Journal of Surgery 2006, 95, 158–161.

[47] Ahmed, A.; Turner, G.; King, B.; et al. Endocrine-Related Cancer 2009, 16, 885–894.

[48] Pape, U. F.; Bohmig, M.; Berndt, U.; et al. Annals of the New York Academy of Science 2004, 1014, 222–233.

[49] Niederle, M. B.; Niederle, B. Oncologist 2011, 16, 602–613.

[50] Arnold, R.; Wilke, A.; Rinke, A.; et al. Clinical Gastroenterology and Hepatology 2008, 6, 820–827.

[51] Korse, C. M.; Bonfrer, J. M. G.; Aaronson, N. K.; et al. Neuroendocrinology 2009, 89, 296–301.

[52] Doerffel, Y.; Wermke, W. Ultraschall in der Medizin 2008, 29, 506–514.

[53] Kaltsas, G.; Rockall, A.; Papadogias, D.; et al. European Journal of Endocrinology 2004, 151, 15–27.

[54] Bader, T. R.; Semelka, R. C.; Chiu, V. C.; et al. Journal of Magnetic Resonance Imaging 2001, 14, 261–269.

[55] Masselli, G.; Polettini, E.; Casciani, E.; et al. Radiology 2009, 251, 743–750.

[56] Kamaoui, I.; De-Luca, V.; Ficarelli, S.; et al. American Journal of Roentgenology 2010, 194, 629–633.

[57] van Tuyl, S. A.; van Noorden, J. T.; Timmer, R.; et al. Gastrointestinal Endoscopy 2006, 64, 66–72.

[58] Belluti, M.; Fry, L. C.; Schmitt, J.; et al. Digestive Diseases and Sciences 2009, 54, 1050–1058.

[59] Kweekkeboom, D. J.; Krenning, E. P. Seminars in Nuclear Medicine 2002, 32, 84–91

[60] Kulaksiz, H.; Eissele, R.; Rossler, D.; et al. Gut 2002, 50, 52–60.

[61] Namwongprom, S.; Wong, F.; Tateishi, U.; et al. Annals of Nuclear Medicine 2008, 22, 237–243.

[62] Haug, A.; Auernhammer, C. J.; Wangler, B.; et al. European Journal of Nuclear Medicine & Molecular Imaging 2009, 36, 765–770.

[63] Eriksson, B.; Kloppel, G.; Krenning, E.; et al. Neuroendocrinology 2008, 87, 8–19.

[64] Garin, E.; Le Jeune, F.; Devillers, A.; et al. Journal of Nuclear Medicine 2009, 50, 858–864.

[65] Belhocine, T.; Foidart, J.; Rigo, P.; et al. Nuclear Medicine Communications 2002, 23, 727–734.

[66] Groth, S. S.; Virnig, B. A.; Al-Refaie, W. B.; et al. Journal of Surgical Oncology 2011, 103, 39–45.

[67] McGory, M. L.; Maggard, M. A.; Kang, H.; et al. Diseases of the Colon & Rectum 2005, 48, 2264–2271.

[68] McCusker, M. E.; Cote, T. R.; Clegg, L. X.; et al. Cancer 2002, 94, 3307–3312.

[69] Plockinger, U.; Couvelard, A.; Falconi, M.; et al. Neuroendocrinology 2008, 87, 20–30.

[70] Toumpanakis, C.; Standish, R. A.; Baishnab, E.; et al. Diseases of the Colon & Rectum 2006, 50, 315–322.

[71] Hubalewska-Dydejczyk, A.; Kulig, A.; Szybiński, P.; et al. European Journal of Nuclear Medicine & Molecular Imaging 2007, 34, 1545–1555.

[72] Adams, S.; Baum, R. P. Quarterly Journal of Nuclear Medicine & Molecular Imaging 2000, 44, 59–67.

[73] Warner, R. R.; O'Dorisio, T. M. Seminars in Nuclear Medicine 2002, 32, 79–83.

[74] Adams, S.; Baum, R. P.; Hertel, A.; et al. Journal of Nuclear Medicine 1998, 39, 1155–1160.

[75] Ohrvall, U.; Westlin, J. E.; Nilsson, S.; et al. Cancer 1997, 80, 2490–2494.

[76] Caplin, M.; Sundin, A.; Nillson, O.; et al. Neuroendocrinology 2012, 95, 88–97.

[77] Ramage, J.; Goretzki, P.; Manfredi, R.; et al. Neuroendocrinology 2008, 87, 31–39.

[78] Matsumoto, T.; Iida, M.; Suekane, H.; et al. Gastrointestinal Endoscopy 1991, 37, 539–542.

[79] Pavel, M.; Baudin, E.; Couvelard, A.; et al. Neuroendocrinology 2012, 95, 157–176.

[80] Steinmuller, T.; Kianmanesh, R.; Falconi, M.; et al. Neuroendocrinology 2008, 87, 47–62.

[81] Rindi, G.; Kloppel, G.; Alhman, H.; et al. Virchows Archiv 2006, 449, 395–401.

[82] Kloppel, G.; Couvelard, A.; Perren, A.; et al. Neuroendocrinology 2009, 90, 162–166.

[83] Perri, M.; Erba, P.; Volterrani, D.; et al. Quarterly Journal of Nuclear Medicine and Molecular Imaging 2008, 52, 323–333.

[84] Patel, C. N.; Chowdhury, F. U.; Scarsbrook, A. F. American Journal of Roentgenology 2008, 190, 815–824.

[85] Prasad, V.; Ambrosini, V.; Hommann, M.; et al. European Journal of Nuclear Medicine and Molecular Imaging 2010, 37, 67–77.

[86] Krausz, Y.; Freedman, N.; Rubinstein, R.; et al. Molecular Imaging and Biology 2011, 13, 583–593.

[87] Ruf, J.; Schiefer, J.; Furth, C.; et al. Journal of Nuclear Medicine 2011, 52, 697–704.

[88] Gabriel, M.; Decristoforo, C.; Kendler, D.; et al. Journal of Nuclear Medicine 2007, 48, 508–518.

[89] Binderup, T.; Knigge, U.; Loft, A.; et al. Clinical Cancer Research 2010, 16, 978–985.

[90] Oberg, K. Annals of the Medical University of Białystok 2005, 50, 62–68.

[91] de Jong, M.; Breeman, W.; Kwekkeboom, D.; et al. Accounts of Chemical Research 2009, 42, 873–880.

4.4.2

THE PLACE OF SOMATOSTATIN RECEPTOR SCINTIGRAPHY AND OTHER FUNCTIONAL IMAGING MODALITIES IN THE SETTING OF PHEOCHROMOCYTOMA AND PARAGANGLIOMA

ALICJA HUBALEWSKA-DYDEJCZYK[1], HENRI J.L.M. TIMMERS[2], AND MALGORZATA TROFIMIUK-MÜLDNER[1]

[1]*Department of Endocrinology with Nuclear Medicine Unit, Medical College, Jagiellonian University, Krakow, Poland*
[2]*Department of Endocrinology, Radboud University Nijmegen Medical Centre, Nijmegen, The Netherlands*

ABBREVIATIONS

CT	computed tomography
FDA	fluorodopamine
FDG	fluoro-2-deoxy-D-glucose
FDOPA	fluoro-L-3,4-dihydroxyphenylalanine
GEP-NET	gastroenteropancreatic neuroendocrine tumor
HNPGL	head and neck paraganglioma
MAX	MYC-associated factor X
MEN-2	multiple endocrine neoplasia type 2

Somatostatin Analogues: From Research to Clinical Practice, First Edition. Edited by Alicja Hubalewska-Dydejczyk, Alberto Signore, Marion de Jong, Rudi A. Dierckx, John Buscombe, and Christophe Van de Wiele.

mIBG   metaiodobenzylguanidine
MRI   magnetic resonance imaging
NF1   neurofibromatosis type 1
PET   positron emission tomography
PGL   paraganglioma
PPGLs   pheochromocytomas and paragangliomas
SDH   succinate dehydrogenase
SDHB   SDH subunit B
SDHD   SDH subunit D
SPECT   single-photon emission computed tomography
SRS   somatostatin receptor scintigraphy
SSTR   somatostatin receptor
T/N   tumor/nontumor
VHL   von Hippel–Lindau syndrome
VMAT-1   vesicular monoamine transporter type 1

PHEOCHROMOCYTOMAS AND PARAGANGLIOMAS

Pheochromocytomas and paragangliomas (PPGLs) are rare neuroendocrine tumors that derive from sympathetic chromaffin tissue in adrenal and extra-adrenal abdominal or thoracic locations [1]. PPGLs arising from the adrenal medulla are commonly referred to as pheochromocytomas. Typical locations for extra-adrenal PPGLs are (1) the Zuckerkandl body, a sympathetic ganglion located at the root of the inferior mesenteric artery; (2) the sympathetic plexus of the urinary bladder, the kidneys, and the heart; and (3) the sympathetic ganglia in the mediastinum. Head and neck paragangliomas (HNPGLs), also called glomus tumors, arise from parasympathetic paraganglia, mainly from the glomus caroticum, glomus (jugulo)tympanicum, and glomus vagale.

The majority of PPGLs produce, metabolize, and secrete catecholamines, whereas HNPGLs usually do not. The main symptoms and signs of catecholamine excess due to PPGLs include headache, palpitations, diaphoresis, and paroxysmal or chronic hypertension. Symptoms and signs of HNPGLs rather relate to the tumor's local space-occupying effects, including cranial nerve damage.

PPGLs occur sporadically or in association with four known familial syndromes: multiple endocrine neoplasia type 2 (MEN-2), von Hippel–Lindau (VHL) syndrome, neurofibromatosis type 1 (NF1), and paraganglioma syndromes associated with mutations of genes encoding subunits of the succinate dehydrogenase (SDH) complex, in particular subunits B (SDHB) and D (SDHD). Reported frequencies of germ line mutations of the aforementioned genes among patients with PPGLs range from 27 to 32% [2] and are likely to increase as further tumor susceptibility genes are identified. Most recently, mutations of genes encoding the SDH complex assembly factor 2, transmembrane protein 127, SDH subunit A, and MYC-associated factor X (MAX) have been identified as further hereditary causes of PPGLs [3–6].

To establish a biochemical diagnosis of PPGL, plasma and/or 24 h urine concentrations of the catecholamines epinephrine and norepinephrine and their

O-methylated metabolites metanephrine and normetanephrine ("metanephrines") are measured. Especially, plasma-free metanephrines and urinary fractionated metanephrines both have an excellent sensitivity (>97%) for detecting PPGL [7].

FUNCTIONAL IMAGING TARGETS

In patients with a biochemically established diagnosis of PPGL, anatomical and functional imaging are critical for primary tumor localization and detection of multiple primary tumors and metastases, guiding the optimal choice between curative surgery and palliative treatment options. CT and MRI provide a high sensitivity and allow precise tumor delineation. Lesions detected by anatomical imaging can be specifically identified as PPGL by functional imaging agents that target the catecholamine synthesis, storage, and secretion pathways of chromaffin tumor cells [8]. These techniques include [$^{123/131}$I]-metaiodobenzylguanidine ($^{123/131}$I-mIBG) single-photon emission computed tomography (SPECT) and 6-[^{18}F]-fluorodopamine (^{18}F-FDA) positron emission tomography (PET). $^{123/131}$I-mIBG and ^{18}F-FDA target the norepinephrine transporter of the PPGL cell membrane and the vesicular monoamine transporters in the membrane of intracellular vesicles. These transporters facilitate the reuptake and storage of catecholamines, respectively. The PET tracers ^{11}C-epinephrine and ^{11}C-hydroxyephedrine are alternatives that accumulate in tumor cells through the same mechanisms but are of limited use for clinical imaging because of their (very) short half-life.

6-[^{18}F]-Fluoro-L-3,4-dihydroxyphenylalanine (^{18}F-DOPA) PET can be used for the imaging of the striatal system and neuroendocrine tumors such as carcinoids but also for PPGL and HNPGL. The target of ^{18}F-DOPA is the large amino acid transporter involved in the uptake of amine precursors.

Other less-specific targets for PPGL imaging are the somatostatin receptors (SSTRs) and glucose transporters. For SSTR-based imaging, ^{111}In-pentetreotide and ^{99m}Tc-HYNIC-TOC/HYNIC-TATE are available for SPECT and ^{68}Ga-DOTATATE/DOTANOC/DOTATOC for PET. 2-[^{18}F]-Fluoro-2-deoxy-D-glucose (^{18}F-FDG) PET provides an index of intracellular glucose metabolism and is taken up by the tumor cell through the glucose transporters [8–10]. Functional imaging (SPECT and PET) are usually combined with CT for attenuation correction and colocalization. A schematic overview of the targets on PPGL tumor cells for the currently available radiopharmaceuticals is given in Figure 4.4.2.1 [11].

SSTR EXPRESSION IN PPGLS

The key phenomenon for applying somatostatin receptor scintigraphy (SRS) in PPGL imaging is the expression of SSTRs on tumor cells, described for the first time in 1992 by Reubi et al. [12]. SSTRs were revealed by autoradiography in 73% of studied pheochromocytomas and 93% PGLs. High receptor density was observed in 36 and 43% of cases, respectively [12]. This discovery was confirmed by Epelbaum et al. [13]. The SSTR status of the examined tumors did not correlate

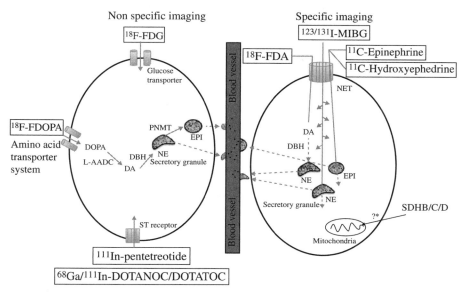

FIGURE 4.4.2.1   Functional imaging targets in PPGL. Adapted from [11], with permission. Nonspecific and specific imaging for pheochromocytoma/paraganglioma. DBH, dopamine-beta-hydroxylase; EPI, epinephrine; ^{18}F-FDA, ^{18}F-fluorodopamine; ^{18}F-FDG, ^{18}F-fluoro-2-deoxy-D-glucose; ^{18}F-FDOPA, ^{18}F-dihydroxyphenylalanine; ^{18}F-FNE, ^{18}F-free norepinephrine; NE, norepinephrine; DOTANOC, DOTA-Nal3-octreotide; DOTATOC, DOTA-Tyr3-octreotide; [123/131]I-mIBG, [123/131]I-metaiodobenzylguanidine; l-AADC, l-aromatic-amino acid decarboxylase; NET, norepinephrine transporter; PNMT, phenylethanolamine-*N*-methyltransferase; SDHB/C/D, succinate dehydrogenase subunit B/C/D; ST receptor, somatostatin receptor; ?*, potential radiopharmaceutical directed at mutations in the mitochondria.

with features such as age, tumor size, tumor location (adrenal vs. extra-adrenal), malignancy, or urinary metanephrine excretion. However, the density of SSTRs was significantly higher in tumors with no uptake of labeled mIBG. It is interesting that mRNAs of all five subtypes of SSTRs (SSTR1–5) were measurable in PPGLs.

In further studies, Mundschenk et al. confirmed the expression of different SSTRs subtypes by immunohistochemistry in up to 94% of examined pheochromocytomas, 48% of which showed the presence of at least two subtypes [14]. In contrast to gastroenteropancreatic neuroendocrine tumors (GEP-NETs), SSTR subtype 3 was the most frequent (90.4% of studied samples), followed by subtype 2A (25%), subtype 5 (15.4%), subtype 4 (10.4%), and subtype 1 (7.7%). The dominant expression of SSTR subtype 3 in benign pheochromocytoma was also confirmed by other authors [15]. However, in a recent study on SSTR mRNA expression in PPGLs, the most frequent receptor subtypes were SSTR subtype 2 and SSTR subtype 1 (100 and 94% of 52 cases included, respectively), with no difference between pheochromocytomas and extra-adrenal PGLs. The level of SSTR subtype 3 and SSTR subtype 5 mRNA was low and detectable only in 53 and 47% of cases, respectively; however, both receptors were more significantly expressed in pheochromocytomas. In 6 of 7 immunohistochemically stained PPGL samples, SSTR2 mRNA was associated with protein expression [16]. No relationship between the SSTR subtype expression pattern and malignant behavior of the tumors has been found; however, SSTR2A presence is more often being observed in extra-adrenal lesions.

From the clinical point of view, it is important that metastatic lesions are also SSTR positive. It should be emphasized that SSTR expression in primary lesions and their metastases, as well as in recurrent lesions, may differ. This phenomenon is well known in neuroendocrine tumors.

EXPRESSION OF SSTRS AND IMAGING DIAGNOSTICS OF PPGLS

The first attempt to apply a labeled somatostatin analogue in localization of tumors derived from chromaffin cells was undertaken over 30 years ago. In the first study on ^{123}I-Tyr-3-octreotide performed by Krenning et al., the only pheochromocytoma patient included was negative on SRS imaging [17]. The next study with the same agent published in 1990 comprised 20 patients with paragangliomas. This time, only two patients were negative on imaging, the tumor size (3–5 mm) being the most probable cause of false-negative results [18]. Reubi et al. correlated the SRS results with autoradiographic tissue studies. They concluded that low density of SSTRs in PPGLs may result in a lack of visualization of the tumor [12]. The group of Krenning was also the first to report the high clinical impact of ^{111}In-octreotide (^{111}In-pentetreotide) imaging on PPGL patient management [19, 20]. The first reports on ^{99m}Tc-labeled somatostatin analogues used in the diagnosis of PPGLs came from the early 2000s [21].

As most of the labeled somatostatin analogues currently used in nuclear diagnostics are highly SSTR2 specific, the membrane expression of this subtype of SSTR is of crucial relevance for imaging results. In a study of Mundschenk et al., SSTR expression was compared with results of preoperatively performed OctreoScan (^{111}In-pentetreotide) scintigraphy. Approximately 50% of patients with positive SRS showed membrane expression of SSTR subtype 2A. If SSTR subtype 2A expression was absent, SSTR subtype 3 was detected. Negative results of SRS were connected with no SSTR subtype 2A or subtype 3 membrane expression or with cytoplasmic presence of SSTR3 only [14].

SENSITIVITY AND SPECIFICITY OF SRS

Due to the relatively high malignant potential of especially extra-adrenal PGLs, and no specific histological or immunohistochemical features of the primary tumor to confirm its malignancy, the detection of distant metastases is of key importance and may influence the choice of the treatment.

It is generally known that SRS is of limited value in nonmetastatic PPGLs. Metastases from malignant PPGLs are more frequently being detected with SRS (Fig. 4.4.2.2). The study comparing ^{123}I-mIBG scintigraphy, ^{18}F-FDA, and ^{111}In-OctreoScan in patients with nonmetastatic and metastatic pheochromocytomas showed equal sensitivity of ^{18}F-FDA and ^{123}I-mIBG (87.5%) and lower of SRS

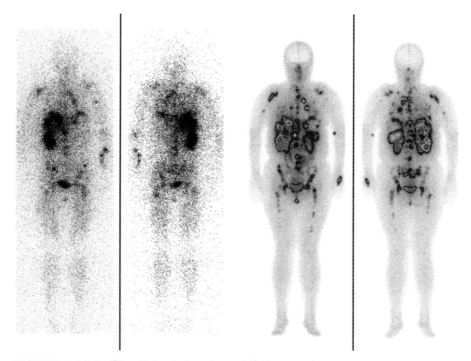

FIGURE 4.4.2.2 ^{131}I-mIBG scintigraphy and ^{99m}Tc-HYNIC-TATE SRS in a 22-year-old woman with metastatic pheochromocytoma.

(28.5%) in patients with nonmetastatic pheochromocytomas [22]. In patients with metastatic tumors, the region-by-region sensitivity was 78.4% for ^{18}F-FDA, 58.9% for ^{123}I-mIBG, and 68.5% for ^{111}In-OctreoScan SRS [22].

The sensitivity of SRS is lower than mIBG scintigraphy in PGLs with exception of HNPGLs, in which it ranges from 86 to 100% for malignant tumors [23, 24]. Kwekkeboom et al. showed the usefulness of SRS with the ^{111}In-labeled somatostatin analogue in the visualization of PGLs already in 1993—the authors were able to detect 94% of previously known lesions and additional foci in 36% investigated patients [20]. Bustillo et al. applied the same tracer and proved high accuracy (90%), sensitivity (94%), and specificity (75%) of the method in PGLs [25].

In their paper, Koopmans et al. showed the superiority of SRS over scintigraphy with labeled mIBG in a group of 29 patients with HNPGLs, in whom SRS revealed additional lesions not detected by other methods. The authors suggested that SRS is likely to contribute to optimal lesion characterization in this patient category (a sensitivity of ^{111}In-octreotide was 95% vs. 44.5% for ^{123}I-mIBG). In view of the low yield of mIBG imaging, this method should not be routinely used in the evaluation of HNPGL [26]. Similar results were obtained for ^{99m}Tc-HYNIC-TOC: the sensitivity of the tracer in the detection of extra-adrenal PGLs was 92.9% for benign and 100% for malignant tumors (the sensitivity of ^{123}I-mIBG scintigraphy was 71.4 and 72.7%, respectively) (Fig. 4.4.2.3) [27].

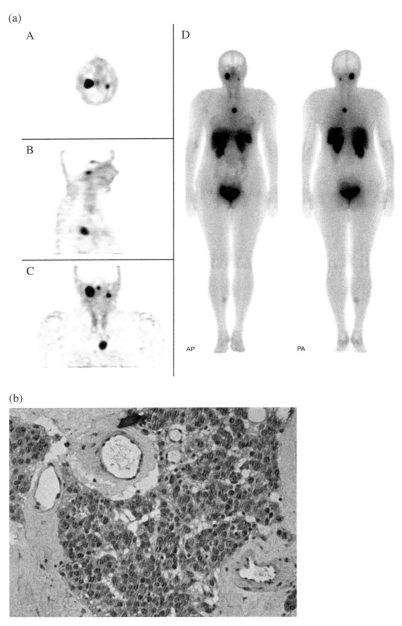

FIGURE 4.4.2.3   (a) ^{99m}Tc-EDDA/HYNIC-TOC SRS in a 37-year-old woman with multiple paragangliomas with prevalent expression of SSTR4 (80% of cells). (A) Axial, (B) sagittal, (C) coronal views, (D) whole body scan. (b) Immunohistochemical staining for SSTR subtype 4 (magn. ×400). Courtesy of Prof. R. Tomaszewska, Dept. of Pathology, Jagiellonian University, Medical College, Krakow. (*See insert for color representation of the figure.*)

In a study assessing various imaging modalities in the detection of SDHx-related HNPGLs, ^{18}F-FDOPA PET proved to be the most efficient in detecting such lesions, followed by CT/MRI (81% lesions detected), ^{18}F-FDG PET/CT (77% of lesions), and ^{111}In-pentetreotide (64% of lesions). ^{18}F-FDA and ^{123}I-mIBG detected less than half of the foci. The authors concluded that SRS or ^{18}F-FDG PET/CT is the second line of imaging in such cases [28].

Implementation of ^{68}Ga-labeled somatostatin analogues in PET has greatly improved the diagnostics of neuroendocrine tumors. Case reports and studies on small groups published so far seem to confirm this phenomenon also for PPGLs [29]. Larger studies are, however, still missing.

There are three available ^{68}Ga-labeled SST analogues: DOTATATE with high affinity to SSTR subtype 2, DOTATOC binding to SSTR2 and 5, and DOTANOC with affinity to SSTR subtypes 2, 3, and 5 [30]. The first report by Fanti et al. showed the suitability of ^{68}Ga-DOTANOC PET in the detection of unknown PGL lesions; however, only 3 patients were presented [31]. Similarly, the available clinical data on malignant pheochromocytomas suggest that the use of ^{68}Ga-DOTATOC PET may be very helpful in the management of patients. In a work published by Kroiss et al., both ^{68}Ga-DOTATOC PET and ^{123}I-mIBG scintigraphy showed 100% sensitivity, when compared with anatomical imaging (CT and MRI). However, in pheochromocytoma patients, on a per-lesion basis, the sensitivity of ^{68}Ga-DOTATOC PET was essentially higher than that of ^{123}I-mIBG (91.7% vs. 63.3%) [32]. The same group highlighted a possible higher sensitivity of ^{68}Ga-DOTATOC PET than ^{123}I-mIBG in the detection of neuroblastoma metastases. There are also limited literature data on ^{68}Ga-DOTATATE PET, with which significantly more lesions with higher tumor/nontumor (T/N) ratio in comparison to ^{123}I-mIBG have been reported [33, 34].

Better visualization of bone metastases from PPGLs with the ^{68}Ga-labeled somatostatin analogue was also noted [35]. Maurice et al. analyzed the results of ^{68}Ga-DOTATATE PET and ^{123}I-mIBG scintigraphy in a retrospective study of 15 patients with PPGLs. The authors concluded that ^{68}Ga-DOTATATE should be considered as a first-line investigation tracer in patients at a high risk of PGLs and metastatic disease, such as in the screening of mutations carriers in familial cases. ^{68}Ga-DOTATATE should also be considered as the preferred tracer in patients with suspected bone metastases [34]. Data comparing the results of ^{68}Ga-labeled somatostatin analogue-based PET with ^{18}F-DOPA, ^{18}F-DA, and ^{18}F-FDG in PPGL patients are not available yet.

Another promising application of labeled somatostatin analogues is radio-guided surgery to improve intraoperative detection of multiple PGLs and to provide radical tumor resection [36, 37].

The choice of the most feasible method of pheochromocytoma and paraganglioma imaging, as well as suitability for radioisotope treatment, may be influenced not only by clinical features (sporadic vs. familial tumors, benign vs. malignant, intra- vs. extra-adrenal) but also by the results of qualitative or quantitative assessment of SSTRs and monoamine transporters. VMAT-1-negative PPGLs do not take up mIBG. Preliminary data have shown that the number of SSTR2A mRNA copies corresponds with positive SRS results [38].

SENSITIVITY OF OTHER FUNCTIONAL IMAGING MODALITIES

The use of ^{123}I-mIBG is preferred over ^{131}I-mIBG because of its higher sensitivity, lower radiation exposure, and improved imaging quality with SPECT [10]. Reported sensitivities of ^{123}I-mIBG scintigraphy for localizing nonmetastatic PPGL vary between 77 and 98% [39, 40], but only 50–79% for metastases [39, 41]. The specificity of ^{123}I-mIBG SPECT approaches 100% [42]. Despite the low sensitivity for detecting PPGL metastases, an advantage of using ^{123}I-mIBG scintigraphy in the setting of metastatic disease is the fact that it might identify patients who possibly benefit from palliative treatment with therapeutic doses of ^{131}I-mIBG [43].

^{18}F-DOPA PET has a high sensitivity for the localization of PPGL [39, 44, 45] and HNPGL [46]. Reported sensitivities vary between 81 and 100%. No specificity data are available. The performance of ^{18}F-DOPA PET is disappointing in case of metastatic PPGL, especially SDHB-related cases [9].

For ^{18}F-FDG PET, sensitivities up to 88% for primary nonmetastatic PPGL were reported. Specificity is ~90% (unpublished results). ^{18}F-FDG PET is highly sensitive for the detection of PPGL metastases, especially SDHB-related cases (region-based sensitivity 97% in reference to CT/MRI) [41] (Fig. 4.4.2.4).

^{18}F-FDA was initially developed at the National Institutes of Health for functional imaging of the sympathetic nervous system and later evaluated as a new imaging tool for PPGL. ^{18}F-FDA PET was shown to have a high sensitivity for both primary

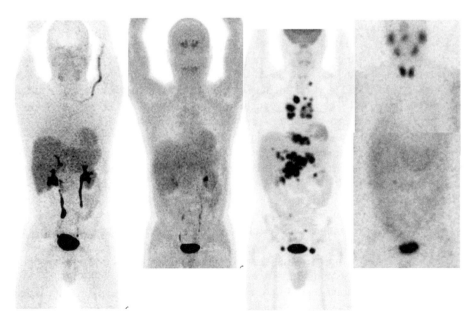

FIGURE 4.4.2.4 Functional imaging results in a 35-year-old male with metastatic SDHB PPGL. Anteriorly reprojected images (respectively, ^{18}F-FDA PET, ^{18}F-DOPA PET, ^{18}F-FDG PET, ^{123}I-mIBG SPECT). Adapted from [39], with permission.

TABLE 4.4.2.1   Sensitivity of functional imaging[a]

(A) Nonmetastatic PPGL (20 patients)[b]

	CT and/or MRI	^{18}F-DOPA	^{18}F-FDA	^{123}I-mIBG	^{18}F-FDG
In ref. to histologically confirmed lesions	100% (26/26)	81% (21/26)	77% (20/26)	77% (20/26)	88% (23/26)

(B) Metastatic PPGL (28 patients)[c]

	CT and/or MRI	^{18}F-DOPA[A]	^{18}F-FDA[B]	^{123}I-mIBG[C]	^{18}F-FDG[D]
In ref. to lesions on CT and/or MRI	—	45% (96/211)	76% (161/211)	57% (106/187)	74% (157/211)

[a]Adapted from [39].
[b]Sensitivities are not significantly different between functional imaging modalities.
[c]A versus B, A versus C, A versus D, B versus C, C versus D: $p < 0.01$. B versus D: $p = 0.760$.

tumors (77–100%) and metastases (77–90%) [39, 42, 47]. The sensitivity of ^{18}F-FDA PET exceeds 90% [42]. So far, ^{18}F-FDA is only available as an experimental imaging agent.

The results of a head-to-head comparison between different functional imaging modalities are presented in Table 4.4.2.1 [39].

IMAGING ACROSS HEREDITARY SYNDROMES

Imaging results appear to be largely determined by the underlying genotypes and related tumor cell characteristics. There is evidence for differential expression of cellular targets for radiopharmaceuticals. For example, it was shown that there is a lower expression of the cell membrane norepinephrine transporter system in VHL-related PPGL cells than in MEN-2-related tumor cells [48]. Considering the higher affinity of ^{18}F-FDA than ^{123}I-mIBG for these transporters, it is no surprise that ^{18}F-FDA PET is superior to ^{123}I-mIBG SPECT in the context of VHL syndrome [49].

There also appears to be a link between tumor biology and imaging [50]. *SDHB* mutations are associated with PPGLs of a particularly high malignant potential. ^{18}F-FDG PET has an excellent sensitivity for *SDHB*-associated metastatic PPGL [41, 51]. ^{18}F-FDG accumulation is an index of increased tissue glucose metabolism, and as a marker of tumor viability, the degree of ^{18}F-FDG uptake usually reflects tumor aggressiveness. ^{18}F-FDG uptake by PPGL does not appear to be merely an indicator of a high metabolic rate due to malignancy per se, but may rather be directly linked to *SDHB*-specific tumor biology [39]. The *SDHB* gene encodes for subunit B of the mitochondrial SDH complex II, a key enzyme in oxidative phosphorylation. *SDHB* mutations can lead to complete loss of SDH enzymatic activity in malignant PPGL, with upregulation of hypoxic–angiogenic-responsive genes [52]. Impairment of mitochondrial function due to loss of *SDHB* function may cause tumor cells to shift from oxidative phosphorylation to aerobic glycolysis, a phenomenon known as the "Warburg effect" [53]. Higher glucose requirement because of a switch to less efficient pathways for cellular energy production may explain the increased ^{18}F-FDG uptake by malignant *SDHB*-related PPGL.

The only study on impact of SRS with ^{111}In-pentetreotide in hereditary *SDHx* gene mutation-related PPGLs has just been published in 2013. The sensitivity of SRS in all PGLs was 69.5% versus 85.4% for MRI and CT and 42.7% for mIBG. SRS performed better than other assessed imaging modalities in thoracic PGL (61.5% vs. 46.2% and 30.8%, respectively) and was more efficient in detecting HNPGL than mIBG, however inferior to CT/MRI (75, 30.6, and 90.4%, respectively). In abdominal PPGLs, SRS was least sensitive [54].

CONCLUSIONS

Imaging in PPGL patients, if chosen appropriately, adds greatly to the patients' management: treatment options, prognosis, follow-up, and early detection of recurrence. Considering the heterogeneous nature of the disease, imaging modality preference should be guided by clinical and genetic background (Fig. 4.4.2.5) [55]. Increasing importance of PET tracers should be stressed.

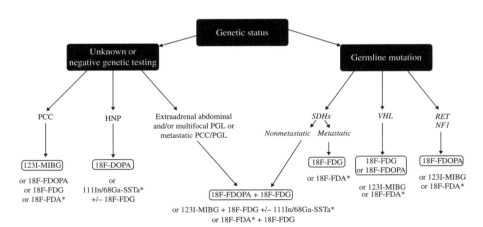

FIGURE 4.4.2.5   Clinical algorithm for imaging investigations in PCC/PGL. Based on the earlier considerations, the following algorithm can be proposed based on the clinical situation. This algorithm should be adapted to the practical situation in each institution and should evolve with time. In bold, first-line imaging procedures according to accessibility of tracers and clinical approvals in European countries. ^{18}F-FDA and ^{68}Ga-DOTA-SSTa (asterisks) are experimental tracers that should be used in the setting of clinical trials. ^{18}F-FDA PET is currently used at the NIH only. ^{68}Ga-DOTA-SSTa is now accessible in many clinical and research centers in Europe [55], with permission.

In apparently sporadic benign adrenal PGLs (pheochromocytomas), ^{123}I-mIBG should be considered the first line of functional imaging (if such is necessary). Only in ^{123}I-mIBG-negative cases or in patients on medications interfering with ^{123}I-mIBG uptake other modalities, particularly ^{18}F-FDOPA PET, should be asked for.

^{18}F-FDOPA PET is currently considered the best method for detecting HNPGL. If not available, SRS either with SPECT or PET traces should be used. ^{18}F-FDOPA PET is also the best functional imaging method for detecting and differential diagnosis of retroperitoneal, extra-adrenal PGLs with unknown genetic background.

Depending on the presence/absence of *SDHB* subunit mutation, ^{18}F-FDG and ^{18}F-FDOPA PET are the best modalities in approaching the patients with metastatic PPGLs. ^{123}I-mIBG and/or SRS are to be performed if further targeted radionuclide therapy is considered.

Currently, the choice of imaging method in PPGL patients should be also guided by the type of mutation causing the disease. ^{18}F-FDOPA is the first line of functional approach in all syndromic patient but VHL-related, *SDHB*-related, and *SDHx*-related metastatic cases, in whom ^{18}F-FDG seems the most appropriate tracer.

SUMMARY

Summing up, the confirmation of SSTRs expression in neuroendocrine tumors has changed modern diagnostic imaging and therapy and starts to influence the approach to PPGLs. *In vitro* studies have revealed SSTR expression, particularly subtypes 2A and 3, in PPGLs, the confinement of which to cell membranes is essential for successful diagnostic use of somatostatin analogues. Scintigraphy with radiolabeled somatostatin analogues is nowadays an approved complementary method for the localization of PPGLs, particularly malignant head and neck PGLs, and, if necessary, for qualification for PRRT and follow-up of the patients.

It seems that labeled analogues with a broader affinity to SSTRs may be a good diagnostic alternative for PPGL patients. Current place of SRS in PPGLs diagnostics may change if tracers labeled with ^{68}Ga and other positron emitters are more profoundly tested.

High tracer accumulation is essential for PRRT in inoperable or disseminated SSTR-positive tumors. Some PPGLs are SRS positive and mIBG negative, suggesting a possible therapeutic role for labeled somatostatin analogues. Promising results of radiotherapy with labeled analogues have been recently announced; however, data concerning this approach are still scarce (Chapter 6.6).

Overall, the prognosis for patients with PPGLs is good. Most pheochromocytoma cases present with the disease limited to the adrenal gland, for whom available imaging methods are sufficient. In some patients, finding markers for the differentiation between pheochromocytoma and other adrenal masses is challenging. The development of nuclear medicine imaging methods is focused on the search for new diagnostic targets to improve the detection of multifocal and malignant primary and metastatic lesions in PPGL cases, as well as to find alternative therapeutic strategies to control tumor growth. The choice of optimal methods of functional imaging including factors like genetic predisposition is also a matter of research.

Somatostatin analogues with affinity to all known SSTR subtypes (e.g., SOM-230) are one of the line of research, as more than 60% of PPGL cells express SSTR subtype 3, some 2 and 5, but solely subtype 4 can also be found. Gastrin, GLP-1, GHRH, LHRH, neuropeptide-Y, and other neuropeptide analogues are also being studied [41, 46].

REFERENCES

[1] Lenders, J. W.; Eisenhofer, G.; Mannelli, M.; Pacak, K. The Lancet 2005, 366, 665–675.

[2] Mannelli, M.; Castellano, M.; Schiavi, F.; et al. Journal of Clinical Endocrinology and Metabolism 2009, 94, 1541–1547.

[3] Bayley, J. P.; Kunst, H. P.; Cascon, A.; et al. Lancet Oncology 2010, 11, 366–372.

[4] Qin, Y.; Yao, L.; King, E. E.; et al. Nature Genetics 2010, 42, 229–233.

[5] Comino-Méndez, I.; Gracia-Aznárez, F. J.; Schiavi, F.; et al. Nature Genetics 2011, 43, 663–667.

[6] Burnichon, N.; Cascón, A.; Schiavi, F.; et al. Clinical Cancer Research 2012, 18, 2828–2837.

[7] Lenders, J. W.; Pacak, K.; Walther, M. M.; et al. Journal of the American Medical Association 2002, 287, 1427–1434.

[8] Ilias, I.; Shulkin, B.; Pacak, K. Trends in Endocrinology and Metabolism 2005, 16, 66–72.

[9] Havekes, B.; Lai, E. W.; Corssmit, E. P.; et al. Quarterly Journal of Nuclear Medicine and Molecular Imaging 2008, 52, 419–429.

[10] Lynn, M. D.; Shapiro, B.; Sisson, J. C.; et al. Radiology 1985, 155, 789–792.

[11] Havekes, B.; King, K.; Lai, E. W.; et al. Clinical Endocrinology (Oxford) 2010, 72, 137–145.

[12] Reubi, J. C.; Waser, B.; Khosla, S.; et al. Journal of Clinical Endocrinology and Metabolism 1992, 74, 1082–1089.

[13] Epelbaum, J.; Bertherat, J.; Prevost, G.; et al. Journal of Clinical Endocrinology and Metabolism 1995, 80, 1837–1844.

[14] Mundschenk, J.; Unger, N.; Schulz, S.; et al. Journal of Clinical Endocrinology and Metabolism 2003, 88, 5150–5157.

[15] Unger, N.; Serdiuk, I.; Sheu, S. Y.; et al. Endocrine Research 2004, 39, 931–934.

[16] Saveanu, A.; Muresan, M.; De Micco, C.; et al. Endocrine-Related Cancer 2011, 18, 287–300.

[17] Krenning, E. P.; Bakker, W. H.; Breeman, W. A.; et al. The Lancet 1989, 8632, 242–244.

[18] Lamberts, S. W. J.; Bakker, W. H.; Reubi J. C.; Krenning E. P. New England Journal of Medicine 1990, 323, 1246–1249.

[19] Krenning, E. P.; Bakker, W. H.; Kooij, P. P.; et al. Journal of Nuclear Medicine 1992, 33, 652–658.

[20] Kwekkeboom, D. J.; van Urk, H.; Pauw, B. K.; et al. Journal of Nuclear Medicine 1993, 34, 873–878.

[21] Płachcińska, A.; Mikołajczak, R.; Maecke, H. R.; et al. European Journal of Nuclear Medicine and Molecular Imaging 2003, 30, 1402–1406.

[22] Ilias, I.; Chen, C. C.; Carrasquillo, J. A.; et al. Journal of Nuclear Medicine 2008, 49, 1613–1619.

[23] Tenenbaum, F.; Lumbroso, J.; Schlumberger, M.; et al. Journal of Nuclear Medicine 1995, 36, 1–6.

[24] Hoefnagel, C. A.; Levington, V. J.; in Ell, P. J.; Gambhir, S. S., ed., *Nuclear Medicine in clinical diagnosis and treatment*, Churchill Livingstone, New York, Vol 1, 2004, pp. 445–457.

[25] Bustillo, A.; Telischi, F. F. Otolaryngology – Head and Neck Surgery 2004, 130, 479–482.

[26] Koopmans, K. P.; Jager, P. L.; Kema, I. P.; et al. Journal of Nuclear Medicine 2008, 49, 1232–1237.

[27] Chen, L.; Li, F.; Zhuang, H.; et al. Journal of Nuclear Medicine 2009, 50, 397–400.

[28] King, K. S.; Chen, C. C.; Alexopoulos, D. K.; et al. Journal of Clinical Endocrinology and Metabolism 2011, 96, 2779–2785.

[29] Win, Z.; Rahman, L.; Murrell, J.; et al. European Journal of Nuclear Medicine and Molecular Imaging 2006, 33, 506.

[30] Cuccurullo, V.; Mansi, L. European Journal of Nuclear Medicine and Molecular Imaging 2012, 39, 1262–1265.

[31] Fanti, S.; Ambrosini, V.; Tomassetti, P.; et al. Biomedicine and Pharmacotherapy 2008, 62, 667–671.

[32] Kroiss, A.; Putzer, D.; Uprimny, C.; et al. European Journal of Nuclear Medicine and Molecular Imaging 2011, 38, 865–873.

[33] Naji, M.; Zao, C.; Welsh, S. J.; et al. Molecular Imaging and Biology 2011, 13, 769–775.

[34] Maurice, J. B.; Troke, R.; Win, Z.; et al. European Journal of Nuclear Medicine and Molecular Imaging 2012, 39, 1266–1270.

[35] Farnsworth, J.; Alsayed, M.; Zerizer, I.; et al. Nuclear Medicine Communications 2011, 32, 426.

[36] Filippi, L.; Valentini, F. B.;Gossetti, B.; et al. Tumori 2005, 91, 173–176.

[37] Martinelli, O.; Irace, L.; Massa, R.; et al. Journal of Experimental and Clinical Cancer Research 2009, 28, 148.

[38] Kolby, L.; Bernhardt, P.; Johanson, V.; et al. Annals of New York Academy of Sciences 2006, 1073, 491–497.

[39] Timmers, H. J.; Chen, C. C.; Carrasquillo, J. A.; et al. Journal of Clinical Endocrinology and Metabolism 2009, 94, 4757–4767.

[40] Van Der Horst-Schrivers, A. N.; Jager, P. L.; Boezen, H. M.; et al. Anticancer Research 2006, 26, 1599–1604.

[41] Timmers, H. J.; Kozupa, A.; Chen, C. C.; et al. Journal of Clinical Oncology 2007, 25, 2262–2269.

[42] Timmers, H. J.; Eisenhofer, G.; Carrasquillo, J. A.; et al. Clinical Endocrinology (Oxford) 2009, 71, 11–17.

[43] Loh, K. C.; Fitzgerald, P. A.; Matthay, K. K.; et al. Journal of Endocrinological Investigation 1997, 20, 648–658.

[44] Fiebrich, H. B.; Brouwers, A. B.; Kerstens, M. N.; et al. Journal of Clinical Endocrinology and Metabolism 2009, 94, 3922–3930.

[45] Hoegerle, S.; Nitzsche, E.; Altehoefer, C.; et al. Radiology 2002, 222, 507–512.

[46] Hoegerle, S.; Ghanem, N.; Altehoefer, C.; et al. European Journal of Nuclear Medicine and Molecular Imaging 2003, 30, 689–694.

[47] Pacak, K.; Eisenhofer, G.; Carrasquillo, J. A.; et al. Hypertension 2001, 38, 6–8.

[48] Huynh, T. T.; Pacak, K.; Brouwers, F. M.; et al. European Journal of Endocrinology 2005, 153, 551–563.

[49] Kaji, P.; Carrasquillo, J. A.; Linehan, W. M.; et al. European Journal of Endocrinology 2007, 156, 483–487.

[50] Timmers, H. J.; Chen, C. C.; Carrasquillo, J. A.; et al. Journal of the National Cancer Institute 2012, 104, 700–708.

[51] Taïeb, D.; Sebag, F.; Barlier, A.; et al. Journal of Nuclear Medicine 2009, 50, 711–717.

[52] Gimenez-Roqueplo, A. P.; Favier, J.; Rustin, P.; et al. Journal of Clinical Endocrinology and Metabolism 2002, 87, 4771–4774.

[53] Warburg, O. Science 1956, 123, 309–314.

[54] Gimenez-Roqueplo, A. P.; Caumont-Prim, A.; Houzard, C.; et al. Journal of Clinical Endocrinology and Metabolism 2013, 98, E162–173.

[55] Taïeb, D.; Timmers, H. J.; Hindié, E.; et al. European Journal of Nuclear Medicine and Molecular Imaging 2012, 39, 1977–1995.

4.4.3

SOMATOSTATIN RECEPTOR SCINTIGRAPHY IN MEDULLARY THYROID CANCER

ANOUK N.A. VAN DER HORST-SCHRIVERS[1],
ADRIENNE H. BROUWERS[2], AND THERA P. LINKS[1]

[1]Departments of Medical Endocrinology, University of Groningen,
University Medical Center Groningen, Groningen, The Netherlands
[2]Department of Nuclear Medicine and Molecular Imaging, University of Groningen, and
University Medical Center Groningen, Groningen, The Netherlands

ABBREVIATIONS

^{18}F-DOPA	^{18}F-dihydroxyphenylalanine
^{18}F-FDG	^{18}F-2-fluoro-2-deoxy-D-glucose
[99m]Tc-V-DMSA	^{99m}Tc-V-dimercapto-sulfuric acid
ATA	American Thyroid Association
CT	computed tomography
FMTC	familial MTC
MEN	multiple endocrine neoplasia
MIBG	metaiodobenzylguanidine
MRI	magnetic resonance imaging
MTC	medullary thyroid cancer
PET	positron emission tomography
PRRT	peptide receptor radionuclide therapy

Somatostatin Analogues: From Research to Clinical Practice, First Edition. Edited by
Alicja Hubalewska-Dydejczyk, Alberto Signore, Marion de Jong, Rudi A. Dierckx,
John Buscombe, and Christophe Van de Wiele.
© 2015 John Wiley & Sons, Inc. Published 2015 by John Wiley & Sons, Inc.

RET   rearranged during transfection
SRS   somatostatin receptor scintigraphy
SSTR   somatostatin receptor

INTRODUCTION

Medullary thyroid cancer (MTC), first described in 1959 [1], is a neuroendocrine tumor originating from the calcitonin-secreting C cells of the thyroid and accounts for 3–10% of the thyroid malignancies [2]. It occurs in a sporadic (75%) and a hereditary form referred to as multiple endocrine neoplasia (MEN) type 2A (MEN-2A), MEN type 2B, or familial MTC (FMTC). The hereditary forms occur due to germ line mutations in the rearranged during transfection (RET) gene [3–6]. Besides a 100% expression of MTC, patients also develop pheochromocytomas (MEN-2A and MEN-2B), hyperparathyroidism (MEN-2A), intestinal ganglioneuromatosis (MEN-2B), and mucosal neuromas (MEN-2B).

MTC metastasizes early during the course of the disease to regional cervical lymph nodes. Up to 20–30% of patients with a primary tumor of <1 cm (T1), up to 50% of patients with a T2 tumor, and almost all patients with a T3 or T4 tumor present with cervical lymph node metastases [7, 8]. Distant metastases are typically found in the mediastinum, lungs, liver, and bone.

Surgery, consisting of a total thyroidectomy and an extensive lymph node dissection, is the only effective curative treatment in primary MTC. Locoregional tumor control may be improved by initial extensive surgery consisting of central, bilateral, and upper mediastinal neck dissection [9].

However, clinically curative surgery resulted in a cure rate between 33 and 61% in groups that routinely employed central and bilateral lymph node dissection [7, 8, 10]. Serum calcitonin is the main biochemical and accurate marker used for the detection of tumor persistence and recurrence and is of importance in the postoperative management of patients with MTC. For patients with recurrent disease and/or lymph node metastases, surgery is the first line of treatment. Patients with distant metastases cannot be cured and have a reduced survival [11]. Systemic treatment modalities such as radiotherapy and chemotherapy have limited success. New molecular targeted therapy such as tyrosine kinase inhibitors shows promising results *in vitro* as well as *in vivo*; a phase III trial has been published recently [12, 13]. The management of patients with MTC is also impaired due to the difficulty of imaging persistent/residual and/or metastatic tumor lesions. Morphological imaging techniques (ultrasonography of the neck, computed tomography (CT), and magnetic resonance imaging (MRI)), scintigraphic imaging techniques (^{99m}Tc-V-dimercapto-sulfuric acid (^{99m}Tc-V-DMSA), ^{111}In-labeled somatostatin receptor scintigraphy (SRS)), and positron emission tomography (PET) labeled with ^{18}F-2-fluoro-2-deoxy-D-glucose (^{18}F-FDG) have complementary values, since they are dependent of different tumor characteristics. Also, they depend on tumor load. Sensitivity for individual imaging techniques can however be disappointing. Newer alternative tracers for PET imaging such as ^{18}F-dihydroxyphenylalanine (^{18}F-DOPA) and ^{68}Ga-labeled analogues of somatostatin (DOTA) are more promising.

Besides the diagnostic application of somatostatin analogues, these have also been used in the treatment of MTC. The focus of this chapter is the diagnostic and therapeutic use of somatostatin in MTC.

SOMATOSTATIN AS AN IMAGING TOOL IN PATIENTS WITH MTC

As previously stated, the imaging of residual and/or recurrent (metastatic) MTC can be very difficult when conventional imaging techniques such as ultrasonography, CT, and/or MRI are used. Patients with postoperative calcitonin levels of <150 pg/ml should undergo ultrasonography of the neck according to the ATA guidelines [14]. However, when calcitonin levels rise above 150 pg/ml, additional imaging techniques are recommended for the detection of distant metastases. For this purpose, $^{123}I(^{131}I)$-metaiodobenzylguanidine (MIBG) and ^{99m}Tc-V-DMSA routinely have been used and were reported to be able to detect residual or recurrent disease after primary surgery, in about 33–70% of patients. After the demonstration of the presence of somatostatin in parafollicular C cells and cells of MTC by immunohistochemistry in the mid-1970s [15–17], several somatostatin receptors (SSTR) (including SSTR2 and SSTR5) have been shown to be present *in vitro* in MTC cells, thus providing the rational for the use of SRS in these patients (Fig. 4.4.3.1) [18–20].

The first report on SRS in three patients with MTC using ^{123}I-labeled tyr-3-octreotide (tyr-3-SMS 201-995, a synthetic derivative of somatostatin) showed no uptake [21]. Since then, several studies have been published regarding the value of SRS in patients with recurrent or residual MTC, but the data do not reveal sensational results. As is shown in Table 4.4.3.1, the number of patients that have been studied is limited, and the sensitivity range varies between 0 and 75%, illustrating the lack of well-designed studies in selected patients [22]. Sensitivities also differ regarding different metastases sites and size of the metastases.

(a)   (b)

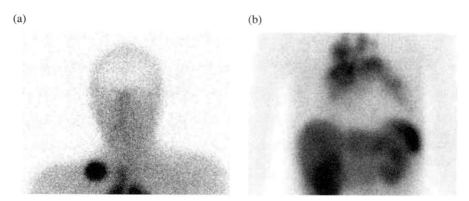

FIGURE 4.4.3.1 ^{111}In-labeled somatostatin receptor scintigraphy, with visible metastases in the supraclavicular lymph nodes and mediastinal and bilateral hili of the lung. (a) Head/neck region anterior view. (b) Chest/abdominal region anterior view.

TABLE 4.4.3.1   Studies regarding somatostatin receptor scintigraphy in recurrent medullary thyroid cancer

	No. of patients	Sensitivity patient based	Remarks
Krenning 1989 [21]	3	0%	^{123}I-labeled tyr-3-octreotide
Frank-Raue 1995 [23]	26	57%	
Krenning 1993 [24]	12	8 out of 12	
Kwekkeboom 1995 [25]	17	65%	No visualization of liver metastases
Krausz 1994 [26]	10	9 out of 10	No visualization of liver metastases
Bernà L 1995 [27]	11	55%	
Bernà 1998 [28]	20	50%	
Celentano L 1995 [29]	14	64%	
Rufini L 1995 [30]	7	72%	
Baudin E 1996 [31]	24	38%	No visualization of small tumor sites
Tisell 1997 [32]	22	50%	Higher CEA and calcitonin in patients with a positive scan
Adams S 1998 [33]	18	29%	No visualization of small tumor sites
Hoegerle 2001 [34]	11	52%	^{18}F-DOPA PET: 63%
Arslan N 2001 [35]	14	79%	Sensitivity combined [99]mT Tc-V-DMSA and SRS: 86%
De Groot JW 2004 [36]	26	41%	Lesion-based sensitivity
Diehl 2001 [37]	24	25%	24 histological confirmed lesions
Kurtaran 1998 [38]	14	71%	For the primary tumor
		0%	For lymph node metastases
Lodish 2012 [39]	11	45%	Pediatric population

Small size: <1 cm. CEA denotes carcinogenic embryonic antigen, ^{18}F-DOPA PET denotes ^{18}F-dihydroxyphenylalanine position emission tomography, SRS denotes somatostatin receptor scintigraphy, and [99]mT Tc-V-DMSA denotes [99]mTc-V-dimercapto-sulfuric acid.

So in current patient practice, sensitivities of SRS but also of ^{123}I-MIBG and [99]mTc-V-DMSA are disappointing, and in clinical practice, the application of these techniques is waning. Other scintigraphic imaging techniques such as ^{18}F-FDG PET, ^{18}F-DOPA PET, and PET imaging with ^{68}Ga-labeled DOTA peptides are upcoming in the diagnostic workup of patients with residual and/or metastatic disease in addition to the conventional imaging. However, the clinical value use of these new tracers has not been established yet. Also PET combined with CT imaging and in the future combined with MRI may be of additional value in these patients.

It has been reported that ^{18}F-FDG PET may be superior in patients with short calcitonin doubling time and in patients with tumor with a Ki67 score of >2.0% [40–43].

This may be in contrast to the ^{18}F-DOPA PET with a reported sensitivity of about 62%, possibly reflecting the more indolent type of MTC. However, these data have to be confirmed in other series [34, 41]. When combining CT imaging with ^{18}F-DOPA PET, the reported sensitivities may increase to 94 and 100% [44, 45]. Although ^{18}F-DOPA PET has less prognostic value, it can assess the extent of the disease in residual MTC [46, 47].

Currently several other tracers for PET imaging binding to somatostatin receptors have become available that could be of value in the detection of MTC: ^{68}Ga-DOTA peptides, DOTA-TOC, DOTA-TATE, and DOTA-NOC, bind with high affinity to SST-2. ^{68}Ga-DOTA-NOC also binds to SST-3 and SST-5 [48]. These new PET traces are very promising, although the clinical experience is limited. Conry et al. investigated 18 patients with recurrent MTC with an overall sensitivity of ^{18}F-FDG PET and ^{68}Ga-DOTA-TATE PET of 77.8 and 72.2%, respectively. On a region-based analysis, ^{18}F-FDG PET was more sensitive [49]. Clearly, more studies with more homogeneous patient groups are needed to evaluate the value of these tracers in patients with MTC.

THERAPEUTIC USE OF SOMATOSTATIN ANALOGUES

Since the most effective treatment of MTC, surgery, does not result in a 100% cure rate, additional therapies are needed in recurrent/persistent and/or metastatic MTC. Radiotherapy is especially used for local tumor control. Chemotherapy has a very limited value.

The first therapeutic intervention with somatostatin analogues in patients with MTC dates from 1987. A somatostatin analogue was prescribed for a 63-year-old man with disseminated MTC and pancreatic nesidioblastosis. The analogue had no effect neither on the calcitonin hypersecretion nor on the growth of the medullary carcinoma [50].

Since then, there have been several trials studying the effect of somatostatin analogues on MTC. Treatment with the current available somatostatin analogues, octreotide and lanreotide (both with a high affinity for SSTR2 and SSTR5), does not seem to have an effect on survival but may control symptoms of flushing and diarrhea in some patients [51–56].

In current guidelines, the use of somatostatin analogues can be considered for symptomatic treatment of diarrhea if other antimotility drugs, such as loperamide, are ineffective [14].

Experience with peptide receptor radionuclide therapy (PRRT) is limited in this patient group and disappointing.

New therapies in the treatment of metastatic MTC use target tyrosine kinase receptors inhibitors that belong to the same family group of proteins as RET. Several TK inhibitors have already been tested *in vitro* and evaluated in mostly phase II clinical trials, and several phase III trials are currently underway and have been published [12, 13].

In summary, the clinical applications of somatostatin analogues in the diagnosis and therapy for patients with MTC are very limited. Possibly, the ^{68}Ga-DOTA-labeled peptides may be diagnostically applicable, but with the scarce data, the specific clinical value in MTC patients must be awaited.

REFERENCES

[1] Hazard, J. B.; Hawk, W. A.; Crile, G. Jr. Journal of Clinical Endocrinology and Metabolism 1959, 19, 152–161.

[2] Hundahl, S. A.; Fleming, I. D.; Fremgen, A. M.; Menck, H. R. Cancer 1998, 832, 638–2648.

[3] Kebebew, E.; Clark, O. H. Current Treatment Options in Oncology 2000, 1, 359–367.

[4] Mulligan, L. M.; Kwok, J. B.; Healey, C. S.; et al. Nature 1993, 363, 458–460.

[5] Hofstra, R. M.; Landsvater, R. M.; Ceccherini, I.; et al. Nature 1994, 367, 375–376.

[6] Donis-Keller, H.; Dou, S.; Chi, D.; et al. Human Molecular Genetics 1993, 2, 851–856.

[7] Moley, J. F.; DeBenedetti, M. K. Annals of Surgery 1999, 229, 880–887; discussion 887–888.

[8] Scollo, C.; Baudin, E.; Travagli, J. P.; et al. Journal of Clinical Endocrinology and Metabolism 2003, 88, 2070–2075.

[9] de Groot, J. W.; Links, T. P.; Sluiter, W. J.; Wolffenbuttel, B. H.; Wiggers, T.; Plukker, J. T. Head and Neck 2007, 29, 857–863.

[10] Dralle, H. British Journal of Surgery 2002, 89, 1073–1075.

[11] de Groot, J. W.; Plukker, J. T.; Wolffenbuttel, B. H.; Wiggers, T.; Sluiter, W. J.; Links, T. P. Clinical Endocrinology (Oxford) 2006, 65, 729–736.

[12] Wells, S. A. Jr.; Robinson, B. G.; Gagel, R. F.; et al. Clinical Oncology 2012, 30, 134–141.

[13] Verbeek, H. H.; Alves, M. M.; de Groot, J. W.; et al. Journal of Clinical Endocrinology and Metabolism 2011, 96, E991–995.

[14] American Thyroid Association Guidelines Task Force; Kloos, R.T.; Eng, C.; et al. Thyroid 2009, 19, 565–612.

[15] Parsons, J. A.; Erlandsen, S. L.; Hegre, O. D.; McEvoy, R. C.; Elde, R. P. Journal of Histochemistry and Cytochemistry 1976, 24, 872–882.

[16] Sundler, F.; Alumets, J.; Håkanson, R.; Björklund, L.; Ljungberg, O. American Journal of Pathology 1977, 88, 381–386.

[17] Reubi, J. C.; Chayvialle, J. A.; Franc, B.; Cohen, R.; Calmettes, C.; Modigliani, E. Laboratory Investigation 1991, 64, 567–573.

[18] Mato, E.; Matías-Guiu, X.; Chico, A.; et al. Journal of Clinical Endocrinology and Metabolism 1998, 83, 2417–2420.

[19] Papotti, M.; Kumar, U.; Volante, M.; Pecchioni, C.; Patel, Y. C. Clinical Endocrinology (Oxford) 2001, 54, 641–649.

[20] Kwekkeboom, D. J.; Reubi, J. C.; Lamberts, S. W.; et al. Journal of Clinical Endocrinology and Metabolism 1993, 76, 1413–1417.

[21] Krenning, E. P.; Bakker, W. H.; Breeman, W. A.; et al. Lancet 1989, 1(8632), 242–244

[22] Koopmans, K. P.; Neels, O. N.; Kema, I. P.; et al. Critical Reviews in Oncology/Hematology 2009, 71, 199–213.

[23] Frank-Raue, K.; Bihl, H.; Dörr, U.; Buhr, H.; Ziegler, R.; Raue, F. Clinical Endocrinology (Oxford) 1995, 42, 31–37.

[24] Krenning, E. P.; Kwekkeboom, D. J.; Reubi, J. C.; et al. Digestion 1993, 54 (Suppl 1), 84–87.

[25] Kwekkeboom, D. J.; Reubi, J. C.; Lamberts, S. W.; et al. Journal of Clinical Endocrinology and Metabolism 1993, 76, 1413–1417.

[26] Krausz, Y.; Ish-Shalom, S.; Dejong, R. B.; et al. Clinical Nuclear Medicine 1994, 19, 416–421.

[27] Bernà, L.; Cabezas, R.; Mora, J.; Torres, G.; Estorch, M.; Carrió, I. Journal of Endocrinology 1995, 144, 339–345.

[28] Bernà, L.; Chico, A.; Matías-Guiu, X.; et al. European Journal of Nuclear Medicine 1998, 25, 1482–1488.

[29] Celentano, L.; Sullo, P.; Klain, M.; Lupoli, G.; Cascone, E.; Salvatore, M. Quarterly Journal of Nuclear Medicine 1995, 39(4 Suppl 1), 131–133.

[30] Rufini, V.; Salvatori, M.; Saletnich, I.; et al. Quarterly Journal of Nuclear Medicine 1995, 39(4 Suppl 1), 140–144.

[31] Baudin, E.; Lumbroso, J.; Schlumberger, M.; et al. Journal of Nuclear Medicine 1996, 37, 912–916.

[32] Tisell, L. E.; Ahlman, H.; Wängberg, B.; et al. British Journal of Surgery 1997, 84, 543–547.

[33] Adams, S.; Baum, R. P.; Hertel, A.; Schumm-Draeger, P. M.; Usadel, K. H.; Hör, G. European Journal of Nuclear Medicine 1998, 25, 1277–1283.

[34] Hoegerle, S.; Altehoefer, C.; Ghanem, N.; Brink, I.; Moser, E.; Nitzsche, E. European Journal of Nuclear Medicine 2001, 28, 64–71.

[35] Arslan, N.; Ilgan, S.; Yuksel, D.; et al. Clinical Nuclear Medicine 2001, 26, 683–688.

[36] de Groot, J. W.; Links, T. P.; Jager, P. L.; Kahraman, T.; Plukker, J. T. Annals of Surgical Oncology 2004, 11, 786–794.

[37] Diehl, M., Risse, J. H.; Brandt-Mainz, K.; et al. European Journal of Nuclear Medicine 2001, 28, 1671–167.

[38] Kurtaran, A.; Scheuba, C.; Kaserer, K.; et al. Journal of Nuclear Medicine 1998, 39, 1907–1909.

[39] Lodish, M.; Dagalakis, U.; Chen, C. C.; et al. Journal of Clinical Endocrinology and Metabolism 2012, 97, E207–212.

[40] Faggiano, A.; Grimaldi, F.; Pezzullo, L.; et al. Endocrine Related Cancer 2009, 16, 225–231.

[41] Koopmans, K. P.; de Groot, J. W.; Plukker, J. T.; et al. Journal of Nuclear Medicine 2008, 49, 524–531.

[42] Marzola, M. C.; Pelizzo, M. R.; Ferdeghini, M.; et al. European Journal of Surgical Oncology 2010, 36, 414–421.

[43] Bogsrud, T. V.; Karantanis, D.; Nathan, M. A.; et al. Molecular Imaging and Biology 2010, 12, 547–553.

[44] Beheshti, M.; Pöcher, S.; Vali, R.; et al. European Radiology 2009, 19, 1425–1434.

[45] Luster, M.; Karges, W.; Zeich, K.; et al. Thyroid 2010, 20, 527–533.

[46] Treglia, G.; Castaldi, P.; Villani, M. F.; et al. European Journal of Nuclear Medicine and Molecular Imaging 2012, 39, 569–580.

[47] Verbeek, H. H.; Plukker, J. T.; Koopmans, K. P.; et al. Journal of Nuclear Medicine 2012, 53, 1863–1871.

[48] Antunes, P.; Ginj, M.; Zhang, H.; Waser, B.; Baum, R. P.; Reubi, J. C.; Maecke, H. European Journal of Nuclear Medicine and Molecular Imaging 2007, 34, 982–993.

[49] Conry, B. G.; Papathanasiou, N. D.; Prakash, V.; et al. European Journal of Nuclear Medicine and Molecular Imaging 2010, 37, 49–57.

[50] Jerkins, T. W.; Sacks, H. S.; O'Dorisio, T. M.; Tuttle, S.; Solomon, S. S. Journal of Clinical Endocrinology and Metabolism 1987, 64, 1313–1319.

[51] Vainas, I.; Drimonitis, A.; Boudina, M.; et al. Hellenic Journal of Nuclear Medicine 2005, 8, 43–47.

[52] Modigliani, E.; Guliana, J. M.; Maroni, M.; et al. Annales d'Endocrinologie (Paris) 1989, 50, 483–488.

[53] Libroia, A.; Verga, U.; Di Sacco, G.; Piolini, M.; Muratori, F. Henry Ford Hospital Medical Journal 1989, 37, 151–153.

[54] Díez, J. J.; Iglesias, P. Journal of Endocrinological Investigations 2002, 25, 773–778.

[55] Vitale, G.; Tagliaferri, P.; Caraglia, M.; et al. Journal of Clinical Endocrinology and Metabolism, 2000; 85, 983–988.

[56] Frank-Raue, K.; Raue, F.; Ziegler, R. Medizinische Klinik (Munich) 1995, 90, 63–66.

4.4.4

SOMATOSTATIN RECEPTOR SCINTIGRAPHY IN OTHER TUMORS IMAGING

MALGORZATA TROFIMIUK-MÜLDNER AND
ALICJA HUBALEWSKA-DYDEJCZYK

*Department of Endocrinology with Nuclear Medicine Unit,
Medical College, Jagiellonian University, Krakow, Poland*

ABBREVIATIONS

CT	computed tomography
DTPA	diethylene triamine pentaacetic acid
FDG	2-deoxy-2-(18F)fluoro-D-glucose
GIST	gastrointestinal stromal tumor
HCC	hepatocellular cancer
HD	Hodgkin's disease
MALT	mucosa-associated lymphoid tissue
MCC	Merkel cell carcinoma
mIBG	metaiodobenzylguanidine
NHL	non-Hodgkin's lymphoma
NSCLC	non-small cell lung cancer
PBMC	peripheral blood mononuclear cells
PET	positron emission tomography
PRRT	peptide receptor radionuclide therapy
RCC	renal cell carcinoma

Somatostatin Analogues: From Research to Clinical Practice, First Edition. Edited by
Alicja Hubalewska-Dydejczyk, Alberto Signore, Marion de Jong, Rudi A. Dierckx,
John Buscombe, and Christophe Van de Wiele.

SCLC   small cell lung cancer
SPECT   single photon emission computed tomography
SRS   somatostatin receptor scintigraphy
SSTR   somatostatin receptor
TIO   tumor-induced osteomalacia

INTRODUCTION

The inhibitory action of the somatostatin occurs through its interaction with the family of specific membrane receptors, expressed in many organs and tissues. High density of somatostatin receptors (SSTRs) has been reported in endocrine and neuro-endocrine cells, particularly in neoplasms originating from those cells. However, the presence, quite often abundant, of functional SSTRs, including type 2, has been also confirmed in neuroendocrine malignancies in broader sense (small cell lung cancer (SCLC), medullary thyroid cancer, pheochromocytomas and paragangliomas, Merkel cell carcinomas (MCC), etc.) and non-neuroendocrine tumors. SSTRs have been found in non-small cell lung, breast, prostate, colon, and many other cancers, not only in neoplastic cells with neuroendocrine features or tumor infiltrating immune-competent cells [1]. This phenomenon implicates the possible auxiliary role of somatostatin receptor scintigraphy (SRS) in clinical management of those neo-plasms. It is also one of the sources of false positive findings, when the "classical" neuroendocrine tumors are searched for.

CENTRAL NERVOUS SYSTEM

The first experience with SRS in pituitary tumors was reported in 1992. The increased uptake of 123I-labeled Tyr3-octreotide in sellar region was noted in 12 of 15 acrome-galic patients studied; the authors also noted that negative scintigraphy results were related to the absence of acute growth hormone response to octreotide administration [2]. However, further studies with ^{111}In-pentetreotide failed to confirm the utility of the SRS in predicting tumor shrinkage or hormonal response to somatostatin analogue therapy in growth hormone producing and non-functioning pituitary adenomas, as well as in detecting the tumor residual mass in non-radically resected cases [3, 4]. So, regardless of the frequently positive SRS in pituitary adenomas, due to its limited added value, the method has not been included in routine pituitary patients' management.

SSTRs are present in leptomeninx. The SSTR type 2 expressing meningiomas have been considered the target for SRS. The increased uptake of the ^{111}In-pentetreotide in those tumors has been consistently confirmed from the very beginning [5–7]. For meningiomas visualization, two things are essential: (1) the expression of the SSTRs and (2) as the SSTR analogues are water-soluble, only the tumors located outside the blood–brain barrier or with disrupted blood–brain barrier are detected [8, 9]. The use of 99mTc-labeled depreotide provided better spatial resolution and enabled the detection of the smaller lesions than with the use of ^{111}In-pentetreotide [10]. The sensitivity of 99mTc-HYNIC-octreotate in meningioma detection is better than the sensitivity of the CT scans (100 vs. 83%, respectively), and the higher tumor/non-tumor ratio correlates with higher meningioma grade [11]. Due to high tumor-to-background ratio, Galium-68-labeled somatostatin analogues for PET/CT have also proved to be useful in meningiomas localization [12, 13]. The sensitivity of the method has been shown to be even better than contrast-enhanced MRI, which detected 92% of 190 meningiomas found by 68Ga-DOTATOC PET/CT[14]. The current role of SRS in meningioma management includes mainly differential noninvasive diagnosis of the intracranial tumors [15], but the new applications, such as therapy, both surgery and radiation, planning and performance improvement, or treatment outcome prognosis, are considered [16–18].

Disturbed blood–brain barrier is also the prerequisite of the positive SRS scans in patients with gliomas [5]. The uptake of the tracer has been shown mostly in high-grade tumors [7, 19], however less intense in comparison to the meningiomas. Although the SRS has not been involved in glioma patient management, the regional injections of 90-Yttrium-labeled somatostatin analogues have been applied in palliative treatment of the grade II and III malignancies—see also Chapter 6.6 [20, 21].

The SRS has also been successfully applied in childhood medulloblastomas [22], primitive neuroectodermal tumors [23], or hemangioblastomas [24].

HEAD AND NECK TUMORS

The SRS has been mostly applied in imaging of the head and neck tumors with neuroendocrine differentiation: carcinoids, small cell cancers, paragangliomas, esthesioneuroblastomas, etc. [25, 26]. In the series of 53 patients with neuroendocrine head and neck tumors, the sensitivity and specificity of the ^{111}In-pentetreotide scintigraphy were 93 and 92%, respectively, with the tumor-to-background ratio depending on the tumor type, the highest in case of paragangliomas and carcinoids [27].

The SSTR type 2 expression in tumor tissues has been confirmed in case reports on positive SRS imaging in esthesioneuroblastoma [28] and juvenile nasopharyngeal angiofibromas [29]. The SRS application in paragangliomas and medullary thyroid carcinoma management has been discussed in detail in Chapters 4.4.2 and 4.4.3.

LUNG

The pulmonary tumors were found to be SSTR positive already in the late 1980s [30]. The expression of SSTR type 1, 2A, 2B, 3, 4, and 5 was observed in 79.7, 96.6, 66.1, 49.1, 5.2, and 0% of typical lung carcinoids, 77.8, 77.8, 77.8, 33.3, 0, and 0% of atypical carcinoids, and 27.6, 69, 24.1, 15.5, 0, and 3.4% of SCLCs, respectively [31]. Frequent expression of SSTR type 2A in those tumors made them the good target for SRS.

The first studies failed to reveal the SSTR expression in non-small cell lung cancer (NSCLC) [32], and the positive results of scintigraphy were explained by the presence of the SSTRs in tumor vessel or in immune-competent cells infiltrating the tumor. However, other authors found SSTR type 2 in NSCLC cells by immunohistochemistry; the higher expression, the better differentiated the tumor. However, no correlation between the SSTR type 2 expression and 99mTc-depreotide uptake was confirmed [33].

Bronchial Carcinoids

Series of case reports stressed the increasing role of SRS in bronchial carcinoids quite early. The ^{111}In-DTPA-octreotide has been shown to be effective in localizing primary tumors, monitoring their growth, and detecting metastatic and/or recurring disease [34]. The radiopharmaceutical was proved to be useful in localizing occult ACTH-secreting bronchial carcinoids [35]. The study including 28 bronchial carcinoid patients showed positive radiolabeled octreotide scans in 71% of the primary tumors; however, CT proved to be more effective in detecting primary tumors as well as intrathoracic recurrence and liver metastases [36]. Kuyumu et al. compared ^{111}In-octreotide with 18F-FDG PET/CT findings in patients with pulmonary carcinoids (both typical and atypical). Sensitivity, specificity, positive predictive value (PPV), and negative predictive value (NPV) of SRS in the detection of primary tumor, lymph nodes, and distant metastasis were 76, 97, 95, and 86%, respectively. PET/CT was performed in 13 of total 21 evaluated patients, with sensitivity and specificity of 85 and 89.4 %, respectively [37].

Other somatostatin-based radiotracers have been proved to be able to detect bronchial carcinoids, including 99mTc-EDDA/HYNIC-TOC (Fig. 4.4.4.1) [38]. One of the PET-dedicated tracers 68Ga-DOTATATE was compared with 18F-FDG in pulmonary neuroendocrine tumors. All typical carcinoids included in the study were 68Ga-DOTATATE positive, whereas 4 of 11 tumors were negative on 18F-FDG PET/CT scans. More than a half of higher grade pulmonary neuroendocrine tumors were 68Ga-DOTATATE negative, while visualized by 18F-FDG.68Ga-DOTATATE was more effective in distinguishing inflammation or collapsed lung from tumor [39]. Similarly the flip-flop phenomenon with decreasing uptake of the labeled somatostatin analogue and increasing uptake of the 18F-FDG in atypical pulmonary carcinoids was shown for 68Ga DOTATOC by Jindal et al. [40].

Small Cell Lung Cancer

Considering its neuroendocrine differentiation, SCLC was an obvious target for imaging with labeled somatostatin analogues. The first analogue available for imaging, 123I-Tyr-3-octreotide, was tested as the SCLC staging agent already in the early 1990s [41, 42]. In the first larger series of 20 patients with histologically confirmed SCLC, iodine-123-Tyr-3-octreotide correctly identified 84% of primary tumors, and 78% of all patients with extensive disease, however with limited ability to detect liver and bone metastases [42].

The first study on ^{111}In-DTPA-octreotide in SCLC showed primary tumor in 13 of 15 evaluated cases, 12 of which with more diffuse disease than it was suspected

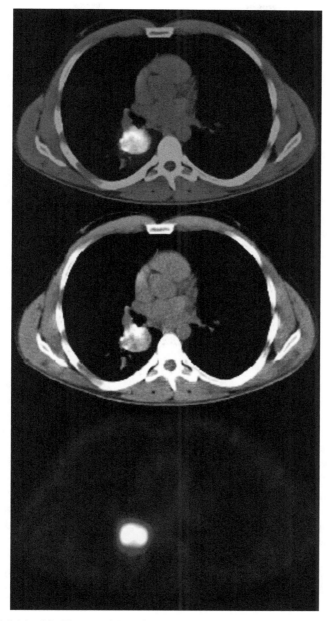

FIGURE 4.4.4.1 99mTc-tectrotide scintigraphy in bronchial carcinoid (SPECT/CT) (Nuclear Medicine Unit, Department of Endocrinology, University Hospital in Krakow, Poland). (*See insert for color representation of the figure.*)

based on other imaging modalities [43]. However, the radiopharmaceutical showed limited ability to detect abdominal, particularly liver, metastases, probably due to high physiological hepatic uptake of the tracer [44]. In series of 21 patients ^{111}In-octreotide detected 86% (48/56) of the lesions already known at the time of scintigraphy, including 94% of mediastinal metastases, 75% of bone metastases, and 71% of abdominal lymph nodes metastases; the technique detected five previously unrecognized SCLC lesions [45]. The next published data were less enthusiastic: both Bohuslavizki et al. and Hochstenbag et al. revealed only limited ability of ^{111}In-DTPA-octreotide to correctly identify distant metastases and stage SCLC [46, 47].

This prompted a larger multicenter study that included 100 SCLC, which confirmed high sensitivity (96%) of the method in visualizing the primary tumor and much poorer performance in detecting the regional and distant metastases (60 and 45%, respectively). The authors concluded that although ^{111}In-DTPA-octreotide is not suitable for staging of the SCLC, the decrease in tumor/background ratio during the chemotherapy noted in patient with remission may be utilized in monitoring the treatment [48].

Non-small Cell Lung Cancer

Octreotide scintigraphy was shown quite early to be able to detect NSCLC [44]. One of the studies compared 99mTc-octreotide with 18F-FDG coincidence imaging (using the same gantry). The studied group of 44 patients included 25 patients with NSCLC. The sensitivity, specificity, PPV, and NPV of 99mTc-octreotide for detecting the primary lesion were 100, 75.7, 90.1, and 100%, respectively, and of 18F-FDG they were 100, 46.1, 83.8, and 100%, respectively. The sensitivity of 99mTc-octreotide for the detection of lung cancer at the primary lesion was comparable with that of 18F-FDG coincidence imaging. SPECT 99mTc-octreotide scintigraphy was less effective in detecting hilar and mediastinal lymph node metastases; however, it proved to be successful in detecting distant metastases [49]. The other radiopharmaceuticals tested in NSCLC included ^{111}In-DOTA-lanreotide [50] and 68Ga-DOTATOC [51].

However, it was 99mTc-P829, later named 99mTc-depreotide, to be the radiotracer most frequently applied in NSCLC [52]. Although it has been mostly applied for the noninvasive assessment of lung nodules, 99mTc-depreotide was also used for NSCLC staging and was compared with other modalities used for that purpose. The largest study published so far comprised data from 166 NSCLC patients. Whole body and SPECT 99mTc-depreotide scintigraphy results were compared with attenuation-corrected 18F-FDG PET. 99mTc-depreotide scintigraphy was equally sensitive as 18F-FDG PET (94% (CI: 88–98%) vs. 96% (CI: 90–98%), respectively), but less specific (51% (CI: 34–68%) and 71% (CI: 54–85%), respectively). The staging accuracy of both methods proved to be similar (45 and 55%, respectively); however, the authors concluded it to be insufficient to correctly assess the extent of the disease [53]. Another study on the assessment of locoregional lymph nodes involvement with 99mTc-depreotide conducted on 86 patients with 204 lymph node stations in total, although revealing high sensitivity of 99% and NPV of 98% of the method, failed to confirm any added value of scintigraphy with equal to CT diagnostic accuracy of 76.4%. The method accurateness did not benefit from quantitative assessment of the tracer uptake [54].

Solitary Pulmonary Nodule

The first experience with 99mTc-depreotide in evaluation of the solitary pulmonary nodules was published in 1999 by Blum et al. [55]. It was followed shortly by the announcement of the results of a multicenter trial, which included 144 patients (88 malignant lung lesions). The sensitivity of the method with chest CT as the reference was 96.6% and specificity was 73.1%, comparable with the performance of 18F-FDG PET assessment carried out at the end of the twentieth century [56].

The sensitivity and specificity of the 99mTc-depreotide scintigraphy in detecting the malignancy in solitary pulmonary nodule varies from 88 to 100% and from 43 to 88%, respectively (Table 4.4.4.1) [55–67]. The results of the meta-analysis by Cronin et al. comparing different methods of cross-sectional imaging in solitary pulmonary nodule differential diagnosis are summed up in Table 4.4.2 [68].

The 99mTc-depreotide in solitary pulmonary lesions was compared head-to-head with other functional imaging modalities. 99mTc-depreotide was proved to be equally sensitive and specific as 201Tl chloride in characterization of the pulmonary lesion, with false positive results being the main disadvantage [63]. Studies assessing 99mTc-depreotide and 18F-FDG usually show greater sensitivity and/or specificity of the latter; however, it has been concluded that 99mTc-depreotide SRS is valuable alternative for 18F-FDG and should be considered if PET is not available [53, 61, 62].

Other labeled somatostatin analogues used in solitary pulmonary nodule assessment were 99mTc-octreotide acetate and 99mTc-EDDA/HYNIC-TOC [49, 69]. The sensitivity of the second tracer was 90% in the largest studies published so far, and true negative results were obtained in 79% of the benign lung nodules [69]. The performance

TABLE 4.4.4.1 **Sensitivity and specificity of 99mTc-depreotide in solitary pulmonary nodule assessment**

References	Number of subjects	Sensitivity	Specificity
Blum (1999) [55]	14	0.93	0.88
Blum (2000) [56]	114	0.97	0.73
Grewal (2002) [57]	39	1.0	0.43
Baath (2004) [58]	28	0.94	0.64
Chcialowski (2004) [59]	31	0.94	0.44
Kahn (2004) [53]	157	0.94	0.51
Martins (2004) [60]	40	0.97	0.63
Halley (2005) [61]	28	0.89	0.80
Ferran (2006) [62]	29	0.84	0.88
Boundas (2007) [63]	33	1.0	0.65
Szalus (2008) [64]	50	0.89	0.60
Axelsson (2008) [65]	99	0.94	0.52
Harders (2012) [66]	140	0.94	0.58
Sobic-Saranovic (2012) [67]	26	0.88	0.85

TABLE 4.4.2   Comparison of different imaging modalities in detecting malignancy in solitary pulmonary nodule[a]

Imaging modality	Sensitivity (95% CI)	Specificity (95% CI)	PPV (95% CI)	NPV (95% CI)	Diagnostic OR (95% CI)	Area under ROC curve (95% CI)
Dynamic CT	0.93 (0.88–0.97)	0.76 (0.68–0.97)	0.80 (0.74–0.86)	0.95 (0.93–0.98)	39.91 (1.21–81.04)	0.93 (0.81–0.97)
Dynamic MR	0.94 (0.91–0.97)	0.79 (0.73–0.86)	0.86 (0.83–0.89)	0.93 (0.90–0.96)	60.59 (5.56–115.62)	0.94 (0.83–0.98)
FDG PET	0.95 (0.93–0.98)	0.82 (0.77–0.88)	0.91 (0.88–0.93)	0.90 (0.85–0.94)	97.31 (6.26–188.37)	0.94 (0.83–0.98)
99mTcdepreotide SPECT	0.95 (0.93–0.97)	0.82 (0.78–0.85)	0.90 (0.83–0.97)	0.91 (0.84–0.98)	84.50 (34.28–134.73)	0.94 (0.83–0.98)

[a] Based on Ref. [68].

of the method was facilitated by semi-quantitative analysis of the tracer uptake [70]. Head-to-head comparison of 99mTc-depreotide and 99mTc-EDDA/HYNIC-TOC showed similar efficacy of two somatostatin analogue–based tracers in distinguishing malignant pulmonary nodules; however, the higher number of false positive results with 99mTc-depreotide was stressed [71]. Similar results were obtained for a two subset of patients, each evaluated with one of two somatostatin-based tracers [67].

BREAST

SSTRs presence, particularly type 2, has been confirmed in breast cancer tissue [30, 72]. Positive correlation between SSTR type 2 mRNA and estrogen and progesterone receptors expression has also been found [73, 74], and the SSTR expression is regulated by the estrogens in cell line studies [75]. These findings led to the clinical application of the SRS as early as in 1994. Positive scintigraphy with ^{111}In-DTPA-D-Phe1-octreotide in 39 of 52 primary breast cancers (75%) was reported, particularly in invasive ductal carcinomas, and nonpalpable cancer-containing lymph nodes were detected in 4 of 13 patients with subsequently histologically proven metastases [76]. Subsequent studies confirmed similar sensitivity of the ^{111}In-pentetreotide scintigraphy in presurgical assessment of breast cancer patients [77, 78]. The ^{111}In-pentetreotide scintigraphy seemed to be less effective in the assessment of axillary lymph nodes involvement [79], which may be due to the weaker expression of the SSTR2 mRNA expression in metastatic breast cancer [74].

The positivity of the ^{111}In-pentetreotide scintigraphy was reported to be related to the SSTR type 2 ($P = 0.025$), as well as SSTR type 5 ($P < 0.001$) expression in cancer tissue [80]. The 99m-Technetium labeled compounds have been also tested in breast cancers. The 99mTc-depreotide value in prediction of the response to hormonal therapy was tested in breast cancer bone metastases—99mTc-depreotide was more specific but less sensitive and accurate than 99mTc-MDP in metastases detection. Depreotide-positive patients remained stable during the follow-up, whereas five of six depreotide-negative patients progressed [81]. In the larger group of patients with advanced breast cancer, the 99mTc-depreotide scintigraphy was performed twice: before and 3 weeks after initiation of the hormonal treatment. The PPV and NPV of baseline 99mTc-depreotide scintigraphy for therapy responsiveness were 73% (8/11) and 100% (7/7), respectively. Sequential scans were always both positive or both negative. The 99mTc-depreotide uptake in subsequent scans decreased in responders and increased in the nonresponders ($P=0.017$). Baseline and follow-up scans combined predicted the responsiveness to the hormonal therapy with 100% accuracy [82]. The sensitivity, specificity, PPV, and NPV of 99mTc-octreotide in the detection of primary breast cancer lesion were 91.8, 22.2, 71.8, and 57.1%, respectively; however, the lymph node metastases may be obscured by the nonspecific breast tissue uptake [83]. 68Ga-DOTATOC PET/CT has been reported to be able to detect both breast metastases from neuroendocrine tumors and primary breast cancer lesions [84].

The radio-guided surgery with 125I-lanreotide was first reported in 1999. The overall accuracy of nodal evaluation with 125I-lanreotide/intraoperative gamma detection was 77% and the NPP of this technique was 97%. False positive results were obtained in 20% of histologically negative axillary lymph nodes. A significant statistical correlation between histology and gamma probe counts (P<0.0001) was found [85]. Intraoperative gamma-probe detection with ^{111}In-octreotide of the axillary lymph node involvement in SRS positive primary breast tumors was shown to be ineffective in microscopic (*in situ*) nodal metastases [86].

OTHER CHEST TUMORS

The use of SRS in thymic malignancies evaluation has been mostly described in case reports. Two patients' series have been published so far. ^{111}In-DTPA-D-Phe1-octreotide scintigraphy performed in a group of 18 patients was effective in the detection of thymic masses larger than 1.5 cm and in differential diagnosis of malignant lesions—thymic hyperplasia was negative on SRS scans [87]. In the second series of 14 cases, the ^{111}In-pentetreotide scintigraphy was positive in 13 cases, whereas in 11 cases, expression of at least one SSTR subtype was confirmed by immunohistochemistry [88].

DIGESTIVE SYSTEM

The experience in SRS in digestive system malignancies other than gastroenteropancreatic neuroendocrine neoplasms is limited to case reports and single small series.

Oesophageal Cancers

99mTc-depreotide has been the only somatostatin analogue tested in oesophageal cancer patients. Although statistically significant difference in tracer uptake between malignant and nonmalignant oesophageal lesions was confirmed, limited sensitivity of the method (76%) makes it unsuitable for the screening and/or primary diagnosis [89]. Although SSTR expression was confirmed by immunohistochemistry mostly in oesophageal adenocarcinoma, the 99mTc-depreotide uptake did not correlate with tumor type and SSTR expression [90].

Gastrointestinal Stromal Tumors

The only published report including SRS in gastrointestinal stromal tumor (GIST) patients showed positive scans in 50% of six patients evaluated with ^{111}In-octreotide [91]. However, tumor cells in primary culture (gastric and small intestinal GIST) specifically bound and internalized 177Lu when incubated with the therapeutic compound 177Lu-octreotate for 4–48 h, which makes the PPRT the possible treatment method in tyrosine kinase inhibitor–resistant patients [91].

Primary Liver Cancers

99mTc-HYNIC-Tyr(3)-octreotide scintigraphy has been proved to detect human hepatocellular cancer (HCC) xenografts in nude mice [92]. The human studies showed positive ^{111}In-octreotide scans in less than 50% of HCC cirrhotic patients—the imaging may be used as the examination qualifying the patient to the subsequent treatment with long-acting somatostatin analogues [93, 94].

Although hepatic cholangiocarcinomas and adenocarcinomas of the gallbladder were proved to take up ^{111}In-DOTA-LAN, the positive scans did not predict the response to the lanreotide treatment [95].

UROGENITAL NEOPLASMS

Kidney

The SRS has been mostly used in renal cell carcinomas (RCC). The largest series published so far included 16 patients with advanced RCC. Although nine of them were positive on ^{111}In-pentetreotide scintigraphy, lesion-based analysis showed only 12.1% sensitivity, proving negligible value of the method in this setting [96].

Prostate

The first published study on hormone-refractory metastatic prostate adenocarcinoma compared ^{111}In-DTPA-D-Phe1-octreotide (OctreoScan) with 99mTc-HDP. Nearly 95% of 31 patients had at least one metastasis positive on OctreoScan, and 37% of all bone metastases were detected with SRS [97]. Small pilot study results suggested that ^{111}In-DTPA-D-Phe1-octreotide scintigraphy might be useful in selecting prostatic cancer patient who benefit from somatostatin analogue therapy [98].

Similar Ga-68-DOTATOC PET/CT sensitivity of 30% in per-lesion analysis was showed in prostatic cancer focal bone metastases. One-third of the patients with super-scan was negative with this imaging method. The maximum Delta SUV(max) between metastases and normal bone was 4.9 (mean =1.6 ± 0.9) and between the prostate and adjacent tissue was 5.9 (mean = 2.8 ± 1.6), suggesting only weak expression of the SSTR type 2 and 5 in tumor tissue [99].

NEUROBLASTOMA

Neuroblastoma is the most common extracranial solid tumor of the childhood. The nuclear medicine, particularly mIBG-scintigraphy, has been established as the valuable tool for detecting, staging, and qualification for the radionuclide therapy [100]. As the SSTR type 2 expression has been shown in some of the neuroblastomas, somatostatin analogue–based imaging and/or treatment offers an alternative management, particularly for the mIBG negative patients [100, 101]. The study comparing ^{123}I-mIBG and ^{111}In-pentetreotide scintigraphy results showed concordant results in 85% of studied patients. The positive SRS scans were connected with more favorable prognosis [102].

The larger study including 88 neuroblastoma patients confirmed better ^{123}I-mIBG sensitivity (94 vs. 64% for ^{111}In-pentetreotide) in tumor detection. However, SRS provided important prognostic information: 4-year survival and 4-year event free–survival were significantly better for ^{111}In-pentetreotide positive than negative children (survival: 90 vs. 48%, log rank P < 0.003; event-free survival: 83 vs. 39%, log rank P < 0.0002) [103].

Peptide receptor radionuclide therapy (PRRT) in neuroblastoma patients is discussed in Chapter 6.6.

MERKEL CELL CARCINOMA

MCC, rare and highly malignant cutaneous tumor, is also considered a neuroendo-crine neoplasia and as such is a potential target for SRS [104]. The first reports of the successful visualization of the MCC larger than 0.5 cm with SRS came from 1992 [105]. In the prospective study in a group of 20 patients, ^{111}In-octreotide sensitivity and specificity for grade I and II tumors were 78 and 96%, respectively. SRS was able to detect the occult tumor lesions; however, the metastases to the organs with physiologic high uptake of the tracer may be omitted [106]. The German group showed in contrast false negative results of ^{111}In-octreotide scintigraphy in 5 of 11 examined patients [107].

A study comparing ^{111}In-pentetreotide with18F-FDG PET/CT showed a better detection rate with the latter agent resulting in frequent upstaging of the patients. None of the octreotide positive patients was negative on 18F-FDG PET/CT, suggest-ing that currently this method is the preferred in the functional evaluation of MCC patient [108]. However, one case report suggested that 68Ga-DOTATATE PET/CT might be more accurate in the evaluation of the tumor burden [109]. Positive SRS in MCC resulted in the first attempts of PRRT in progressive patients [110].

LYMPHOMA

Both somatostatin and SSTRs are detected in the immune system. In human studies, somatostatin mRNA has been found in the thymic epithelial cells and spleen [111], and SSTRs expression in various lymphoid organs [112]. The SSTR type 3 mRNA was shown to be consistently expressed in normal human lymphocytes, their activation resulted in upregulation of SSTR type 5 mRNA. SSTR type 2 expression in peripheral blood mononuclear cells (PBMC) has been regarded to be low in comparison to the cell lines and PBMC from leukemic patients, and to increase after activation with phytohemagglutinin [113, 114]. However, in the quantitative assessment by RT-PCR, the SSTR2 mRNA expression was found to be low, corresponding to the low binding of [(125)I-Tyr(3)]octreotide and absent immunoreactivity for SSTR type 2 in immu-nohistochemistry [115]. It was concluded that lymphomas are not the good candidates for PRRT with labeled somatostatin analogues, which have been however tried in imaging to stage the tumors and to assess therapy response.

The successful attempts to detect the lymphoma tissue by SRS were reported in early 1990s [116, 117]. The sensitivity of ^{111}In-DTPA-D-Phe-1-octreotide in Hodgkin's disease (HD) varied from 70 to 98% (with lowest detection rate for the disease localized in abdomen) [118–121]. For stage I–II HD SRS resulted in upstaging in18% of the patients [122].

In non-Hodgkin's lymphoma (NHL) patients, the sensitivity of the method has been considered to be lower and varied between 35 and 85% (29–35% for low-grade and 43–44% for high-grade disease) [118, 120, 121]. 99mTc-depreotide has also been tested in NHL as the potential targeting agent, although high bone marrow uptake made it unsuitable for that purpose [123]. ^{111}In-octreotide scintigraphy was found to be able to distinguish between gastric and non-gastric primary mucosa-associated lymphoid tissue (MALT) lymphoma [124]. SRS effectiveness in staging of cutaneous lymphomas is very limited—43% of B-cell and none of the T-cell cutaneous lymphomas were detected [125].

In SRS effectiveness studies, anatomical imaging (CT, ultrasound) was used as a standard. Only a few studies comparing SRS with other functional method have been published so far. The head to head comparison of 111In labeled octreotide with 18F-FDG was published in 1993. In a group of 22 malignant lymphoma patients, metabolic imaging yielded a higher rate of detection of lymphoma manifestations (92% vs. 64%) and better tumor contrast [126]. The study comparing 67Ga citrate, ^{111}In-DOTA-dPhe(1)-Tyr(3)-octreotide (^{111}In-DOTA-TOCT), and ^{111}In-DOTA-lanreotide (^{111}In-DOTA-LAN) scintigraphy in patients with proven MALT-type lymphoma showed similar detection rate (63, 60, and 64%, respectively) and false positive cases. Whereas SRS was better in detecting infradiaphragmatic lesions, 67Ga-citrate was more sensitive in detecting supra-diaphragmatic disease [127].

TUMOR-INDUCED OSTEOMALACIA (TIO)

After a series of case reports on SRS in detection of the hypophosphatemia causing neoplasia, a few larger studies has been published. 68Ga-DOTATATE PET/CT was able to find FGF-23-secreting tumors in all six patients included in the multicenter Australian study [128]. The largest study published so far included 183 patients with hypophosphatemia, 80 of whom were considered to have TIO. The sensitivity of 99mTc-HYNIC-TOC was 86.3%, specificity 99.1%, and the overall accuracy 93.4% [129].

ACKNOWLEDGMENTS

Authors would like to thank Ms. Monika Tomaszuk and Mr. Boguslaw Glowa for their help in preparing the manuscript.

REFERENCES

[1] Volante, M.; Bozzalla-Cassione, F.; Papotti, M. Endocrine Pathology 2004, 15, 275–291.

[2] Ur, E.; Mather, S. J.; Bomanji, J.; et al. Clinical Endocrinology (Oxford) 1992, 36, 147–150.

[3] Plöckinger, U.; Reichel, M.; Fett, U.; et al. Journal of Clinical Endocrinology and Metabolism 1994, 79, 1416–1423.

[4] Plöckinger, U.; Bäder, M.; Hopfenmüller, W. et al. European Journal of Endocrinology 1997, 136, 369–376.

[5] Scheidhauer, K.; Hildebrandt, G.; Luyken, C.; et al. Hormone and Metabolism Research Supplement Series 1993, 27, 59–62.

[6] Maini, C. L.; Sciuto, R.; Tofani, A.; et al. Nuclear Medicine Communications 1995, 16, 756–766.

[7] Schmidt, M.; Scheidhauer, K.; Luyken, C.; et al. European Journal of Nuclear Medicine 1998, 25, 675–686.

[8] Haldemann, A. R.; Rösler, H.; Barth, A.; et al. Journal of Nuclear Medicine 1995, 36, 403–410.

[9] Nathoo, N.; Ugokwe, K.; Chang, A. S.; et al. Journal of Neuro-oncology 2007, 81, 167–174.

[10] Hellwig, D.; Samnick, S.; Reif, J.; et al. Clinical Nuclear Medicine 2002, 27, 781–784.

[11] Wang, S.; Yang, W.; Deng, J.; et al. Journal of Neuro-oncology 2013, 113, 519–526.

[12] Henze, M.; Schuhmacher, J.; Hipp, P.; et al. Journal of Nuclear Medicine 2001, 42, 1053–1056.

[13] Henze, M.; Dimitrakopoulou-Strauss, A.; Milker-Zabel, S.; et al. Journal of Nuclear Medicine 2005, 46, 763–769.

[14] Afshar-Oromieh, A.; Giesel, F. L.; Linhart, H. G.; et al. European Journal of Nuclear Medicine and Molecular Imaging 2012, 39, 1409–1415.

[15] Valotassiou, V.; Leondi, A.; Angelidis, G; et al. Scientific World Journal 2012, 2012, 412580.

[16] Wang, S.; Yang, W.; Deng, J.; et al. Nuclear Medicine Communications 2013, 34, 249–253.

[17] Nyuyki, F.; Plotkin, M.; Graf, R.; et al. European Journal of Nuclear Medicine and Molecular Imaging 2010, 37, 310–318.

[18] Nicolato, A. Quarterly Journal of Nuclear Medicine and Molecular Imaging 2004, 48, 26–32.

[19] Luyken, C.; Hildebrandt, G.; Scheidhauer, K.; et al. Acta Neurochirgica (Wien) 1994, 127, 60–64.

[20] Merlo, A.; Hausmann, O.; Wasner, M.; et al. Clinical Cancer Research 1999, 5, 1025–1033.

[21] Schumacher, T.; Hofer, S.; Eichhorn, K.; et al. European Journal of Nuclear Medicine and Molecular Imaging 2002, 29, 486–493.

[22] Müller, H. L.; Frühwald, M. C.; Scheubeck, M.; et al. Journal of Neuro-oncology 1998, 38, 27–40.

[23] Frühwald, M. C.; Rickert, C. H.; O'Dorisio, M. S.; et al. Clinical Cancer Research 2004, 10, 2997–3006.

[24] Chowdhury, F. U.; Scarsbrook, A. F. Clinical Nuclear Medicine 2008, 33, 294–296.

[25] Kau, R.; Arnold, W. Acta Oto-laryngologica 1996, 116, 345–349.

[26] Whiteman, M. L.; Serafini, A. N.; Telischi, F. F.; et al. AJNR American Journal of Neuroradiology 1997, 18, 1073–1080.

[27] Myssiorek, D.; Tronco, G. Laryngoscope 2005, 115, 1707–1716.

[28] Rostomily, R. C.; Elias, M.; Deng, M.; et al. Head and Neck 2006, 28, 305–312.

[29] Kukwa, W.; Andrysiak, R.; Kukwa, A.; et al. Head and Neck 2011, 33, 1739–1746.

[30] Reubi, J. C.; Krenning, E.; Lamberts, S. W.; Kvols L. Journal of Steroid Biochemistry and Molecular Biology 1990, 20, 1073–1077.

[31] Tsuta, K.; Wistuba, I. I.; Moran, C.A. Pathology—Research and Practice 2012, 208, 470–474.

[32] Reubi, J. C.; Waser, B.; Sheppard, M.; Macaulay, V. International Journal of Cancer 1990, 45, 269–274.

[33] Herlin, G.; Kölbeck, K. G.; Menzel, P. L.; et al. Acta Radiologica 2009, 50, 902–908.

[34] Musi, M.; Carbone, R. G.; Bertocchi, C.; et al. Lung Cancer 1998, 22, 97–102.

[35] Tabarin, A.; Valli, N.; Chanson, P.; et al. Journal of Clinical Endocrinology and Metabolism 1999, 84, 1193–1202.

[36] Granberg, D.; Sundin, A.; Janson, E. T.; et al. Clinical Endocrinology (Oxford) 2003, 59, 793–799.

[37] Kuyumcu, S.; Adalet, I.; Sanli, Y.; et al. Annals of Nuclear Medicine 2012, 26, 689–697.

[38] Płachcińska, A.; Mikołajczak, R.; Maecke, H. R.; et al. European Journal of Nuclear Medicine and Molecular Imaging 2004, 31, 1005–1010.

[39] Kayani, I.; Conry, B. G.; Groves, A. M.; et al. Journal of Nuclear Medicine 2009, 50, 1927–1932.

[40] Jindal, T.; Kumar, A.; Venkitaraman, B.; et al. Cancer Imaging 2011, 11, 70–75.

[41] Kwekkeboom, D. J.; Krenning, E. P.; Bakker, W. H.; et al. Journal of Nuclear Medicine 1991, 32, 1845–1848.

[42] Leitha, T.; Meghdadi, S.; Studnicka, M.; et al. Journal of Nuclear Medicine 1993, 34, 1397–1402.

[43] Maini, C. L.; Tofani, A.; Venturo, I.; et al. Nuclear Medicine Communications 1993, 14, 962–968.

[44] Kwekkeboom, D. J.; Kho, G. S.; Lamberts, S. W.; et al. European Journal of Nuclear Medicine 1994, 21, 1106–1113.

[45] Bombardieri, E.; Crippa, F.; Cataldo, I.; et al. European Journal of Cancer 1995, 31A, 184–188.

[46] Bohuslavizki, K. H.; Brenner, W.; Günther, M.; et al. Nuclear Medicine Communications 1996, 17, 191–196.

[47] Hochstenbag, M. M.; Heidendal, G. A.; Wouters, E. F.; ten Velde, G. P. Clinical Nuclear Medicine 1997, 22, 811–816.

[48] Reisinger, I.; Bohuslavitzki, K. H.; Brenner, W.; et al. Journal of Nuclear Medicine 1998, 39, 224–227.

[49] Wang, F.; Wang, Z.; Yao, W.; et al. Journal of Nuclear Medicine 2007, 48, 1442–1448.

[50] Traub, T.; Petkov, V.; Ofluoglu, S.; et al. Journal of Nuclear Medicine 2001, 42, 1309–1315.

[51] Dimitrakopoulou-Strauss, A.; Georgoulias, V.; Eisenhut, M.; et al. European Journal of Nuclear Medicine and Molecular Imaging 2006, 33, 823–830.

[52] Virgolini, I.; Leimer, M.; Handmaker, H., et al. Cancer Research 1998, 58, 1850–1859.

[53] Kahn, D.; Menda, Y.; Kernstine, K.; et al. Chest 2004, 125, 494–501.

[54] Danielsson, R.; Bååth, M.; Svensson, L.; et al. European Journal of Nuclear Medicine and Molecular Imaging 2005, 32, 925–931.

[55] Blum, J. E.; Handmaker, H.; Rinne, N. A. Chest 1999, 115, 224–232.

[56] Blum, J.; Handmaker, H.; Lister-James, J.; Rinne, N. Chest 2000, 117, 1232–1238.

[57] Grewal, R. K.; Dadparvar, S.; Yu, J. Q.; et al. The Cancer Journal 2002, 8, 400–404.

[58] Bååth, M.; Kölbeck, K. G.; Danielsson, R. Acta Radiologica 2004, 45, 833–839.

[59] Chciałowski, A.; Dziuk, E.; From, S.; et al. Polskie Archiwum Medycyny Wewnętrznej 2004, 112, 1031–1038.

[60] Martins, T.; Lino, J. S.; Ramos, S., Oliveira, L. Cancer Biotherapy and Radiopharmaceuticals 2004, 19, 253–259.

[61] Halley, A.; Hugentobler, A.; Icard, P.; et al. European Journal of Nuclear Medicine and Molecular Imaging 2005, 32, 1026–1032.

[62] Ferran, N.; Ricart, Y.; Lopez, M.; et al. Nuclear Medicine Communications 2006, 27, 507–514.

[63] Boundas, D.; Karatzas, N.; Moralidis, E.; et al. Nuclear Medicine Communications 2007, 28, 533–540.

[64] Szaluś, N.; Chciałowski, A.; From, S.; et al. Polski Merkuriusz Lekarski 2008, 25, 330–334.

[65] Axelsson, R.; Herlin, G.; Bååth, M.; et al. Acta Radiologica 2008, 49, 295–302.

[66] Harders, S. W.; Madsen, H. H.; Hjorthaug, K.; et al. British Journal of Radiology 2012, 85, e307–313.

[67] Sobic-Saranovic, D. P.; Pavlovic, S. V.; Artiko, V. M.; et al. Clinical Nuclear Medicine 2012, 37, 14–20.

[68] Cronin, P.; Dwamena, B. A.; Kellu, A. M.; Carlos, R. C. Radiology 2008, 246, 772–782.

[69] Płachcińska, A.; Mikołajczak, R.; Maecke, H.; et al. Cancer Biotherapy and Radiopharmaceuticals 2004, 19, 613–620.

[70] Płachcińska, A.; Mikołajczak, R.; Kozak, J.; et al. Nuclear Medicine Review—Central and Eastern Europe 2004, 7, 143–150.

[71] Płachcińska, A.; Mikołajczak, R.; Kozak, J.; et al. Nuclear Medicine Review—Central and Eastern Europe 2006, 9, 24–29.

[72] Evans, A. A.; Crook, T.; Laws, S. A.; et al. British Journal of Cancer 1997, 75, 798–803.

[73] Kumar, U.; Grigorakis, S. I.; Watt, H. L.; et al. Breast Cancer Research and Treatment 2005, 92, 175–186.

[74] Orlando, C.; Raggi, C. C.; Bianchi, S.; et al. Endocrine-Related Cancer 2004, 11, 323–332.

[75] Van Den Bossche, B.; D'haeninck, E.; De Vos, F.; et al. European Journal of Nuclear Medicine and Molecular Imaging 2004, 31, 1022–1030.

[76] van Eijck, C. H.; Krenning, E. P.; Bootsma, A.; et al. The Lancet 1994, 343(8898), 640–643.

[77] Bajc, M.; Ingvar, C.; Palmer, J. Journal of Nuclear Medicine 1996, 37, 622–626.

[78] Chiti, A.; Agresti, R.; Maffioli, L. S.; et al. European Journal of Nuclear Medicine 1997, 24, 192–196.

[79] Albérini, J. L.; Meunier, B.; Denzler, B.; et al. Breast Cancer Research and Treatment 2000, 61, 21–32.

[80] Schulz, S.; Helmholz, T.; Schmitt, J.; et al. Breast Cancer Research and Treatment 2002, 72, 221–226.

[81] Van Den Bossche, B.; D'haeninck, E.; De Winter, F.; et al. Nuclear Medicine Communications 2004, 25, 787–792.

[82] Van Den Bossche, B.; Van Belle, S.; De Winter, F.; et al. Journal of Nuclear Medicine 2006, 47, 6–13.

[83] Wang, F.; Wang, Z.; Wu, J.; et al. Nuclear Medicine and Biology 2008, 35, 665–671.

[84] Elgeti, F.; Amthauer, H.; Denecke, T.; et al. Nuklearmedizin 2008, 47, 261–265.

[85] Cuntz, M. C.; Levine, E. A.; O'Dorisio, T. M.; et al. Annals of Surgical Oncology 1999, 6, 367–372.

[86] Skånberg, J.; Ahlman, H.; Benjegård, S. A.; et al. Breast Cancer Research and Treatment 2002, 74, 101–111.

[87] Lastoria, S.; Vergara, E.; Palmieri, G.; et al. Journal of Nuclear Medicine 1998, 39, 634–639.

[88] Ferone, D.; Montella, L.; De Chiara, A.; et al. Frontiers in Bioscience, Landmark 2009, 14, 3304–3309.

[89] Herlin, G.; Idestrom, L.; Lundell, L.; et al. International Journal of Molecular Imaging 2011, 2011, 279345

[90] Herlin, G.; Lundell, L.; Ost, A.; et al. Radiology Research and Practice 2012, 2012, 415616

[91] Arne, G.; Nilsson, B.; Dalmo, J.; et al. Acta Oncologica 2013, 52, 783–792.

[92] Li, Y.; Si, J. M.; Zhang, J.; et al. World Journal of Gastroenterology 2005, 7, 3953–3957.

[93] Dimitroulopoulos, D.; Xinopoulos, D.; Tsamakidis, K.; et al. World Journal of Gastroenterology 2007, 13, 3164–3170.

[94] Nguyen-Khac, E.; Ollivier, I.; Aparicio, T.; et al. Cancer Biology and Therapy 2009, 8, 2033–2039.

[95] Fiebiger, W. C.; Scheithauer, W.; Traub, T.; et al. Scandinavian Journal of Gastroenterology 2002, 37, 222–225.

[96] Freudenberg, L. S.; Gauler, T.; Görges, R.; et al. Nuklearmedizin 2008, 47, 127–131.

[97] Nilsson, S.; Reubi, J. C.; Kalkner, K. M.; et al. Cancer Research 1995, 55(23 Suppl), 5805s–5810s.

[98] Kälkner, K. M.; Nilsson, S.; Westlin, J. E. Anticancer Research 1998, 18, 513–516.

[99] Luboldt, W.; Zöphel, K.; Wunderlich, G.; et al. Molecular Imaging and Biology 2010, 12, 78–84.

[100] Mueller, W. P.; Coppenrath, E.; Pfluger, T. Pediatric Radiology 2013, 43, 418–427.

[101] Manil, L.; Edeline, V.; Lumbroso, J.; et al. Journal of Nuclear Medicine 1996, 37, 893–896.

[102] Kropp, J.; Hofmann, M.; Bihl, H. Anticancer Research 1997, 17, 1583–1588.

[103] Schilling, F. H.; Bihl, H.; Jacobsson, H.; et al. Medical and Pediatric Oncology 2000, 35, 688–691.

[104] Nguyen, B. D.; McCullough, A. E. Radiographics 2002, 22, 367–376.

[105] Kwekkeboom, D. J.; Hoff, A. M.; Lamberts, S. W.; et al. Archives of Dermatology 1992, 128, 818–821.

[106] Guitera-Rovel, P.; Lumbroso, J.; Gautier-Gougis, M. S.; et al. Annals of Oncology 2001, 12, 807–811.

[107] Durani, B. K.; Klein, A.; Henze, M.; et al. British Journal of Dermatology 2003, 148, 1135–1140.

[108] Lu, Y.; Fleming, S. E.; Fields, R. C.; et al. Clinical Nuclear Medicine 2012, 37, 759–762.

[109] Epstude, M.; Tornquist, K.; Riklin, C.; et al. Clinical Nuclear Medicine 2013, 38, 283–284.

[110] Salavati, A.; Prasad, V.; Schneider, C. P.; et al. Annals of Nuclear Medicine 2012, 26, 365–369.

[111] Dalm, V. A.; Van Hagen, P. M.; de Krijger, R. R.; et al. Clinical Endocrinology (Oxford) 2004, 60, 625–629.

[112] Reubi, J. C.; Waser, B.; Horisberger, U.; et al. Blood 1993, 82, 2143–2151.

[113] Tsutsumi, A.; Takano, H.; Ichikawa, K.; et al. Cellular Immunology 1997, 181, 44–49.

[114] Ferone, D.; Hofland, L. J.; Colao, A.; et al. Annals of Oncology 2001, 12(suppl. 2), S125–S130.

[115] Dalm, V. A.; Hofland, L. J.; Mooy, C. M.; et al. Journal of Nuclear Medicine 2004, 45, 8–16.

[116] Reubi, J. C.; Waser, B.; van Hagen, M.; et al. International Journal of Cancer 1992, 50, 895–900.

[117] Vanhagen, P. M.; Krenning, E. P.; Reubi, J. C.; et al. British Journal of Haematology 1993, 83, 75–79.

[118] Lipp, R. W.; Silly, H.; Ranner, G., et al. Journal of Nuclear Medicine 1995, 36, 13–18.

[119] van den Anker-Lugtenburg, P. J.; Krenning, E. P.; Oei, H. Y., et al. British Journal of Haematology 1996, 93, 96–103.

[120] van den Anker-Lugtenburg, P. J.; Löwenberg, B.; Lamberts, S. W.; Krenning, E. P. Metabolism 1996, 45 (8 Suppl 1), 96–97.

[121] Ivancević, V.; Wörmann, B.; Nauck, C.; et al. Leukemia and Lymphoma 1997, 26, 107–114.

[122] Lugtenburg, P. J.; Krenning, E. P.; Valkema, R.; et al. British Journal of Haematology 2001, 112, 936–944.

[123] Bushnell, D. L.; Menda, Y.; Madsen, M. T.; et al. Nuclear Medicine Communications 2004, 25, 839–843.

[124] Raderer, M.; Traub, T.; Formanek, M.; et al. British Journal of Cancer 2001, 85, 1462–1466.

[125] Valencak, J.; Trautinger, F.; Raderer, M.; et al. Journal of European Academy of Dermatology and Venereology 2010, 24, 13–17.

[126] Bares, R.; Galonska, P.; Dempke, W.; et al. Hormone and Metabolism Research. Supplement Series 1993, 27, 56–58.

[127] Li, S.; Kurtaran, A.; Li, M; et al. European Journal of Nuclear Medicine and Molecular Imaging 2003, 30, 1087–1095.

[128] Clifton-Bligh, R. J.; Hofman, M. S.; Duncan, E.; et al. Journal of Clinical Endocrinology and Metabolism 2013, 98, 687–694.

[129] Jing, H.; Li, F.; Zhuang, H.; et al. European Journal of Radiology 2013 May 27. pii: S0720-048X(13)00201-5. doi: 10.1016/j.ejrad.2013.04.006. [Epub ahead of print].

4.4.5

SOMATOSTATIN RECEPTOR SCINTIGRAPHY IN INFLAMMATION AND INFECTION IMAGING

ALBERTO SIGNORE[1], LUZ KELLY ANZOLA FUENTES[2], AND MARCO CHIANELLI[3]

[1]*Nuclear Medicine Unit, Department of Medical-Surgical Sciences and of Translational Medicine, Faculty of Medicine and Psychology, "Sapienza" University of Rome, Rome, Italy*
[2]*Nuclear Medicine, ClinicaColsanitas, Bogotà, Colombia*
[3]*Endocrinology Unit, "Regina Apostolorum" Hospital, Albano (Rome), Italy*

ABBREVIATIONS

99mTc-HYNIC-TOC	99mTc-HYNIC-[D-Phe1,Tyr3]-octreotide
anti-TG	antibodies against thyroglobulin
anti-TPO	antibodies against thyroid peroxidase
anti-TSHr	antibodies against thyroid-stimulating hormone receptor
IBD	inflammatory bowel disease
SRS	somatostatin receptor scintigraphy
SS	Sjögren syndrome
TSH	thyroid-stimulating hormone
TSHr	thyroid-stimulating hormone receptor

Somatostatin Analogues: From Research to Clinical Practice, First Edition. Edited by Alicja Hubalewska-Dydejczyk, Alberto Signore, Marion de Jong, Rudi A. Dierckx, John Buscombe, and Christophe Van de Wiele.
© 2015 John Wiley & Sons, Inc. Published 2015 by John Wiley & Sons, Inc.

INTRODUCTION

Inflammatory diseases are a heterogeneous class of diseases characterized by chronic inflammation of the target organ, often relapsing, invalidating, and requiring lifelong treatment. The so-called aseptic chronic inflammatory diseases include autoimmune diseases, graft rejection, sarcoidosis, vasculitis, atherosclerosis, and some degenerative diseases. In these patients, it is very important to achieve specific immune suppression to extinguish the immune process with the aim of stopping the disease, preventing or delaying complications, and avoiding disease relapse. It is important that while attempting to improve the quality of life of these patients by means of anti-inflammatory drugs, side effects are reduced to a minimum via the use of specific immune therapies that block as selectively as possible the pathologic mechanisms responsible for the disease.

New therapeutic options are being developed for specific targeted therapies. Several trials are being performed to assess the efficacy and safety of new approaches. All of them, however, rely mostly on the clinical assessment of the patients to evaluate the effect of treatment. It would be important to use an objective and reliable method to highlight directly the immune process underlying the individual disease; specific diagnostic tests, furthermore, may allow the selection of patients to be treated.

Nowadays, nuclear medicine techniques are not often used for the diagnosis of chronic inflammatory diseases but greatly contribute to the management and prognosis of the disease. Most importantly, somatostatin receptor scintigraphy (SRS) has been proposed for the evaluation of the state of activity of some inflammatory diseases and to early evaluate therapy efficacy. This is particularly important because new molecular therapeutic biological agents that specifically target and block inflammatory reactions are continuously being developed. The referring physician not only obtains information on the activity of the disease but also on the nature of the process and can, therefore, decide which treatment to start, when to start it, and when to stop it or modify it.

SOMATOSTATIN RECEPTOR EXPRESSION IN INFLAMMATORY DISEASES

Many hormones and some neuropeptides and neurotransmitters play a key role in regulating lymphoid cells; somatostatin, in particular, is involved in numerous regulating mechanisms of cell activities in the immune system. The expression of somatostatin receptors in the thymus in man has prompted the hypothesis that this hormone participates in the maturation process of T lymphocytes.

Somatostatin receptors are expressed on both activated lymphocytes and inflamed vascular endothelium. SRS holds important information not only by demonstrating the presence of inflammation but also by providing a rationale, in positive patients, for the use in selected cases, of unlabelled somatostatin for the treatment of the disease. The expression of receptors for somatostatin has been investigated in patients with autoimmune diseases and cancer [1].

Hyperexpression of the somatostatin receptor (SSTR) has been found in intestinal samples from patients with active ulcerative colitis and Crohn's disease. SSRs were localized in intramural veins and were not detected in noninflamed control intestine [2]. SSRs were reported *in vitro* in patients with active rheumatoid arthritis [3].

SRS is applicable in imaging of chronic inflammation but is unsuitable for visualization of acute infectious diseases [4]. Ten Bokum et al. showed in 2002 accumulation of In-111 DTPA-octreotide in the thymus and the pituitary of normal Balb/c mice and of nonobese diabetic mice, a strain prone to autoimmune type 1 diabetes. They were unable, however, to show any uptake in the inflamed pancreas of prediabetic animals [5].

Radiolabeled somatostatin analogues have been extensively used for the study of neuroendocrine tumors [6], particularly in gastroenteropancreatic tumors, where the presence of receptors for somatostatin has been demonstrated [7]. The most commonly used analogue is ^{111}In-[D-Phe^1]-pentetreotide (OctreoScan), a small octapeptide that binds with high affinity to the somatostatin type 2 receptor expressed on the cell membrane of the target tissues. It has no side effects. Several analogues have been synthetized and are currently used in routine clinical practice. A recent analogue that has been used in autoimmune diseases is ^{99m}Tc-HYNIC-[D-Phe1,Tyr3]-octreotide (^{99m}Tc-HYNIC-TOC), which has extensively been used in neuroendocrine tumors and in inflammatory diseases [8]. ^{99m}Tc-Depreotide is another analogue that has been used to localize sites of inflammation in patients with viral myocarditis [9] and in patients with bone infection [10, 11].

STUDY OF SJÖGREN SYNDROME BY SRS

Sjögren syndrome (SS) is characterized by dry mouth and dry eyes (sicca syndrome) as a result of autoimmune destruction of salivary and lacrimal glands. Specific auto-antibodies (anti-SSA and anti-SSB) are detectable in the peripheral blood, but the diagnosis of the disease is based on biopsy and/or salivary gland hypofunction as detected by Schirmer test (AECG criteria). The role of salivary gland scintigraphy (scialoscintigraphy) is still a matter of debate and new imaging modalities are indeed required to demonstrate the presence of lymphocytic infiltration in the salivary glands. New immunological treatments are being tested for SS (infliximab, rituximab), and it would be useful to have a diagnostic test capable of detecting infiltrated glands that could be used for therapy selection and monitoring (Fig. 4.4.5.1).

A recent study has described the use of the new somatostatin analogue ^{99m}Tc-HYNIC-Tyr(3)-octreotide for the diagnosis of the state of activity in patients with rheumatoid arthritis and secondary SS before and after treatment with infliximab. Results showed that inflamed parotid glands could be diagnosed by this radiopharmaceutical. Inflamed joints were also detected in patients with active rheumatoid arthritis [8]. Interestingly, after treatment with infliximab, normalization of the uptake in most inflamed joints was noted but not in salivary glands, probably reflecting the different nature of the two diseases.

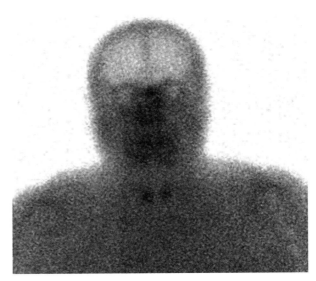

FIGURE 4.4.5.1   SRS in a patient with Sjögren syndrome showing pathological uptake of labeled octreotide in major salivary glands. A minor uptake is also detectable in the thyroid gland. SRS—somatostatin receptor scintigraphy.

STUDY OF THYROID DISEASES BY SRS

Autoimmune thyroid diseases, including Graves' disease, primary myxedema, and Hashimoto's thyroiditis, appear related in certain aspects of their pathogenesis and clinical course. Evidence of humoral immunity is provided in all of these disorders by the presence of antibodies against thyroid peroxidase (formerly known as microsomal antigen) and often against thyroglobulin. Antibody titles tend to be highest in Hashimoto's disease and lowest in primary hypothyroidism at the time it is diagnosed. More specific to Graves' disease are circulating autoantibodies that are capable of binding to the thyroid-stimulating hormone receptor (TSHr) on the surface of thyroid cells and stimulate cell growth and hormone production. Graves' disease, primary myxedema, and Hashimoto's thyroiditis all share evidence of cell-mediated immunity against thyroid antigens and are characterized by a varying degree of infiltration by lymphocytes and plasma cells. The infiltrating cells collect in aggregates, forming lymphoid follicles with germinal centers [12, 13].

Exophthalmos is a frequent manifestation of Graves' disease that may lead to severe complications. It is caused by muscle hypertrophy and lymphocytic infiltration of the retro-orbital space, which may eventually turn into fibrosis. Exophthalmos is usually treated with corticosteroids and/or cyclosporin or by local X-ray therapy. It would be extremely important to be able to diagnose the state of activity of the disease and to differentiate between active infiltration and fibrosis because of the difference in their treatment.

Current diagnosis of thyroid autoimmunity is based on the detection of autoantibodies (anti-TSHr, anti-TPO, and anti-TG) and clinical signs and symptoms.

However, the relationship between autoantibodies and disease activity is still unclear, and it is generally believed that the activity of the autoimmune process is determined by the intensity of intrathyroidal lymphocytic infiltration. *In vivo* measurement of thyroid cellular infiltration, particularly in patients with undetectable serum thyroid autoantibodies, would be ideal for evaluating the disease activity, determining the need for therapy, and monitoring the efficacy of treatment (Fig. 4.4.5.2).

Forster and colleagues described the use of SRS in patients with Graves' ophthalmopathy and stressed the importance of SPECT acquisition although with high interoperator variability [14]. Very elegantly, Savastano and colleagues demonstrated the TSHr dependence of Graves' ophthalmopathy in a patient with negative orbital SRS that became positive after the administration of recombinant TSH [15]. Krassas et al. postulated in their review that the accumulation of the radionuclide is most probably due to the presence in the orbital tissue of activated lymphocytes bearing somatostatin receptors; alternative explanations are binding to receptors on other cell types (myoblasts, fibroblasts, or endothelial cells) and local blood pooling due to venous stasis by the autoimmune orbital inflammation [16].

Krassas et al. in 1997 used ^{111}In-octreotide scan to select patients with thyroid eye disease to be treated with lanreotide and to follow them up to evaluate the response to the treatment [17]. Diaz and colleagues in 1994 demonstrated, in 40 patients, the diagnostic value of somatostatin receptor scan not only for detecting and quantifying the inflammation severity in the retrocular space but in the follow-up of the disease [18]. This same observation was made by Moncayo et al. who observed the inflammation in the retrobulbar tissue with ^{111}In-octreotide and used it for evaluating the response to the therapy [19]. Likewise, in another study, Colao et al. used

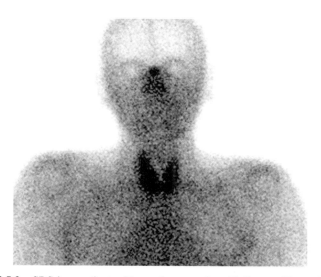

FIGURE 4.4.5.2 SRS in a patient with autoimmune thyroid disease (Graves' disease with mild orbitopathy) showing pathological uptake of labeled octreotide in the thyroid gland. No activity is detectable in salivary glands. SRS—somatostatin receptor scintigraphy.

[111]In-octreotide to predict the clinical response to corticosteroid treatment in Graves' ophthalmopathy and suggested using it as a useful approach to select patients for the proper treatment [20].

[111]In-pentetreotide has been used in imaging of Graves' disease, obtaining different contrasting results: a few studies have reported that this radiopharmaceutical accumulates in thyroid and in the retro-orbital space in patients with exophthalmos and there is a positive correlation with the activity of disease [21, 22], in disagreement with other authors [23, 24]. Differences between authors might be explained by the possible mechanisms of accumulation of octreotide: uptake occurs in the early phases of Graves' ophthalmopathy when active infiltration is present; in the later stage of the disease, there is fibroblastic activity with subsequent fibrosis in the retro-orbital region without expression of somatostatin receptors [25, 26].

A study was performed with ^{111}In-pentetreotide orbital scintigraphy on patients with severe ophthalmopathy caused by Graves' disease, Hashimoto's thyroiditis, and Means' syndrome. Activated lymphocytes express somatostatin receptor during the active phase of the disease, permitting ^{111}In-pentetreotide scintigraphy. Authors concluded that ^{111}In-pentetreotide scintigraphy allows to select patients for octreotide therapy, which seems to be adequate in active, moderately severe thyroid eye disease, especially when it involves soft tissues [27–30].

Finally, Galuska et al. studied patients with Graves' orbitopathy with [99mTc]-depreotide with comparable results as for the other somatostatin analogues but showing excellent SPECT images [31, 32].

STUDY OF SARCOIDOSIS BY SRS

Octreotide is not suitable for the imaging of experimental abscesses [4], but SSRs were observed *in vitro* in multiple confluent granulomas in patients with active sarcoidosis. In this case, SSRs are not expressed on the surface of lymphocytes but are located in the areas containing epithelial cells (Fig. 4.4.5.3). In patients successfully treated with steroids with complete sclerosis of the granulomatous lesion, SSRs were not found. These studies are in agreement with studies in patients with tuberculosis [33]. A recent study by Lebtahi et al. compared the use of ^{111}In-pentetreotide to that of ^{67}Ga-citrate in patients with sarcoidosis, showing similar diagnostic accuracy [34]. A study by Migliore et al. assessed the role of SRS using [99mTc]-HYNIC-Tyr(3)-octreotide in a patient with systemic sarcoidosis. The study showed that the technique was able to detect pulmonary and extrapulmonary localization of sarcoidosis. It was also possible to select the best therapeutic options. After treatment with infliximab, the patient showed normalization of the scan that correlated with improvement of clinical status [35]. Kwekkeboom et al., in a study of 46 patients with sarcoidosis, demonstrated uptake of 111In-pentetreotide in 36 of 37 patients with known mediastinal, hilar, and interstitial disease. They postulated that somatostatin receptor imaging can demonstrate active granulomatous disease in patients with sarcoidosis [36]. The possible role of SRS in diagnosis, staging, and follow-up of patients suffering from sarcoidosis was reviewed by Dalm et al. [37].

Shorr et al. used [99mTc]-depreotide in patients with sarcoidosis [38].

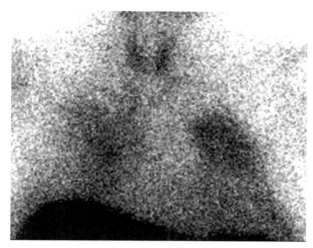

FIGURE 4.4.5.3 SRS in a patient with pulmonary sarcoidosis showing diffuse uptake in both lungs. Little activity is detectable also in the thyroid. SRS—somatostatin receptor scintigraphy.

STUDY OF IBD BY SRS

Crohn's disease is characterized by a chronic mononuclear cell infiltration of the intestinal wall and hypertrophy of local lymphoid tissues [39]. Immune erosion of the intestinal wall may lead to severe complications of the affected bowel such as stenosis and ulceration, which may require surgical resection. In over 70% of patients, relapse of the disease is noted within 1 year after the intervention. In the early relapse phase, the symptoms are infrequent and nonspecific, and conventional X-ray examinations are negative. Since effective therapies are available, early diagnosis of the relapse might allow prompt initiation of therapy to prevent the onset of complications and the need for further surgical resection [40].

Hyperexpression of the somatostatin receptor has been found in intestinal samples from patients with active ulcerative colitis and Crohn's disease. SSR were localized in intramural veins and were not detected in noninflamed control intestine [2]. The use of SRS in inflammatory bowel diseases (IBD) has been explored so far.

STUDY OF RHEUMATOID ARTHRITIS BY SRS

Rheumatoid arthritis is a chronic autoimmune disease characterized by severe short- and long-term complications of the joints. Chronic mononuclear cell infiltration of the synovial membrane and subsequent erosion of cartilage and bone lead to joint ankylosis. The typical hemodynamic changes of acute inflammation and the persistence of the chronic infiltrate are both present.

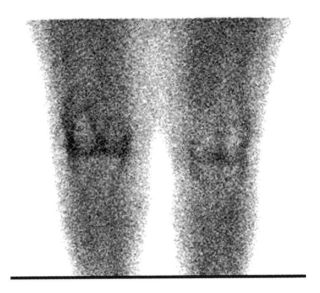

FIGURE 4.4.5.4   SRS in two patients with rheumatoid arthritis showing pathological uptake in knees (upper panel) and wrist and interphalangeal joints (lower panel). It is interesting to note that the left knee shows more uptake of labeled ^{99m}Tc-HYNIC-TOC compared to the right knee, indicating a different activity of the disease in the two joints. SRS—somatostatin receptor scintigraphy.

Specific and nonspecific signs of inflammation are normally used for the clinical diagnosis and follow-up of the disease. Systemic treatment with anti-inflammatory drugs (steroidal and nonsteroidal) is commonly employed for relief of symptoms and to delay disease progression. Treatment is usually lifelong and is accompanied by several side effects; local therapy is also used and has the advantage of higher local concentrations and fewer side effects. It would be very useful for the prevention of disease progression to diagnose affected joints before they become clinically evident, and local therapies could be applied before complications develop (Fig. 4.4.5.4). Rheumatoid arthritis has been extensively studied by nuclear medicine techniques, and all radiopharmaceuticals tested showed accumulation in the inflamed joints [41].

SSRs were reported *in vitro* in patients with active rheumatoid arthritis [42]. Van Hagen et al. in 1994 in a sample of 14 patients with active AR and 4 with severe osteoarthritis showed uptake in 76% of the painful and swollen joints of AR group positive findings [33].

CONCLUSIONS

SRS has extensively been studied in several chronic inflammatory diseases and offers a valuable diagnostic tool that cannot be obtained by conventional diagnostic imaging. The radiopharmaceutical is commercially available, and the scintigraphy is easy to perform and may be used in all departments.

The main indications of SRS in inflammation are the selection of patients with active disease to be treated with immunomodulating therapies and monitoring of their efficacy, in particular where conventional diagnostic imaging lack to offer valuable information, such as Graves' ophthalmopathy. It may also offer advantages for the study of systemic diseases, such as sarcoidosis or rheumatoid arthritis; for the staging of the disease; and for the study of associated pathologies.

REFERENCES

[1] Ferone, D.; Lombardi, G.; Colao, A. Minerva Endocrinology 2001, 26, 165–173.

[2] Reubi, J.; Mazzucchelli, L.; Laissue, J. Gastroenterology 1994, 106, 951–959.

[3] Van Hagen, P.; Markusse, H.; Lamberts, S.; et al. Arthritis and Rheumatism 1994, 37, 1521–1527.

[4] Oyen, W. J. G.; Boerman, O. C.; Claessens, R. A. M. J.; et al. Nuclear Medicine Communications 1994, 15, 289–293.

[5] Ten Bokum, A. M.; Rosmalen, J. G.; Hofland, L. J.; et al. Nuclear Medicine Communications 2002, 23, 1009–1017.

[6] Krenning, E. P.; Kwekkeboom, D. J.; Pauwels, S.; et al. Nuclear Medicine Annual 1995, 150.

[7] Cascini, G. L.; Cuccurullo, V.; Mansi, L. Journal of Nuclear Medicine and Molecular Imaging 2010, 54, 24–36.

[8] Chianelli, M.; Martin, S.; Signore, A.; et al. European Journal of Nuclear Medicine 2005, 32, S61.

[9] Moralidis, E.; Mantziari, L.; Gerasimou, G.; et al. Journal of Nuclear Medicine 2012, 15, 144–146.

[10] Papathanasiou, N. D.; Rondogianni, P. E.; Pianou, N. K.; et al. Nuclear Medicine Communications 2008, 29, 239–246.

[11] Spyridonidis, T.; Patsouras, N.; Alexiou, S.; et al. Journal of Nuclear Medicine 2011, 14, 260–263.

[12] Weetman, A. P.; McGregor, A. M. Endocrine Review 1984, 5, 309–315.

[13] DeGroot, L. J.; Quintans, J. Endocrine Review 1989, 10, 537–562.

[14] Förster, G. J.; Krummenauer, F.; Nickel, O.; et al. Cancer Biotherapy and Radiopharmaceuticals 2000, 15, 517–525.

[15] Savastano, S.; Pivonello, R.; Acampa, W.; et al. Journal of Clinical Endocrinology and Metabolism 2005, 90, 2440–2444.

[16] Krassas, G. E.; Kahaly, G. J. European Journal of Endocrinology 1999, 140, 373–375.

[17] Krassas, G. E.; Kaltsas, T.; Dumas, A.; et al. European Journal of Endocrinology 1997, 136, 416–422.

[18] Diaz, M.; Kahaly, G.; Mühlbach, A.; et al. Rofo 1994, 161, 484–488.

[19] Moncayo, R.; Baldissera, I.; Decristoforo, C.; et al. Thyroid 1997, 7, 21–29.

[20] Colao, A.; Lastoria, S.; Ferone, D.; et al. Journal of Clinical Endocrinology and Metabolism 1998, 83, 3790–3794.

[21] Postema, P. T. E.; Wijnggaarde, R.; Vandenbosch, W. A.; et al. Journal of Nuclear Medicine 1995, 203P.

[22] Diaz, M.; Bokisch, A.; Kahaly, G.; et al. European Journal of Nuclear Medicine 1993,(abstract)844

[23] Eberhardt, J. U.; Oberwohrmann, S.; Clausen, M.; et al. European Journal of Nuclear Medicine 1993, (abstract) 844.

[24] Bohuslavizki, K. H.; Oberworhmann, S.; Brenner, W.; et al. Nuclear Medicine Communications 1995, 16, 912–916.

[25] Bahn, R.; Heufelder, A. New England Journal of Medicine 1993, 329, 1468–1475.

[26] Hurley, J. Journal of Nuclear Medicine 1994, 35, 918–920.

[27] Krassas, G. E.; Dumas, A.; Pontikides, N.; et al. Clinical Endocrinology 1995, 42, 571–580.

[28] Nocaudie, M.; Bailliez, A.; Itti, E.; et al. European Journal of Nuclear Medicine 1999, 26, 511–517.

[29] Kung, A. W.; Michon, J.; Iai, K. S.; et al. Thyroid 1996, 6, 381–384.

[30] Ozata, M.; Bolu, E.; Sengul, A.; et al. Thyroid 1996, 6, 283–288.

[31] Galuska, L.; Leovey, A.; Szucs-Farkas, Z.; et al. Nuclear Medicine Communications 2005, 26, 407–414.

[32] Galuska, L.; Nagy, E.; Szucs-Farkas, Z.; et al. Orvosi Hetilap 2003, 144, 2017–2022.

[33] Van Hagen, P.; Krenning, E.; Reubi, J.; et al. European Journal of Nuclear Medicine 1994, 21, 497–502.

[34] Lebtahi, R.; Crestani, B.; Belmatoug, N.; et al. Journal of Nuclear Medicine 2001, 42, 21–26.

[35] Migliore, A.; Signore, A.; Capuano, A.; et al. European Review for Medical and Pharmacological Sciences 2008, 12, 127–130.

[36] Kwekkeboom, D. J.; Krenning, E. P.; Kho, G. S.; et al. European Journal of Nuclear Medicine 1998, 25, 1284–1292.

[37] Dalm, V. A.; van Hagen, P. M.; Krenning, E. P. Journal of Nuclear Medicine 2003, 47, 270–278.

[38] Shorr, A. F.; Helman, D. L.; Lettieri, C. J.; et al. Chest 2004, 126, 1337–1343.

[39] Podolsky, D. K. New England Journal of Medicine 1991, 325, 928–938.

[40] Podolsky, D. K. New England Journal of Medicine 1991, 325, 1008–1018.

[41] De Bois, M. H. W.; Pauwels, E. K. J.; Breedveld, F. C. European Journal of Nuclear Medicine 1995, 22, 1339–1346.

[42] Duet, M.; Liote, F. Joint Bone Spine 2004, 71, 530–535.

5

SOMATOSTATIN ANALOGUES IN PHARMACOTHERAPY: INTRODUCTION

WOUTER W. DE HERDER

Department of Internal Medicine, Erasmus MC, Sector of Endocrinology, Rotterdam, The Netherlands

Somatostatin is a cyclic peptide, which is present in the mammalian circulation in two bioactive forms: somatostatin-14 and somatostatin-28 [1, 2]. Somatostatin-14 was detected accidentally during studies of the distribution of growth hormone-releasing factor in the hypothalamus of rats [3, 4]. Subsequent studies showed that somatostatin is present and plays an inhibitory role in the regulation of several organ systems and tissues in man and other mammals, like the intestinal tract, the exocrine and endocrine pancreas, the central nervous system, the hypothalamus and the pituitary gland, the immune system, the retina, and the blood vessels. Somatostatin inhibits a variety of physiological functions in the gastrointestinal tract, like gastrointestinal motility, gastric acid production, pancreatic enzyme secretion, and bile and colonic fluid secretion. It inhibits the secretion of pancreatic and intestinal hormones like insulin, glucagon, secretin, and vasoactive intestinal polypeptide. In view of the ability of somatostatin to inhibit such a variety of physiological processes, it was predicted that this peptide might be of therapeutic value in clinical conditions involving hyperfunction or hypersecretion of the organ systems mentioned earlier. However, the multiple simultaneous effects of pharmacological concentrations of somatostatin in different organs, the need for intravenous administration, the short duration of action (a half-life in the circulation of less than 3 min), and the

Somatostatin Analogues: From Research to Clinical Practice, First Edition. Edited by
Alicja Hubalewska-Dydejczyk, Alberto Signore, Marion de Jong, Rudi A. Dierckx,
John Buscombe, and Christophe Van de Wiele.
© 2015 John Wiley & Sons, Inc. Published 2015 by John Wiley & Sons, Inc.

postinfusion rebound hypersecretion of hormones considerably hampered the initial enthusiasm, as well as its clinical use [5].

Octreotide acetate (Sandostatin®, SMS 201-995) was the first octapeptide somatostatin analogue that was synthesized. Its elimination half-life after subcutaneous administration is two hours, and rebound hypersecretion of hormones does not occur [6]. Somatostatin and its commercially available analogues octreotide and lanreotide (Somatuline®, BIM 23014) exert their effects through interaction with somatostatin receptor, which are expressed on the cells. Somatostatin binds with high affinity to all somatostatin subtypes 1 through 5 (sst_{1-5}), whereas octreotide and lanreotide bind only with a high affinity to sst_2 and sst_5 [5, 7]. Expression of somatostatin receptors by endocrine tumors is essential for the control of hormonal hypersecretion by the octapeptide somatostatin analogues. In sst_2- or sst_5-positive patients, clinical symptomatology related to hormonal hypersecretion can be controlled by the chronic administration of one of the currently available octapeptide somatostatin analogues [5, 8–11]. These drugs may also exert antiproliferative actions in these patients [9, 12–15].

REFERENCES

[1] Reichlin, S. New England Journal of Medicine 1983, 309, 1556–1563.

[2] Reichlin, S. New England Journal of Medicine 1983, 309, 1495–1501.

[3] Krulich, L.; Dhariwal, A. P.; McCann, S. M. Endocrinology 1968, 83, 783–790.

[4] Brazeau, P.; Vale, W.; Burgus, R.; et al. Science 1973, 179, 77–79.

[5] Lamberts, S. W.; van der Lely, A. J.; de Herder, W. W.; Hofland, L. J. New England Journal of Medicine 1996, 334, 246–254.

[6] Bauer, W.; Briner, U.; Doepfner, W.; et al. Life Sciences 1982, 31, 1133–1140.

[7] Patel, Y. C. Frontiers in Neuroendocrinology 1999, 20, 157–198.

[8] Kvols, L. K.; Moertel, C. G.; O'Connell, M. J.; Schutt, A. J.; Rubin, J.; Hahn, R. G. New England Journal of Medicine 1986, 315, 663–666.

[9] Colao, A.; Cappabianca, P.; Caron, P.; et al. Clinical Endocrinology (Oxford) 2009, 70, 757–768.

[10] Feelders, R. A.; Hofland, L. J.; van Aken, M. O.; et al. Drugs 2009, 69, 2207–2226.

[11] Oberg, K.; Kvols, L.; Caplin, M.; et al. Annals of Oncology 2004, 15, 966–973.

[12] Shojamanesh, H.; Gibril, F.; Louie, A.; et al. Cancer 2002, 94, 331–343.

[13] Rinke, A.; Muller, H. H.; Schade-Brittinger, C.; et al. Journal of Clinical Oncology 2009, 27, 4656–4663.

[14] Colao, A.; Pivonello, R.; Auriemma, R. S.; et al. Journal of Clinical Endocrinology and Metabolism 2008, 93, 3436–3442.

[15] Colao, A.; Auriemma, R. S.; Rebora, A.; et al. Clinical Endocrinology (Oxford) 2009, 71, 237–245.

5.1

SOMATOSTATIN ANALOGUES IN PHARMACOTHERAPY

WOUTER W. DE HERDER

Department of Internal Medicine, Erasmus MC, Sector of Endocrinology, Rotterdam, The Netherlands

In the early 1980s, several somatostatin analogues were developed including SMS 201-995 (octreotide acetate, Sandostatin®; Novartis, Basel, Switzerland), RC-160 (Vapreotide, Sanvar®, Octastatin), BIM-23014 (lanreotide, Somatuline®; Ipsen, Paris, France), and MK 678 (Seglitide). These new drugs are more resistant to biological degradation in the body than native somatostatin [1, 2]. As a result, their half-lives and biological activities are considerably longer than that of native somatostatin (1.5–2 h vs. 1–2 min). At present, only octreotide and lanreotide are still clinically used.

While native somatostatin binds with high affinity to all somatostatin receptor subtypes (sst_{1-5}), octreotide and lanreotide only bind with a high affinity to sst_2 and sst_5 (Table 5.1.1) [3]. Octreotide acetate (Sandostatin®) was the first octapeptide somatostatin analogue developed for clinical use [4]. Another advantage of this drug over native somatostatin is that rebound hypersecretion of hormones does not occur [4]. Octreotide has to be administered two to three times daily as a subcutaneous formulation (in single doses ranging from 100 to 500 μg) or can be administered as a continuous intravenous infusion [5]. The development of an intramuscular depot formulation of octreotide, Sandostatin® long-acting repeatable (LAR®) (Novartis, Basel, Switzerland), which can be administered up to 30–90 mg once every 3–4 weeks, has to a large extent abolished the need for daily injections. Thirty milligrams of

Somatostatin Analogues: From Research to Clinical Practice, First Edition. Edited by
Alicja Hubalewska-Dydejczyk, Alberto Signore, Marion de Jong, Rudi A. Dierckx,
John Buscombe, and Christophe Van de Wiele.

TABLE 5.1.1 Properties of somatostatin receptor subtypes

	Binding affinity: IC_{50} value in nM (mean ± SEM)				
Compound	sst_1	sst_2	sst_3	sst_4	sst_5
SS-14	0.93–2.3	0.2–0.3	0.6–1.4	1.5–1.8	0.3–1.4
Lanreotide	180–2330	0.5–0.8	14–107	230–2100	5.2–17
Octreotide	280–1140	0.4–0.6	7.1–34.5	>1000	6.3–7.0
Pasireotide	9.3	1.0	1.5	>100	0.2

lanreotide (Somatuline® PR; Ipsen, Paris, France) has to be administered intramuscularly every 10–14 days and roughly has an equal efficacy to octreotide [6]. A slow-release depot preparation of lanreotide, lanreotide Autogel® (Ipsen, Paris, France), has to be administered deep subcutaneously (s.c.) in dosages ranging from 60 to 120 mg once every 3–6 weeks [7, 8].

Expression of somatostatin receptors by endocrine tumors is essential for the control of hormonal hypersecretion by somatostatin analogues. In sst_2- and/or sst_5-positive patients, clinical symptomatology can be controlled by the chronic administration of one of these currently available octapeptide somatostatin analogues [1, 9]. Octreotide (Sandostatin®) and lanreotide (Somatuline®) have been registered in most countries for the control of hormonal symptoms in patients with well-differentiated neuroendocrine tumors of the digestive tract (carcinoids) and pancreas and in patients with acromegaly [2, 5, 10]. These drugs may also exert antiproliferative effects on tumors in these patients [11, 12]. In patients with well-differentiated neuroendocrine tumors of the digestive tract (carcinoids) and pancreas, treatment with very high doses of somatostatin analogues might induce more antiproliferative effects than relatively low doses [10, 13, 14]. In the past, treatment of these patients with ultrahigh-dose octreotide pamoate (Onco-LAR®; Novartis, Basel, Switzerland), of which 160 mg had to be administered intramuscularly every 2–4 weeks, did show promising results [15]. However, the development of this drug was discontinued by the manufacturing company.

Pasireotide (SOM 230) is a somatostatin analogue that binds to all somatostatin receptor subtypes, except sst_4 (Table 5.1.1) [16]. The drug is currently produced as a short-acting formulation that has to be administered s.c. and a long-acting intramuscular LAR formulation. These drugs currently undergo phase III study programs in Cushing's disease, acromegaly, and well-differentiated neuroendocrine tumors of the digestive tract (carcinoids) and the pancreas [17–21]. Pasireotide is generally well tolerated, although impaired glucose tolerance and hyperglycemia can occur.

New fundamental insights in receptor physiology also opened the concept of multireceptor family crosstalk, like between somatostatin and dopamine receptors. Focus has, therefore, been directed toward the development of new drugs interacting with these phenomena [22]. BIM-23A760 (Ipsen, Paris, France) is a chimeric molecule that binds to sst_2 and sst_5 and dopamine receptor 2 [23]. However, in a phase IIb study in patients with acromegaly, this drug showed strong dopaminergic activity but only very weak somatostatinergic activity. On the basis of these preliminary data, the manufacturing company decided to discontinue the development of this drug.

REFERENCES

[1] Lamberts, S. W.; van der Lely, A. J.; de Herder, W. W.; Hofland, L. J. New England Journal of Medicine 1996, 334, 246–254.

[2] Feelders, R. A.; Hofland, L. J.; van Aken, M. O.; et al. Drugs 2009, 69, 2207–2226.

[3] Patel, Y. C. Frontiers in Neuroendocrinology 1999, 20, 157–198.

[4] Bauer, W.; Briner, U.; Doepfner, W.; et al. Life Sciences 1982, 31, 1133–1140.

[5] Oberg, K.; Kvols, L.; Caplin, M.; et al. Annals of Oncology 2004, 15, 966–973.

[6] Caron, P.; Cogne, M.; Gusthiot-Joudet, B.; Wakim, S.; Catus, F.; Bayard, F. European Journal of Endocrinology 1995, 132, 320–325.

[7] Chanson, P.; Boerlin, V.; Ajzenberg, C.; et al. Clinical Endocrinology (Oxford) 2000; 53, (5), 577–586.

[8] O'Toole, D.; Ducreux, M.; Bommelaer, G.; et al. Cancer 2000, 88, 770–776.

[9] Kvols, L. K.; Moertel, C. G.; O'Connell, M. J.; Schutt, A. J.; Rubin, J.; Hahn, R. G. New England Journal of Medicine 1986, 315, 663–666.

[10] Eriksson, B.; Oberg, K.; Andersson, T.; Lundqvist, G.; Wide, L.; Wilander, E. Scandinavian Journal of Gastroenterology 1988, 23, 508–512.

[11] Shojamanesh, H.; Gibril, F.; Louie, A.; et al. Cancer 2002, 94, 331–343.

[12] Rinke, A.; Muller, H. H.; Schade-Brittinger, C.; et al. Journal of Clinical Oncology 2009, 27, 4656–4663.

[13] Eriksson, B.; Renstrup, J.; Imam, H.; Oberg, K. Annals of Oncology 1997, 8, 1041–1044.

[14] Imam, H.; Eriksson, B.; Lukinius, A.; et al. Acta Oncologica 1997, 36, 607–614.

[15] Welin, S. V.; Janson, E. T.; Sundin, A.; et al. European Journal of Endocrinology 2004, 151, 107–112.

[16] Bruns, C.; Lewis, I.; Briner, U.; Meno-Tetang, G.; Weckbecker, G. European Journal of Endocrinology 2002, 146, 707–716.

[17] Kvols, L.; Oberg, K.; de Herder, W.; et al. Journal of Clinical Oncology (Meeting Abstracts) 2005, 23 (16 suppl), Abstract 8024.

[18] van der Hoek, J.; van der Lelij, A. J.; Feelders, R. A.; et al. Clinical Endocrinology (Oxford) 2005, 63, 176–184.

[19] Feelders, R. A.; de Bruin, C.; Pereira, A. M.; et al. New England Journal of Medicine 2010, 362, 1846–1848.

[20] Schmid, H. A. Molecular and Cellular Endocrinology 2008, 286, 69–74.

[21] Petersenn, S.; Schopohl, J.; Barkan, A.; et al. Journal of Clinical Endocrinology and Metabolism 2010, 95(6), 2781–2789.

[22] Rocheville, M.; Lange, D. C.; Kumar, U.; Patel, S. C.; Patel, R. C.; Patel, Y. C. Science 2000, 288, 154–157.

[23] Florio, T.; Barbieri, F.; Spaziante, R.; et al. Endocrine-Related Cancer 2008, 15, 583–596.

5.2

PITUITARY TUMOR TREATMENT WITH SOMATOSTATIN ANALOGUES

ALICJA HUBALEWSKA-DYDEJCZYK, ALEKSANDRA GILIS-JANUSZEWSKA, AND MALGORZATA TROFIMIUK-MÜLDNER

Department of Endocrinology with Nuclear Medicine Unit, Medical College, Jagiellonian University, Krakow, Poland

ABBREVIATIONS

ACTH   adrenocorticotrophic hormone, corticotrophin
AE   adverse event
ATG   Autogel
CAB   cabergoline
CI   confidence interval
D2R   dopamine type 2 receptor
FSH   follicle-stimulating hormone
fT3   free triiodothyronine
fT4   free thyroxine
GH   growth hormone
GHRH   growth hormone-releasing hormone
HDL-C   high-density lipoprotein cholesterol
i.m.   intramuscular
IGF-1   insulin-like growth factor type 1
LAN   lanreotide
LAR   long-acting release

Somatostatin Analogues: From Research to Clinical Practice, First Edition. Edited by Alicja Hubalewska-Dydejczyk, Alberto Signore, Marion de Jong, Rudi A. Dierckx, John Buscombe, and Christophe Van de Wiele.

LDL-C   low-density lipoprotein cholesterol
LH   luteinizing hormone
NFA   nonfunctioning adenomas
NFPT   nonfunctioning pituitary tumors
OCT   octreotide
OR   odds ratio
PPARγ   peroxisome proliferator-activated receptor gamma
PRL   prolactin
s.c.   subcutaneously
SMR   standardized mortality rate
SR   slow release
SSAs   somatostatin analogues
SST   somatostatin
SSTR   somatostatin receptor
T-C   total cholesterol
TSH   thyroid-stimulating hormone, thyrotropin
UFC   urinary free cortisol

INTRODUCTION

Pituitary tumors, mainly adenomas, are one of the most frequent brain neoplasias. Their prevalence varies depending on the population and method of the assessment. On the autopsy and radiological examination, the small pituitary tumors including clinically nonsignificant tumors (incidentalomas) are present in one in every six people [1]. The prevalence of clinically significant pituitary lesions in a large cross-sectional study performed in Liege, Belgium, was 1 per 1064 individuals [2]. In this study, the prolactinomas, null cell adenomas, somatotropinomas, and corticotropinomas constituted 60, 14.7, 13.2, and 5.9% of all pituitary tumors, respectively [2]. Manifestations of clinically apparent pituitary adenomas are related to the hormone oversecretion and/or mass effect (hypopituitarism included).

Treatment options for patients with pituitary tumors are neurosurgery, radiotherapy, and pharmacotherapy, alone or in combination. Neurosurgery, usually via transsphenoidal approach performed by an experienced neurosurgeon, is chosen to alleviate the compressive mass effect symptoms or to provide control of hormonal hypersecretion in tumors not suitable or resistant to the medical treatment, particularly if curative resection is possible and patient is willing to. Radiotherapy is indicated for persistent hormonal hypersecretion or residual mass after surgery or when surgical resection of compressive mass is contraindicated. It should be also considered in aggressively growing or recurring tumors [3]. Medical therapy with dopamine agonists (DAs) is considered the first-line treatment for prolactin (PRL)-secreting adenomas; however, the pharmacotherapy role in managing patients not suitable for surgical treatment, in preparing for tumor debulking, or in hypersecretion control is rapidly increasing.

Somatostatin (SST) has been initially identified as a factor inhibiting pituitary growth hormone (GH) secretion [4], but it also plays a role in the regulation of secretion of various pituitary hormones. SST effects are mediated by five membrane receptors (SSTR1–5) belonging to the G-protein-coupled receptor family [5]. The pattern of tissue SSTR expression and interaction between SSTR subtypes determine the physiological action of SST. In the normal pituitary gland, SSTR types 1, 2, 3, and 5 are expressed, the last being the predominant subtype [6, 7]. The SSTR subtype 4 is present only in the human fetal anterior pituitary [8]. The expression of SSTR subtypes in pituitary adenomas is presented in Table 5.2.1 [7]. Pituitary tumors may also express SSTR variants, for example, truncated forms of SSTR5—sst5TND5 and sst5TMD4—have been identified in the cytoplasm of NFA, corticotropinomas, somatotropinomas, and prolactinoma. Those isoforms are not detected in normal anterior pituitary cells and in spite of intracellular localization are functional [9].

SST seems rather to acutely decrease pituitary hormone exocytosis rather than the synthesis. GH secretion is inhibited by SST via SSTR2, SSTR5, and, to some extent, SSTR1; thyroid-stimulating hormone (TSH) secretion via SSTR2 and SSTR5; and PRL secretion via SSTR2. The exact role of SST in regulating corticotrophin (adrenocorticotrophic hormone (ACTH)) release has not been elucidated yet, although the main role of SSTR5 is postulated. The inhibiting effect of SST on luteinizing hormone (LH) and follicle-stimulating hormone (FSH) secretion is modest, and the responsible mechanism has not been discovered yet [7, 10]. The ability of SST and somatostatin analogues (SSAs) to decrease pituitary hormone secretion may also be altered by the presence of truncated isoforms of SSTR, particularly SSTR5, which (viz., sst5TMD4) is negatively correlated with the ability of SSA to inhibit GH release [11].

Inhibition of pituitary/pituitary tumor cell growth usually accompanies the inhibition of the hormone secretion. It seems that this effect is due to induction of apoptosis or cell senescence rather than to mitosis rate decrease, and it is still disputed which SSTR is mediating the process: SSTR2 or/and SSTR5 [7].

The short life of natural SST ($t\frac{1}{2} \leq 3$ min), which makes it unsuitable for clinical use, has led to the development of SSAs with longer half-life. The first clinically used SMS 201-995—octreotide (OCT)—was prepared as acetate salt solution for frequent subcutaneous or intravenous injections [12]. As the continuous subcutaneous administration of OCT resulted in better control of GH levels in acromegalic patients than subcutaneous injections a few times a day [13], the long-acting formulations of SSA (octreotide long-acting release (OCT LAR), lanreotide slow release (LAN SR), or lanreotide Autogel (LAN ATG)) have been manufactured, resulting in better clinical outcome and improved quality of life of patients [14]. Modifications of the SST structure increasing the molecules half-life have changed their affinity to SSTR subtypes (Table 5.2.2) [15]. OCT and LAN bind with the greatest affinity to SSTR2. The novel multireceptor-targeted SSA—pasireotide (SOM230)—has a 39-, 30-, and 5-fold higher binding affinity for SSTR5, SSTR1, and SSTR3, respectively, and 2.6 times lower affinity for SSTR2 compared with OCT. Pasireotide has a two-fold higher binding affinity for SSR5 than endogenous SST (Table 5.2.2). Pasireotide exhibits also greater metabolic stability than OCT because of the presence of cysteine–cysteine bridge that protects the stability of the amide bond in the cyclic ring, which may translate into a

TABLE 5.2.1 SSTR expression in human pituitary adenomas

	SRIF receptor subtype expression						
	Positive tumors/total tumors tested (%)						
	SSTR1	SSTR2	SSTR3	SSTR4	SSTR5	Detection method	Number of tumors studied
GH	27/44 (61)	95/108 (88)	24/55 (44)	2/48 (4)	92/104 (88)	A, B, C, D, E, F	111
ACTH	17/27 (63)	20/27 (74)	3/26 (11)	7/26 (27)	42/56 (75)	A, B, C, D	56
PRL	24/27 (89)	17/27 (63)	3/15 (20)	0/15 (0)	20/25 (80)	A, B, C, D	30
NFA	8/32 (25)	18/32 (56)	14/31 (45)	0/19 (0)	15/31 (48)	A, B, C, E, F	58
TSH	2/2 (100)	2/2 (100)	0/2 (0)	0/2 (0)	1/2 (50)	C	2

Adapted from [7], with permission.

TABLE 5.2.2   Binding affinities of somatostatin (SRIF-14), pasireotide, octreotide, and lanreotide to the five human SSTRs [15][a]

Compound	SSTR1	SSTR2	SSTR3	SSTR4	SSTR5
Somatostatin (SRIF-14)	1.0–2.3	2.0–1.3	0.3–1.6	0.3–1.8	0.2–0.9
Lanreotide	180 to >1000	0.5–1.8	14–107	66 to >1000	0.6–17
Octreotide	280 to >1000	0.4–2.1	4.4–34.5	>1000	5.6–32
Pasireotide	9.3	1.0	1.5	>1000	0.16

[a] Courtesy of BioMed Central.

prolongedvpharmacologic effect compared with OCT [5, 16]. More detailed information about the currently used SSA is presented in Chapter 5.1.

This chapter is focused on medical therapy of pituitary adenomas with SSAs.

PITUITARY TUMORS PRODUCING GH: ACROMEGALY

Acromegaly, due to increased levels of GH and insulin-like growth factor type 1 (IGF-1), is related to cardiovascular, metabolic, and respiratory morbidities and premature death [1, 17–20].

Acromegaly is also related to higher incidence of cancers [21]. The mortality risk in patients with acromegaly is 2.4- to 4.8-fold higher than in the general population; 60, 25, and 15% of acromegalic patients, respectively, will die from cardiovascular and respiratory disease and cancers [22–24]. A meta-analysis published in 2008, in which 16 studies on mortality in acromegaly were included, has confirmed increased all-cause mortality risk in acromegalic patients (a weighted mean of standardized mortality ratio (SMR) of 1.82 (95% confidence interval (CI), 1.12–1.56), even treated with transsphenoidal surgery (a weighted mean of SMR of 1.32 (95% CI, 1.12–1.56) in studies including at least 80% of patients treated surgically) [25].

The goals of the therapy are to control biochemical indices of activity, control tumor size and prevent local mass effect, reduce signs and symptoms of the disease, and eliminate morbidity and restore mortality rates to normal age- and sex-adjusted rates [26, 27]. The normalization of serum GH and IGF-1 restores acromegalic patients' mortality to normal level (SMR of 1.1 (95% CI, 0.9–1.4) vs. 1.9 (95% CI, 1.5–2.4) in patients with random GH <2.5 and >2.5 µg/l, respectively; SMR of 1.1 (95% CI, 0.9–1.4) vs. 2.5 (95% CI, 1.6–4.0) in patients with IGF-1 levels within sex- and age-adjusted normal range and without IGF-1 normalization) [28].

As in other pituitary adenomas, the treatment options for acromegaly includes neurosurgery, radiation, and medical. The optimal use of each modality is the issue of still ongoing discussion [29]. The curative neurosurgery rates in intrasellar micro-adenomas reach 75–95%, whereas in noninvasive macroadenomas, they drop to 40–68% [29]. Surgery remission rate in patients harboring adenomas larger than 20 mm is as low as 20% [30]. Another factor limiting therapy success is high GH level [30]. Radiation is usually considered the last line of therapy in acromegaly.

Conventional radiotherapy results in normalization of GH and IGF-1 levels in over 60% of patients; however, usually the maximum response is seen even 15 years after its administration. A faster response is seen when Gamma Knife or linear accelerator are applied, maybe because they are used in patients with smaller tumors [29]. The other issue is radiation safety: a substantial rate of hypopituitarism, risk of visual deterioration, secondary brain tumors, and radiation vasculopathy [29]. Excess mortality in patients undergoing conventional radiotherapy has also been postulated [23].

Three types of medical therapy are currently available in acromegaly treatment: SSAs, DAs, and GH receptor agonists, alone or in combination [29]. SSAs are the most widely used medical therapy to control the disease [31]. Approximately 90% of somatotrophs of GH-secreting adenomas express SSTR2 and SSTR5. SSTR ligands cause decrease secretion of GH and finally IGF-1 synthesis. It has been also demonstrated that SSTR ligands can also influence the peripheral action of GH through binding to the SSTRs present on hepatocytes and directly inhibiting liver IGF-1 secretion [32]. The ability of levodopa to reduce GH levels in acromegalic patients was observed already in the 1970s [33]. However, bromocriptine normalizes serum IGF-1 levels only in 10% of acromegalic patients and cabergoline in about 39%, particularly with lower pretreatment IGF-1 levels [34, 35]. Pegvisomant, a GH receptor antagonist, normalizes IGF-1 levels in a dose-dependent manner in up to 90% of patients. Currently, it is recommended as the second-line medical treatment when SSAs fail to achieve adequate biochemical control of acromegaly [35].

The first ever report on inhibiting GH release in acromegalic patients with SST was already published in 1974 [36]. However, only SST infusions gave satisfactory, from the clinical point of view, results [37]. In 1984, OCT was announced as the first SSA suitable for clinical management of acromegaly [38]. As it has been already mentioned, the introduction of SSA of modified release has impacted routine clinical practice.

SSAs in acromegaly are used as adjunctive (secondary) or primary therapy, as well as a presurgical pretreatment. Up to 75% of patients treated for 12–36 months with OCT LAR as adjuvant therapy achieved control of GH or GH/IGF-1 levels [39]: 47–75% (mean 56%) of patients achieved GH levels <2.5 µg/l, and 41–75% (mean 66%) IGF-1 normalization. In patients treated with LAN SR as adjuvant therapy, 14–78% (mean 49%) of patients achieved control of GH levels <2.5 µg/l, and 30–63% (mean 47%) IGF-1 normalization [39]. In meta-analysis by Freda et al. including patients on secondary and primary treatment with long-acting SSAs, OCT LAR was more effective than LAN SR in providing biochemical control in unselected population (GH efficacy criteria met in 54 and 48% of patients, respectively; IGF-1 normalization obtained in 63 and 42% of subjects, respectively). If the patients were preselected—assessed for SSA responsiveness before entering the study—the difference in efficacy between those two analogues no longer existed [40]. Treatment results are usually less satisfactory in unselected populations. In a prospective, multicenter study on OCT LAR as the primary therapy after 48 weeks, GH level below 2.5 µg/l was observed in 44% of patients, and IGF-1 normalization in 34% [41]. A summary of results from other studies on OCT LAR as the first-line therapy is presented in Table 5.2.3 [15]. Retrospective head-to-head comparison of OCT LAR and LAN SR as the primary therapy did not reveal statistically significant

TABLE 5.2.3   Summary of results from studies of first-line therapy with OCT LAR in patients with acromegaly

Reference	No of points	Duration of treatment	Patients meeting criterion for GH control (%)	Patients with IGF-1 normalization (%)	Mean tumor shrinkage (%)	% of patients with significant tumor shrinkage (definition of significant)
Colao et al. [42]	15	12–24 months	73	53	53	80 (>20%)
Amato et al. [43]	8	24 months	50	50	34.8	100 (>10%)
Ayuk et al. [44]	25	48 weeks	62	64	NR	NR
Jallad et al. [45]	28	6–24 months	NR	43	NR	76 (>25%)
Colao et al. [46]	34	6 months	61	45.5	54 (median)	74 (>30%)
Cozzi et al. [47]	67	6–108 months	69	70	62	82 (>25%)
Mercado et al. [41]	68	48 weeks	44	34	39	75 (>20%)
Colao et al. [48]	56	24 months	86	84	68	NR
Colao et al. [49]	67	12 months	52	58	49	85 (>25%)
Colao et al. [50]	40	48 weeks	NR	NR	35	73 (>20%)

NR, not reported.

difference in disease control, in tumor shrinkage, or in improvement of cardiovascular risk markers [51]. A randomized, open-label, multicenter study including 104 patients indicated that the outcome of the primary treatment with OCT LAR did not significantly differ from the surgery [50]. In meta-analysis of the effects of SST analogues on tumor volume, OCT was shown to induce tumor shrinkage in 53% (95% CI, 45–61%) of treated patients, and its LAR formulation—in 66% (95% CI, 57–74%). The mean reduction of tumor size was 37.4% (95% CI, 22.4–52.4%) and 50.6 (95% CI, 42.7–58.4%), respectively [52].

SSA as the primary therapy is suitable also for a long-term first-line treatment. A prospective long-term study (up to 108 month) comprising 67 de novo acromegalic patients with macroadenoma treated with individually tailored OCT LAR confirmed high efficacy of such approach. In this study, 68.7% of patients achieved safe GH level (<2.5 µg/l), 70.1% IGF normalization, and 82% tumor shrinkage by more than 25% of initial volume [47]. The nadir of GH and IGF-1 levels may be obtained even 10 years after the SSA treatment has been started [53]. Tailoring of the SSA dose according to the GH and/or IGF-1 level may increase the efficacy of treatment, as demonstrated in the study by Colao et al. [48].

Lanreotide Autogel (LAR ATG) is a long-acting aqueous preparation in prefilled syringe, suitable and approved for self-administration. In a 3-month study in 107 patients, the GH normalization rate with LAN SR was 48 and 56% with the use of LAN ATG. Normal IGF-1 was observed in 45 and 48% of patients treated with SR or ATG LAN formulation, respectively [54]. In the one-year extension of this study, dose titration of the LAN ATG improved GH and IGF-1 control beyond that achieved by fixed dose [55]. In the 3-year extension of the previous study in 14 patients on LAN ATG, the frequency of normal GH increased from 36 to 77% and normal IGF-1 from 36 to 54% [56]. In a meta-analysis by Roelfsema et al. published in 2008, treatment with LAN SR and LAN ATG normalized GH and IGF-1 concentrations in about 50% of acromegalic patients [57]. The efficacy of 120 mg LAN LTG on GH and IGF-1 was comparable with that of 20 mg OCT LAR. There were no differences in the improvement of cardiac function, decrease in beta cell pancreatic function, and side effects between both SSA formulations [57].

The role of preoperative medical therapy with SSAs has been intensively studied. The preoperative treatment with SSAs might affect the quality of the tumor and therefore improve the effectiveness of the surgery [58]. In the study by Abe and Ludecke from 2001, higher rates of GH and IGF-1 normalization were observed in patients pretreated with OCT before surgery compared to those who were medically naive preoperatively [58]. Also in the multicenter study conducted in Norway, the 6-month pretreatment with OCT LAR (20 mg monthly) resulted in surgical and biochemical remission (defined as normal IGF-1 level) in 50% of patients with macroadenomas in comparison to 16% of those who underwent surgery without pretreatment. However, if biochemical remission was defined as a GH level lower than 1 ng/ml after glucose suppression test, the difference between the pretreatment and no pretreatment group was no longer statistically significant [59]. Colao et al. in 67 patients treated preoperatively with OCT LAR for 12 months reported GH control in 52%, IGF-1 normalization in 58%, and tumor shrinkage of more than 25% in 85%

of patients [49]. In contrary, the difference in surgery outcome (surgical remission and complication rate) in retrospective analysis was not significant between patients pretreated with OCT or LAN and matched, medically naive controls [60].

The potential role of preoperative treatment with SSAs in the decrease of cardiopulmonary comorbidities and decrease of anesthesia-related risk of surgical treatment has been discussed. Treatment with SSAs improves cardiac function and reduces the incidence of cardiac dysrhythmias, which can improve the surgery outcomes. It is also related to the reduction of soft tissue swelling, which can result in the resolution of sleep apnea and reduction of intubation-related complications. These issues however need further examination for higher quality of evidence.

Increasing use of SSA as a primary therapy brings a question of long-term efficacy, particularly mortality reduction. In a retrospective cohort study including 438 consecutive acromegalic patients, Bogazzi et al. compared the effects of different therapies (curative neurosurgery, adenomectomy with SSA therapy, and primary therapy) on survival. In whole studied population, the following risk factors were associated with excessive mortality: age, physical status, macroadenoma, hypopituitarism, and uncontrolled disease. Patients treated with adenectomy, adenectomy plus SSA, and primary SSA therapy had similar risk of death (HR of 1, 0.3 (95% CI, 0.04–2.36), and 3.24 (95% CI, 0.61–17.33)). Primary SSA therapy harbored increased risk of death when compared with neurosurgery (regardless adjunctive treatment with SSA; HR of 5.52 (95% CI, 1.06–28.77); however, increased mortality was observed only in diabetic patients (HR of 21.94 (95% CI, 1.56–309.04) vs. 1.3 (95% CI, 0.04–38.09) for diabetic and nondiabetic patients, respectively) [61]. Good biochemical disease control, not the way of achieving it, determines improvement in glucose tolerance and insulin sensitivity parameters [62]. Patients with acromegaly after curative surgery and on SSA therapy were shown to harbor a similar risk of deterioration of glucose tolerance in long-term follow-up [63]. Even partial biochemical control may result in amelioration of cardiovascular risk factors: a significant reduction in systolic and diastolic blood pressure, glucose, insulin, HbA1c, total cholesterol (T-C) and low-density lipoprotein cholesterol (LDL-C), and triglyceride levels and a significant increase in apoA1, high-density lipoprotein cholesterol (HDL-C), and insulin sensitivity compared to pretreatment levels [64].

The multireceptor binding profile of pasireotide suggests its possible use in resistant or refractory to OCT and/or LAN acromegalic patients. As it was demonstrated in the phase II study, pasireotide effectively controlled GH and IGF-1 as well as significantly reduced the tumor size in patients with de novo or resistant/refractory to OCT acromegaly [65]. In an open-label, single-arm, open-ended extension of phase II study, biochemical control was achieved in 6, IGF-1 normalization in 13, and proper GH control in 12 of 26 patients evaluable at month 6. Significant tumor size reduction was observed in 17 (9 in core study, 8 at extension) of 29 patients with MRI data [66].

Results of the phase III study performed in 358 patients with active acromegaly who were de novo diagnosed with a visible adenoma on MRI or medically naïve (no previous medical therapy, but prior pituitary surgery) showed that pasireotide LAR was significantly more effective at inducing full biochemical control (GH ≤2.5 µg/l and/or IGF-1 upper limit of normal) compared to the current standard medical therapy (OCT LAR i.m. injections). The study met its primary end point, with significantly more patients treated with pasireotide LAR (31.3%) experiencing full control of their disease than those taking OCT LAR (19.2%). Patients treated with pasireotide LAR were 63% more likely to achieve control of their disease than those on OCT LAR. Tumor volume reduction and relief in clinical symptoms were similar in both treatment groups [67].

The tolerability of SSAs in most patients is good, most of the adverse events (AEs) are transient with mild or moderate severity. Discontinuation of the treatment due to the side effects is rare. The most common AEs of SSAs are injection site discomfort, erythema, and gastrointestinal problems like diarrhea, abdominal pain, flatulence, steatorrhea, nausea, vomiting, and gallstone formation [39]. The most common AEs with pasireotide LAR versus OCT LAR were diarrhea (39.3% vs. 45%), cholelithiasis (25.8% vs. 35.6%), headache (18.5% vs. 26.1%), and hyperglycemia (28.7% vs. 8.3%) [67].

Based on clinical research, the Acromegaly Consensus Group updated guidelines for acromegaly management published in 2009 has recommended the following use of SSA [29]: (a) as the first-line therapy when the probability of surgical cure is low, (b) after surgery has failed to achieve biochemical control, (c) before surgery to improve severe comorbidities that prevent or could complicate immediate surgery, and (d) to provide disease control, at least partial, in the time between radiation therapy and the onset of maximum benefit from that treatment.

New formulations of the OCT have been recently tested. The oral formulation of OCT—Octreolin—uses transient permeability enhancer technology, which enables intestinal absorption of the peptide by reversible opening of the intestinal epithelial cell tight junctions. Octreolin administration in human subjects resulted in dose-dependent increased plasma octreotide concentrations, with observed rate of plasma decay similar to parenteral administration. A single dose of 20 mg of Octreolin resulted in similar pharmacokinetic parameters to the injection of 0.1 mg of OCT; it also suppressed both basal and GHRH-stimulated GH levels by 49 and 80%, respectively [68]. The study on efficacy and safety of Octreolin for acromegaly (NCT01412424) is expected to be finished in December 2014 [69].

The research on prolonged-release i.m. formulation of OCT (octreotide C2L) has been terminated, in spite of encouraging results, due to commercial reasons (NCT00642421) [70].

The Combination Therapy

The mechanism by which DAs inhibit GH release from somatotrophs is not known; however, the addition of the DAs cabergoline to SSAs may improve the response to SSA in patients not fully controlled with SSA or in mixed tumors cosecreting GH and PRL [71]. The combination of SSA and DA is effective in long-term treatment— the addition of cabergoline in 19% not properly controlled on OCT LAR alone resulted in normalization of IGF-1 levels in 37% of them during a mean 18-month follow-up [72]. The additive role of DA is independent of PRL cosecretion and is observed even in patients with basal PRL levels within the normal range [71, 73].

Observations from *in vitro* and clinical studies on enhanced efficacy of SSAs and DAs in controlling GH secretion, as well as the discovery of SSTRs and DA heterodimerization, have led to the creation of hybrid SST/DA molecules [74, 75]. One of the hybrid molecules tested was BIM 23A760.The initial reports from *in vitro* studies suggested higher efficacy and longer duration of GH suppression with this molecule [76]; however, additive effects were vanishing with prolonged administration, probably due to interfering metabolite accumulation [77]. Due to lack of expected efficacy in suppressing GH and IGF-1 in acromegalic patient, the clinical trial NCT00994214 with BIM-23A760 has been terminated [78].

More detailed information on dopastatins is presented in Chapter 9.

Another option of combination treatment is cotherapy with SSA and pegvisomant— a GH receptor antagonist. Although pegvisomant is very efficient in achieving biochemical control, there is some risk of liver failure and/or an increase in pituitary adenoma size [35], particularly after SSA discontinuation [79].

Trainer et al. in a randomized controlled trial proved that in patients suboptimally controlled on OCT LAR, adding pegvisomant to SSA is equally efficient in optimizing acromegaly control as switching to the GH receptor antagonist alone [80]. A 7-month treatment with LAN ATG and pegvisomant resulted in the normalization of IGF-1 in 57.9% of patients previously biochemically uncontrolled on SSA or pegvisomant, particularly in nondiabetic and older ones [81]. A study by Neggers et al. showed that long-term combination treatment with SSA and pegvisomant is safe—the most common side effect of therapy was transient liver enzymes elevation, to which patients with diabetes mellitus were particularly prone (odds ratio (OR), 2.28 (95% CI, 1.16–9.22)) [82]. In patients well controlled on SSA, the addition of a small dose of pegvisomant allows the reduction of analogue dose by 50% [83]. During combine treatment, pegvisomant injection may be effectively administered once weekly [84].

PITUITARY TUMORS PRODUCING ACTH: CUSHING'S DISEASE

Cushing's disease is a rare condition [85] associated with chronic hypercortisolemia-related increased mortality, mainly due to cardiovascular complications like coronary heart disease, congestive heart disease, and cerebrovascular disease [86]. The main aim of the Cushing's disease treatment is to reverse clinical symptoms, to normalize cortisol level, and to achieve long-term disease control with no recurrence. The first-line therapy in Cushing's disease is surgery. If curative, it usually normalizes the increased mortality (SMR of 0.31 (95% CI, 0.01–1.72) in Danish patients successfully treated with surgery vs. SMR of 5.06 (95% CI, 1.86–11.0) in patients with persistent hypercortisolism after initial neurosurgery) [85]. The second-line therapies include repeated surgery, radiotherapy, medical therapy, and bilateral adrenalectomy [87]. Unfortunately, the outcomes of neurosurgery and radiotherapy in Cushing's disease are not satisfactory; the 10-year cure rates of both are 77 and 65%, respectively [88]. Bilateral adrenalectomy provides immediate control of the hypercortisolism; however, patient requires lifelong substitution of gluco- and mineralocorticoids. About one quarter of adrenalectomized patients will develop aggressive pituitary tumor secreting large quantities of ACTH—the Nelson's syndrome [89]. The medical therapies used in Cushing's disease are not curative, but usually adjuvant to neurosurgery or radiation. There are (a) adrenal-directed therapies (i.e., ketoconazole and metyrapone), which lower cortisol levels by inhibiting its adrenal production; (b) glucocorticoid receptor antagonist (mifepristone), which blocks the peripheral action of cortisol; and (c) the new pituitary-directed therapies aimed at the oversecretion of ACTH by tumor cells. Several pituitary targeting agents are being investigated, for example, peroxisome proliferator-activated receptor gamma (PPARγ) agonists, DAs, and SSAs [90].

The rationale for SSA use in Cushing's disease is the expression of the various SSTR subtypes—mainly SSTR5—on corticotrophs [91]. The studies on OCT in Cushing's disease showed however its ineffectiveness probably due to high affinity of OCT to SSTR2 and moderate to SSTR5. It can also be explained by downregulation of SSTR2 in tumor cells [92].

The high affinity of pasireotide to SSTR5 predominately expressed in corticotrophs has made it more suitable for further testing in Cushing's disease [5]. In *in vitro* studies, pasireotide has caused significant inhibition of basal and stimulated ACTH secretion by cultured human corticotroph tumors, as well as suppression of their proliferation [93, 94]. In studies on murine corticotroph tumor cells, similar effect was observed [95]. Pasireotide has also shown significant inhibition of corticotrophin-releasing hormone (CRH)-induced ACTH release in rats [96].

The results of *in vitro* studies on pasireotide prompted further clinical research in humans. In a multicenter phase II open-label study in 29 patients with de novo or persistent Cushing's disease receiving 600 μg of pasireotide self-administered twice a day (b.i.d) subcutaneously (s.c.) for 15 days, urinary free cortisol (UFC) levels decreased in 22/29 (76%) patients, of whom five had normalized UFC. Serum cortisol and plasma ACTH levels were also reduced [97]. Results from the extension of this study showed reduction of UFC in 56% of patient eligible for efficacy analysis and UFC normalization in 22% at month 6. Of the four patients who remained on pasireotide for 2 years, one had normalized UFC. The most frequent AEs were diarrhea, nausea, hyperglycemia, and abdominal pain [98].

Results of a 12-month phase III study (6 months of double-blind treatment followed by 6 months open-label treatment) of pasireotide in Cushing's disease has been published in 2012. In this study, 162 adults with Cushing's disease and UFC levels of at least 1.5 times the upper limit of the normal range were randomly assigned to receive pasireotide at a dose of 600 μg (82 patients) or 900 μg (80 patients) s.c. b.i.d. Patients with UFC not exceeding two times the upper limit of the normal range and not exceeding the baseline level at month 3 continued to receive their randomly assigned dose; all others received an additional 300 μg b.i.d. The primary end point was a UFC level at or below the upper limit of the normal range at month 6 without an increased drug dose. 12 of the 82 patients in the 600 μg group and 21 of the 80 patients in the 900 μg group met the primary end point. The median UFC level decreased by approximately 50% by month 2 and remained stable in both groups. A normal UFC level was achieved more frequently in patients with baseline levels not exceeding five times the upper limit of the normal range than in patients with higher baseline levels. Serum and salivary cortisol and plasma corticotropin levels decreased, and clinical signs and symptoms of Cushing's disease diminished. Pasireotide was associated with hyperglycemia-related AEs in 118 of 162 patients; other AEs were similar to those associated with other SSAs [99].

Treatment with DA is based on evidence of the dopamine receptor D2 expression in more than 75% of corticotrophs [100]. Cabergoline, a dopamine receptor subtype 2 (D2R) agonist, is also able to normalize UFC level; however, this effect is often not maintained during prolonged treatment [101].

The use of mixed medical treatment, "a pharmacological cocktail" with three different drugs targeting SSTR5, D2R, and steroidogenic enzymes in the adrenal cortex, was studied in a prospective, open-label, multicenter trial in 17 patients with Cushing's disease. This stepwise approach with pasireotide as the initial form of treatment and the sequential addition of cabergoline and low-dose ketoconazole resulted in biochemical control in nearly 90% of patients: 29% received pasireotide (100–250 µg three times a day), 24% pasireotide and cabergoline, and 35% pasireotide–cabergoline–ketoconazole cocktail [102].

The potential for interaction between SST and dopamine receptors to achieve greater suppression of ACTH levels has been explored with the chimeric agent BIM 23A760 with high affinity to SSTR2 and D2R and moderate affinity to SSTR5 [76]; however, its clinical evaluation in Cushing's disease has not yet been done.

Pasireotide may be also effective in targeting aggressive corticotroph tumor growth in the Nelson's syndrome [103]. Case reports on significant clinical and biochemical response to pasireotide administration have already been published [104]. Patients are currently recruited for the open-label, longitudinal study of the effects of subcutaneous pasireotide therapy on adrenocorticotrophic hormone and tumor volume in patients with Nelson's syndrome (NCT01617733) [105].

PITUITARY TUMORS PRODUCING TSH: THYROTROPINOMAS

TSH-producing pituitary tumors (TSH-secreting adenomas) account for 0.5–1.1% of all pituitary tumors [106]. The diagnosis is usually delayed as the clinical symptoms are very heterogeneous and vary from the symptoms of hyperthyroidism to neurological ones such as headache and visual disturbances—the results of the mass effect [107]. Surgery is the first-line treatment; however, due to the usually large size of the tumor and local invasion, the prevalence of persistent hyperthyroidism after neurosurgery is reported to be more than 50% of cases [108]. In such cases or in inoperable masses, medical therapy is usually needed.

Due to the rarity of the disease, the experience with pharmacotherapy of TSH-secreting pituitary adenomas comes from a small series or case reports. The expression of SSTR2 and SSTR5 on tumor cells prompted the use of SSA [107, 109]. In approximately 90 and 75% of patients with TSH-secreting adenomas treated with OCT, TSH secretion has been reported to be decreased and normalized, respectively. Circulating hormone levels were normalized in 96% of cases and goiter volume diminished in 20% of them [107, 108, 110–112]. LAN SR and OCT LAR treatment has also been shown to decrease TSH, free thyroxine (fT4), and free triiodothyronine (fT3) levels [111, 112]. Colao et al. also reported efficient normalization of TSH and fT4 and fT3 after 3–6 months of treatment with low dose of LAN, OCT, and OCT LAR without any significant side effects [113].

PITUITARY TUMORS PRODUCING PRL: PROLACTINOMAS

Prolactinomas are the most common pituitary tumors accounting for approximately 40–60% of them. The aim of the prolactinoma treatment is to normalize PRL level, restore gonadal function and fertility, and diminish the size of the tumor. In the majority of patients, medical therapy with the use of DAs is sufficient. However, in some cases resistant to DAs, other methods of treatment like neurosurgery and radiotherapy might be necessary. The other medical therapy for DA resistant tumors is also searched for [114].

Shimon et al. demonstrated the inhibition of the PRL secretion with SSTR5-selective analogues *in vitro* in cultured prolactinomas [115]. Similar effect was observed *in vitro* when pasireotide was applied [116]. In selected prolactinoma case reports or in experimental settings, SSAs have been used to date. However, SSAs have not been used in prolactinomas in the routine clinical practice [114].

The hybrid molecules possessing selective agonist activity to SSTR2 and D2R have been experimentally used in prolactinomas. As it was mentioned earlier, SSTR2 and D2R after heterodimerization in the presence of the appropriate ligands show enhanced adenyl cyclase inhibitory activity [74]. In *in vitro* study in primary cultures of prolactinomas responsive and nonresponsive to DA, the chimeric SST/DA molecule BIM-23A760 was compared with cabergoline in suppressing PRL secretion. In this study, BIM 23A760 and cabergoline produced a similar partial inhibition of PRL secretion [117].

NONFUNCTIONING PITUITARY TUMORS

Nonfunctioning pituitary tumors (NFPT) are a very heterogeneous group of tumors, representing up to 25% of all pituitary tumors. Most of NFPTs are producing small amounts of glycoproteins (FSH, LH, and their α and β subunits). Less commonly, they are nonfunctioning somatotroph, lactotroph, or corticotroph adenomas. Due to the lack of the clinical syndrome of hormonal activity, these tumors are diagnosed with a variable delay usually at the time of the presence of the neurological symptoms related to mass effects [118]. The main presenting symptoms of NFPTs are visual impairment, headache, and hypogonadism. As in hormone-overproducing tumors, the first-line treatment in these tumors is neurosurgery. However, due to the tumor extension, frequently, it is nonradical with the residual tissue left. Reported "gross total removal" rate ranges from 27 to 87%. The radiotherapy is used to prevent tumor recurrence, although less and less frequently due to associated side effects like hypopituitarism and early cerebrovascular mortality attributed particularly to the conventional technique. No routine medical therapy of NFPTs has been established so far [118].

The presence of the different subtypes of the SSTRs and dopamine receptors on cell membranes gave the rationale for the medical treatment of NFPTs with SSAs and DAs or the combination of both. In NFPT, the dominant receptor subtypes expressed are SSTR3 followed by the SSTR2. SSTR1, SSTR4, and SSTR5 are detected less commonly [119]. Pawlikowski et al. [120] demonstrated inhibition of chromogranin A (CgA) and α subunit secretion by NFPTs cells incubated *in vitro* with the native SST and SSA selective to SSTR3 and SSTR2. The expression of SSTR2 and SSTR5 was associated with NFPTs cell viability reduction when exposed to the SSA [121]. The use of pasireotide (SOM230) completely abolished the promoting effects of vascular endothelial growth factor on the NFPTs cell viability [122].

There are only a few clinical trials that have been conducted to examine the effects of OCT in NFPTs summarized by Colao et al. [123]. Tumor shrinkage was seen in 12% of treated cases. Most of the treated patients had stable remnant tumor. Only in 5% of the patients treated with OCT that the increase in the tumor size was observed. Visual field improvement was reported in 32% (27 of 84) of patients, and its compromise in 8% (7 of 84) of OCT-treated patients (the daily dose varied from 100 to 1500 μg s.c.) [123]. The strength of the data on preventive effect of OCT treatment in NFPT recurrence prevention is limited by the short-term follow-up.

In reports of Warnet et al. [124, 125], the use of OCT in NFPT patients caused rapid improvement in visual impairment and headache with no change in the tumor size probably explained by the direct effect on the retina and the optic nerve [126].

The experience with the use of the combination treatment, with SSA and DA in NFPT is very limited. Andersen et al. observed more than 10% shrinkage of the tumor in 60% of the six NFPTs patients treated for 6 months with OCT (200 μg three times daily) and cabergoline (0.5 mg a day) [127]. Colao et al. found a significant shrinkage of the tumor associated with improved mean defect at visual perimetry with the use of LAN (60 mg every fortnight) and CAB 0.5 mg every alternate day for six months [113]. However, due to the lack of placebo-controlled longitudinal studies on the SSA treatment and combination treatment with SSAs and DAs, such therapies are still not evidence based.

REFERENCES

[1] Ezzat, S.; Asa, S. L.; Couldwell, W. T.; et al. Cancer 2004, 101, 613–619.

[2] Daly, A. F.; Rixhon, M.; Adam, C.; et al. Journal of Clinical Endocrinology and Metabolism 2006, 91, 4769–47756.

[3] Buchfelder, M.; Schlaffer, S. Best Practice and Research Clinical Endocrinology and Metabolism 2009, 23, 677–692.

[4] Brazeau, P.; Vale, W.; Burgus, R.; et al. Science 1973, 179, 77–79.

[5] Schmid, H. A.; Schoeffter, P. Neuroendocrinology 2004, 80 (Suppl 1), 47–50.

[6] Neto, L. V.; Machado Ede, O.; Luque, R. M.; et al. Journal of Clinical Endocrinology and Metabolism 2009, 94, 1931–1937.

[7]   Ben-Shlomo, A.; Melmed, S. Trends in Endocrinology and Metabolism 2010, 21, 123–133.

[8]   Panetta, R.; Patel, Y. C. Life Sciences 1995, 56, 333–342.

[9]   Durán-Prado, M.; Gahete, M. D.; Martínez-Fuentes, A. J.; et al. Journal of Clinical Endocrinology and Metabolism 2009, 94, 2634–2643.

[10] Hofland, L. J.; Feelders, R. A.; de Herder, W. W.; Lamberts, S. W. Molecular and Cellular Endocrinology 2010, 15, 89–98.

[11] Durán-Prado, M.; Saveanu, A.; Luque, R. M.; et al. Journal of Clinical Endocrinology and Metabolism 2010, 95, 2497–2502.

[12] Bauer, W.; Briner, U.; Doepfner, W.; et al. Life Sciences 1982, 31, 1133–1140.

[13] Christensen, S. E.; Weeke, J.; Orskov, H.; et al. Clinical Endocrinology (Oxford) 1987, 27, 297–306.

[14] Stewart, P. M.; Kane, K. F.; Stewart, S. E.; et al. Journal of Clinical Endocrinology and Metabolism 1995, 80, 3267–3272.

[15] Biller, M. K.; Colao, A. M.; Petersenn, S.; et al. BMC Endocrine Disorders 2010, 10, 11–14.

[16] Bruns, C.; Lewis, I.; Briner, U.; et al. European Journal of Endocrinology 2002, 146, 707–716.

[17] Ezzat, S.; Forster, M. J.; Berchtold, P.; et al. Medicine (Baltimore) 1994, 73, 233–240.

[18] Melmed, S. New England Journal of Medicine 2006, 355, 2558–2573.

[19] Ben-Shlomo, A.; Melmed, S. Endocrinology and Metabolism Clinics of North America 2008, 37, 101–122.

[20] Rajasoorya, C.; Holdaway, I. M.; Wrightson, P.; et al. Clinical Endocrinology (Oxford) 1994, 41, 95–102.

[21] Melmed, S. Journal of Clinical Endocrinology and Metabolism 2001, 86, 2929–2934.

[22] Holdaway, I. M.; Rajasoorya, R. C.; Gamble, G. D. Journal of Clinical Endocrinology and Metabolism 2004, 89, 667–674.

[23] Kauppinen-Mäkelin, R.; Sane, T.; Reunanen, A.; et al. The Journal of Clinical Endocrinology and Metabolism 2005 90, 4081–4086.

[24] Melmed, S. Journal of Clinical Investigation 2009, 119, 3189–3202.

[25] Dekkers, O. M.; Biermasz, N. R.; Pereira, A. M.; et al. Journal of Clinical Endocrinology and Metabolism 2008, 93, 61–67.

[26] Katznelson, L.; Atkinson, J. L.; Cook, D. M.; et al. Endocrine Practice 2011, 17, 636–646.

[27] Melmed, S.; Sternberg, R.; Cook, D.; et al. Journal of Clinical Endocrinology and Metabolism 2005, 90, 4405–4410.

[28] Holdaway, I. M.; Bolland, M. J.; Gamble, G. D. European Journal of Endocrinology 2008, 159, 89–95.

[29] Melmed, S.; Colao, A.; Barkan, A.; et al. Journal of Clinical Endocrinology and Metabolism 2009, 94, 1509–1517.

[30] Shimon, I.; Cohen, Z. R.; Ram, Z.; Hadani, M. Neurosurgery, 2001, 48, 1239–1243; discussion 1244–1245.

[31] Melmed, S.; Vance, M. L.; Barkan, A. L.; et al. Pituitary 2002, 5, 185–196.

[32] Murray, R. D.; Kim, K.; Ren, S. G.; et al. Journal of Clinical Investigation 2004 114, 349–356.

[33] Dunn, P. J.; Donald, R. A.; Espiner, E. A. Clinical Endocrinology (Oxford) 1976, 5, 167–174.

[34] Abs, R.; Verhelst, J.; Maiter, D.; et al. Journal of Clinical Endocrinology and Metabolism 1998, 83, 374–378.

[35] Chanson, P.; Salenave, S.; Kamenicky, P. et al. Best Practice and Research. Clinical Endocrinology and Metabolism 2009, 23, 555–574.

[36] Yen, S. S.; Siler, T. M.; DeVane, G. W. New England Journal of Medicine 1974, 290, 935–938.

[37] Christensen, S. E.; Nerup, J.; Hansen, A. P.; Lundbaek, K. Journal of Clinical Endocrinology and Metabolism 1976, 42, 839–845.

[38] Plewe, G.; Beyer, J.; Krause, U.; et al. Lancet 1984 2(8406), 782–784.

[39] Freda, P. U. Journal of Clinical Endocrinology and Metabolism 2002, 87, 3013–3018.

[40] Freda, P. U.; Katznelson, L.; van der Lely, A. J.; et al. Journal of Clinical Endocrinology and Metabolism 2005, 90, 4465–4473.

[41] Mercado, M.; Borges, F.; Bouterfa, H.; et al. Clinical Endocrinology (Oxford), 2007, 66, 859–868.

[42] Colao, A.; Ferone, D.; Marzullo P.; et al. Journal of Clinical Endocrinology and Metabolism 2001, 86, 2779-2786.

[43] Amato, G.; Mazziotti, G.; Rotondi, M.; et al. Clinical Endocrinology (Oxford) 2002, 56, 65–71.

[44] Ayuk, J.; Steward, S. E.; Steward, P. M.; Sheppard, M. C. Clinical Endocrinology (Oxford) 2004, 60, 2779–2786.

[45] Jalld, R. S.; Musolino, N. R.; Salgado, L. R.; Bronstein, M. D. Clinical Endocrinology (Oxford) 2005, 63, 168–175.

[46] Colao, A.; Pivonello, R.; Rosato, F.; et al. Clinical Endocrinology (Oxford) 2006, 64, 342–351.

[47] Cozzi, R.; Montini, M.; Attanasio, R.; et al. Journal of Clinical Endocrinology and Metabolism 2006, 91, 1397–1403.

[48] Colao, A. M.; Pivonello, R.; Auriemma, R. S.; et al. European Journal of Endocrinology 2007, 157, 579–587.

[49] Colao, A.; Pivonello, R.; Auriemma, R. S.; et al. The Journal of Clinical Endocrinology and Metabolism 2008, 93, 3436–3442.

[50] Colao, A.; Cappabianca, P.; Caron, P.; et al. Clinical Endocrinology (Oxford) 2009, 70, 757–768.

[51] Auriemma, R. S.; Pivonello, R.; Galdiero, M.; et al. Journal of Endocrinological Investigation 2008, 31, 956–965.

[52] Giustina, A.; Mazziotti, G.; Torri, V.; et al. PLoS One 2012, 7, e36411.

[53] Maiza, J. C.; Vezzosi, D.; Matta, M.; et al. Clinical Endocrinology (Oxford) 2007, 67, 282–289.

[54] Caron, Ph. A.; Beckers, D. R.; Cullen, M. I.; et al. Journal of Clinical Endocrinology and Metabolism 2002, 87, 99–104.

[55] Caron, Ph.; Bex, M.; Cullen, D. R.; et al. Clinical Endocrinology 2004, 60, 734–740.

[56] Caron, Ph.; Cogne, M.; Raingeard, I.; et al. Clinical Endocrinology (Oxford) 2006, 64, 209–214.

[57] Roelfsema, F.; Biermasz, N. R.; Pereira, A. M.; et al. Biologics: Targets & Therapy 2008, 2, 463–479.

[58] Abe, T.; Ludecke, P. K. European Journal of Endocrinology 2001, 145, 137–145.

[59] Carlsen, S. M.; Lund-Johansen, M.; Schreiner, T.; et al. Journal of Clinical Endocrinology and Metabolism 2008, 93, 2984–2990.

[60] Losa, M.; Mortini, P.; Urbaz, L.; et al. Journal of Neurosurgery 2006, 6, 899–906.

[61] Bogazzi, F.; Colao, A.; Rossi, G.; et al. European Journal of Endocrinology 2013, 169, 367–376.

[62] Giordano, C.; Ciresi, A.; Amato, M. C.; et al. Pituitary 2012, 15, 539–551.

[63] Colao, A.; Auriemma, R. S.; Galdiero, M.; et al. Journal of Clinical Endocrinology and Metabolism 2009, 94, 528–537.

[64] Delaroudis, S. P.; Efstathiadou, Z. A.; Koukoulis, G. N.; et al. Clinical Endocrinology (Oxford) 2008, 69, 279–284.

[65] Petersenn, S.; Schopohl, J.; Barkan, A.; et al. Journal of Clinical Endocrinology and Metabolism 2010, 95, 2781–2789.

[66] Petersenn, S.; Farrall, A. J.; Block, C.; et al. Pituitary 2013, Mar 26. [Epub ahead of print].

[67] Colao, A.; Bronstein, M.; Freda, P.; et al. Endocrine Abstracts 2012, 29, OC1.1.

[68] Tuvia, S.; Atsmon, J.; Teichman, S. L.; et al. Journal of Clinical Endocrinology and Metabolism 2012, 97, 2362–2369.

[69] http://clinicaltrials.gov/ct2/show/NCT01412424?term=octreolin&rank=1.

[70] http://www.clinicaltrials.gov/ct2/show/NCT00642421?term=octreotide+C2L&rank=1.

[71] Cozzi, R.; Attanasio, R.; Lodrini, S.; et al. Clinical Endocrinology (Oxford) 2004, 62, 209–215.

[72] Mattar, P.; Alves Martins, M. R.; Abucham, J. Neuroendocrinology 2010, 92, 120–127.

[73] Jallad, R. S.; Bronstein, M. D. Neuroendocrinology 2009, 90, 82–92.

[74] Rocheville, M.; Lange, D. C.; Kumar, U.; et al. Science 2000, 288, 154–157.

[75] Culler, M. D. Hormones and Metabolism Research 2011, 43, 854–857.

[76] Jaquet, P.; Gunz, G.; Saveanu, A.; et al. Journal of Endocrinological Investigations 2005, 28(11 Suppl International), 21–27.

[77] Gruszka, A.; Ren, S. G.; Dong, J.; et al. Endocrinology 2007, 148, 6107–6114.

[78] http://clinicaltrials.gov/ct2/show/NCT00994214?term=BIM&rank=2.

[79] Jimenez, C.; Burman, P.; Abs, R.; et al. European Journal of Endocrinology 2008, 159, 517–523.

[80] Trainer, P. J.; Ezzat, S.; D'Souza, G. A.; et al. Clinical Endocrinology (Oxford) 2009, 71, 549–557.

[81] van der Lely, A. J.; Bernabeu, I.; Cap, J.; et al. European Journal of Endocrinology 2011, 164, 325–333.

[82] Neggers, S. J.; de Herder, W. W.; Janssen, J. A.; et al. European Journal of Endocrinology 2009, 160, 529–533.

[83] Madsen, M.; Poulsen, P. L.; Orskov, H.; et al. Journal of Clinical Endocrinology and Metabolism 2011, 96, 2405–2413.

[84] Feenstra, J.; de Herder, W. W.; ten Have, S. M.; et al. Lancet 2005, 365, 1644–1646.

[85] Lindholm, J.; Juul, S.; Jørgensen, J. O.; et al. Journal of Clinical Endocrinology and Metabolism 2001, 86, 117–123.

[86] Atkinson, A. B., Kennedy, A., Ivan Wiggam M., et al. Clinical Endocrinology 2005, 63, 549–559.

[87] Biller, B. M.; Grossman, A. B.; Stewart, P. M.; et al. Journal of Clinical Endocrinology and Metabolism 2008, 93, 2454–2462.

[88] Sonino, N.; Zielezny, M.; Fava, G. A.; et al. Journal of Clinical Endocrinology and Metabolism 1996, 81, 2647–2652.

[89] Assié, G.; Bahurel, H.; Coste, J.; et al. Journal of Clinical Endocrinology and Metabolism 2007, 92, 172–179.

[90] Gadelha, M. R.; Neto, L. V. Clinical Endocrinology (Oxford) 2014, 80, 1–12.

[91] Ueberberg, B.; Tourne, H.; Redmann, A.; et al. Hormone and Metabolic Research 2005, 37, 722–728.

[92] Petersen S., Rasch, A. C., Presch, S.; et al. Molecular and Cellular Endocrinology 1999, 25, 75–85.

[93] Hofland, L. J.; van der Hoek, J.; Feelders, R.; et al. European Journal of Endocrinology 2005, 152, 645–654.

[94] Batista, D. L.; Zhang, X.; Gejman, R.; et al. Journal of Clinical Endocrinology and Metabolism 2006, 91, 4482–4488.

[95] van der Hoek, J.; Waaijers, M.; van Koetsveld, P. M.; et al. American Journal Physiology Endocrinology and Metabolism 2005, 289, E278–E287.

[96] Silva, A. P.; Schoeffter, P.; Weckbecker, G.; et al. European Journal of Endocrinology 2005, 153, R7–R10.

[97] Boscaro, M.; Ludlam, W. H.; Atkinson, B.; et al. Journal of Clinical Endocrinology and Metabolism 2009, 94, 115–122.

[98] Boscaro, M.; Bertherat, J.; Findling, J.; et al. Pituitary 2013 Aug 14. [Epub ahead of print].

[99] Colao, A.; Petersenn, S.; Newell-Price, J.; et al. New England Journal of Medicine 2012, 366, 914–924.

[100] Pivonello, R.; Ferone, D.; de Herder, W.W.; et al. Journal of Clinical Endocrinology and Metabolism 2004, 89, 2452–2462.

[101] Pivonello, R.; De Martino, M. C.; Cappabianca, P.; et al. Journal of Clinical Endocrinology and Metabolism 2009, 94, 223–230.

[102] Feelders, R. A.; de Bruin, C.; Pereira, A. M.; et al. New England Journal of Medicine 2010, 362, 1846–1848.

[103] Barber, T. M.; Adams, E.; Ansorge, O.; et al. European Journal of Endocrinology 2010, 163, 495–507.

[104] Katznelson, L. Journal of Clinical Endocrinology and Metabolism 2013, 98, 1803–1807.

[105] http://clinicaltrials.gov/ct2/show/NCT01617733?term=pasireotide&rank=18.

[106] Socin, H.V.; Chanson, P.; Delemer, B.; et al. European Journal of Endocrinology 2003, 148, 433–442.

[107] Beck-Peccoz, P.; Persani, L. Pituitary 2002, 5, 83–88.

[108] Beck-Peccoz, P.; Persani, L.; Mantovani, S.; et al. Metabolism 1996, 45, 75–79.

[109] Hofland, L. J.; Lamberts, S. W. Frontiers in Hormones Research 2004, 32, 235–252.

[110] Chanson, P.; Weintraub, B. D.; Harris, A. G. Annals of Internal Medicine 1993, 119, 236–240.

[111] Caron, P.; Arlot, S.; Bauters, C.; et al. Journal of Clinical Endocrinology and Metabolism 2001, 86, 2849–2853.

[112] Kuhn, J. M.; Arlot, S.; Lefebvre, H.; et al. Journal of Clinical Endocrinology and Metabolism 2000, 85, 1487–1491.

[113] Colao, A.; Filippella, M.; Di Somma, C.; et al. Endocrine 2003, 20, 279–283.

[114] Colao, A. Best Practice and Research. Clinical Endocrinology and Metabolism 2009, 23, 575–596.

[115] Shimon, I.; Yan, X.; Taylor, J. E.; et al. Journal of Clinical Investigation 1997, 100, 2386–2392.

[116] Hofland, L. J.; van der Hoek, J.; van Koetsveld, P. M.; et al. Journal of Clinical Endocrinology and Metabolism 2004, 89, 1577–1585.

[117] Fusco, A.; Gunz, G.; Jaquet, P.; et al. European Journal of Endocrinology 2008, 158, 595–603.

[118] Greeman, Y.; Stern, N. Best Practice and Research Clinical Endocrinology and Metabolism 2009, 23, 625–638.

[119] Taboada, G. F.; Luque, R. M.; Bastos, W.; et al. European Journal of Endocrinology 2007, 156, 65–74.

[120] Pawlikowski, M.; Lawnicka, H.; Pisarek, H.; et al. Journal of Physiology and Pharmacology 2007, 58, 179–188.

[121] Padova, H.; Rubinfeld, H.; Hadani, M.; et al. Molecular and Cellular Endocrinology 2008, 14, 214–218.

[122] Zatelli, M. C.; Piccin, D.; Vignali, C.; et al. Endocrine-Related Cancer 2007, 14, 91–102.

[123] Colao, A.; Di Somma, C.; Pivonello, R.; et al. Endocrine-Related Cancer 2008, 15, 905–915.

[124] Warnet, A.; Harris, A. G.; Renard, E.; et al. Neurosurgery 1997, 41, 786–795.

[125] de Bruin, T. W.; Kwekkeboom, D. J.; Van't Verlaat, J. W.; et al. Journal of Clinical Endocrinology and Metabolism 1992, 75, 1310–1317.

[126] Lamberts, S. W. J.; de Herder, W. W.; van der Lely, A. J.; Hofland, L. J. Endocrinologist 1995, 5, 448–451.

[127] Andersen, M.; Bjerre, P.; Schrøder, H. D.; et al. Clinical Endocrinology (Oxford) 2001, 54, 23–30.

5.3

SOMATOSTATIN ANALOGUES IN PHARMACOTHERAPY OF GASTROENTEROPANCREATIC NEUROENDOCRINE TUMORS

FRÉDÉRIQUE MAIRE AND PHILIPPE RUSZNIEWSKI

Service de Gastroentérologie-Pancréatologie, Hôpital Beaujon, Clichy and Université Paris Denis-Diderot, Paris, France

ABBREVIATIONS

ACTH	adrenocorticotropic hormone
cAMP	cyclic adenosine monophosphate
EGF	epidermal growth factor EGF
IGF-1	insulin-like growth factor
PDGF	platelet-derived growth factor PDGF
pre-proSST	somatostatin amino acid precursor
PTHrp	parathyroid hormone-related peptide
SST	somatostatin
SSTR	somatostatin receptor
VEGF	vascular endothelial growth factor
VIP	vasoactive intestinal peptide

The discovery of somatostatin and the synthesis of a variety of analogues constituted a major therapeutic advance in the treatment of gastroenteropancreatic neuroendocrine tumors.

Somatostatin Analogues: From Research to Clinical Practice, First Edition. Edited by Alicja Hubalewska-Dydejczyk, Alberto Signore, Marion de Jong, Rudi A. Dierckx, John Buscombe, and Christophe Van de Wiele.
© 2015 John Wiley & Sons, Inc. Published 2015 by John Wiley & Sons, Inc.

BIOLOGY

Natural Somatostatin

The somatostatin (SST) neuropeptide family comprises peptides that originate from different posttranslational processing of a 116-amino-acid precursor (pre-proSST), which is encoded by a single gene located in humans on chromosome 3q28. Pre-proSST is processed to proSST (96 amino acids), which is further cleaved to produce two bioactive proteins, the predominant but functionally less active SST molecule consisting of 14 amino acids (SST-14) and a larger more potent molecular form (SST-28) [1]. The SSTs have a very short circulation half-life of 1.5–3 min, rendering analysis of their physiological activity difficult. SST is a pan-inhibitory agent for all known gastrointestinal tract hormones.

SST Receptors

Twenty years after the discovery of SST in 1972, molecular cloning facilitated the identification of its receptor structure. Subsequently, it has become apparent that in mammals, SST mediates its inhibitory effects through binding to at least G-protein-coupled membrane receptors. Five different receptor subtypes have been cloned and characterized (SSTR1–5). Each receptor consists of a single polypeptide chain with seven transmembrane-spanning domains with the extracellular domains exhibiting the ligand-binding sites and the intracellular sites providing linkage to second messenger activation [1].

The diverse inhibitory effects of SST on neurotransmission, motor and cognitive functions, smooth muscle contractility, glandular and exocrine secretions, intestinal motility, and absorption of nutrients and ions are all mainly mediated by cyclic adenosine monophosphate (cAMP) (a second messenger that is important in many biological processes and used for intracellular signal transduction) and Ca^{2+} reduction with activation of protein phosphatases [1].

All five SST receptors (SSTR) have been identified throughout the gastrointestinal tract and endocrine and exocrine glands, as well as on inflammatory and immune cells. Abundant expression of SSTR has been identified in endocrine tumors of the gastrointestinal tract and pancreas. SSTR expression varies between patients and between tumors, but SSTR2 predominance is found in more than 80% of these tumors. The expression of SSTR by human endocrine tumors has important clinical implications, such as inhibition of tumoral peptide hormone secretion and inhibition of tumoral growth by SST analogues.

SST Analogues

The clinical utility of native SST was clearly limited given its rapid blood clearance. The development of stable and potent analogues therefore became necessary for therapeutic efficacy. Several structural analogues of SST have been synthesized (Fig. 5.3.1). Compared with natural SST, octreotide contains three substituted amino

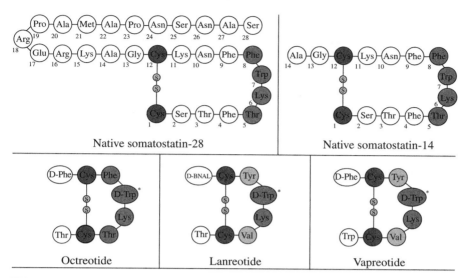

FIGURE 5.3.1 The structure of native somatostatin-28 and somatostatin-14 and the somatostatin analogues octreotide, lanreotide, and vapreotide.

acids that make it resistant to metabolic degradation and increase its *in vivo* half-life. SST analogues have restricted receptor affinity profiles (high affinity to SSTR2 and SSTR5, low affinity to SSTR3, and no affinity to SSTR1 and SSTR4). Octreotide and lanreotide are synthetic octapeptide with similar binding profiles. Both are available either in a rapid action form administrated subcutaneously or intravenously (octreotide, Sandostatin®, Novartis, and lanreotide, Somatuline®, Ipsen) or in a slow-release intramuscular/deep subcutaneous depot formulation (octreotide, Sandostatin LAR®, Novartis, and lanreotide, Somatuline LAR®, Ipsen).

SYMPTOMATIC EFFECTS

Carcinoid Syndrome

Pooled data of octreotide and lanreotide trials from the past 20 years, including more than 400 patients, show a mean symptomatic response rate of 73% (range of 50–100%), with similar results for immediate-release and long-acting formulations [2]. Disappearance of flushing episodes is observed in approximately 60% of patients, while the frequency and/or the severity of the flushing periods can be reduced to less than 50% in more than 85% of patients [3, 4]. Disappearance of diarrhea is observed in more than 30%, and more than 50% improvement is seen in more than 75% of patients. The antisecretory effect results in a reduction of biochemical markers (chromogranin A and urinary 5-hydroxyindoleacetic acid) in up to 40–60% of patients. Tolerance to SST analogues and efficacy should be tested individually by initiating therapy with short-acting analogues (100 μg subcutaneously 2–3 times daily).

Thereafter, depot formulations, usually octreotide LAR (20–30 mg) or lanreotide Autogel (90 mg) every 4 weeks, can be started and should be individually titrated [5]. The efficacy of lanreotide and octreotide is comparable [6]. To prevent carcinoid crisis during interventional procedures, SST analogues should be given intravenously. Moreover, carcinoid heart disease secondary to serotonin production of the tumor has been significantly reduced by the introduction of SST analogue therapy.

The duration of the effect of SST varies and can be limited due to tachyphylaxis or desensitization, which can be temporarily circumvented by an increase in dose. If the treatment is insufficient, it can also be proposed to decrease interval time between two injections or to add short-acting SST analogues. Another cause of diarrhea must also be considered, such as bacterial overgrowth, large small bowel resection, diarrhea linked to biliary salts, exocrine pancreatic insufficiency, and specific treatment required.

Other Hypersecretion Syndromes

SST analogues are also recommended for patients with other functioning tumors.

In patients with glucagonoma, SST analogues are effective in controlling the necrolytic migratory erythema but less effective in the management of weight loss and diabetes mellitus.

Vipoma syndrome often requires intensive intravenous supplementation of fluid losses and careful correction of electrolyte and acid–base abnormalities. SST analogues reduce tumoral vasoactive intestinal peptide (VIP) secretion and thus inhibit intestinal water and electrolyte secretion. By this mechanism, these drugs control the secretory diarrhea in more than 70% of patients.

In patients with insulinoma, SST analogues are of limited use because less than 40% of insulinomas express SSTR subtypes that bind these drugs. Moreover, caution must be exercised in SST analogue therapy in these patients, since hypoglycemia may worsen due to a more profound suppression of counterregulatory hormones, such as glucagon. Diazoxide is the most effective drug for controlling hypoglycemia.

In patients presenting gastrinoma, gastric acid hypersecretion should always first be treated medically with high doses of proton pump inhibitors. SST analogues are also effective in reducing both gastrin and acid secretion in Zollinger–Ellison patients; however, they are rarely used because proton pump inhibitor is the drug of choice (high potencies, long duration of action, oral medication with once-a-day or twice-a-day dosing) [7].

Cushing's syndrome in patients with ectopic adrenocorticotropic hormone (ACTH) production can be controlled by SST analogues in combination with ketoconazole, metyrapone, and etomidate or by biadrenalectomy.

Hypercalcemia in patients with paraneoplastic parathyroid hormone-related peptide (PTHrp) can be controlled by SST analogues and bisphosphonates.

Side Effects

Minor side effects include pain at injection site, abdominal discomfort, bloating, and sometimes steatorrhea. Gallstones can develop in 20–50% of cases; however, virtually all remain asymptomatic. Moreover, patients with diabetes mellitus should be monitored closely when therapy is initiated and adjustment of antidiabetic therapy might be necessary.

ANTITUMORAL EFFECT

Physiopathology

SST analogues have been implicated in tumor proliferation in some nonendocrine and endocrine tumors. Several mechanisms involving SSTR-mediated control of cell proliferation have been described: inhibition of the mitogenic signal of serum growth factors (epidermal growth factor (EGF), platelet-derived growth factor (PDGF), insulin-like growth factor (IGF-1)) and direct inhibition of the proliferation by regulating tyrosine kinase, tyrosine phosphatase, nitric oxide synthase, and cyclic guanosine 3′,5′-cyclic monophosphate-dependent protein kinase. Signaling via SSTR2 or SSTR3 has also been shown to result in apoptosis via p53-dependent pathways [8]. Several *in vitro* and *in vivo* studies support the observation that SST analogues inhibit angiogenesis. Although SST analogues may exert an angiogenic effect directly on SSTR-positive cells, such as endothelial and monocyte cells, they may also have indirect effects via inhibition of the production of angiogenic factors, such as vascular endothelial growth factor (VEGF) [2].

Results

Since the introduction of SST analogues, multiple retrospective series and phase II trials have demonstrated that SST analogue treatment is associated with prolonged survival and disease stabilization. But these studies are difficult to interpret for several reasons: heterogeneity of tumor types (site, histology, tumor load), use of different formulations and doses, lack of objective tumor progression prior to treatment with SST analogues, and complete lack of randomized study examining the question. Although objective radiological response rates have been rare (generally <5%), the rate of tumor stabilization observed in most studies has ranged from 40 to 60%, with a response duration of 10–25 months [9, 10]. Recently, a review article by Modlin et al. [1] reported overall tumor response (stable disease and partial response) in the entire patient population treated with octreotide of 57.4% (36.5–100%), octreotide LAR 69.8% (47.0–87.5%), lanreotide 46.6% (32.0–75.0%), and long-acting lanreotide 64.4% (48.0–87.0%) (Fig. 5.3.2).

To prove or to disprove an antiproliferative effect of octreotide LAR 30 mg, the PROMID study was initiated [11]. This randomized, double-blind, placebo-controlled phase III trial reports the effect of octreotide LAR in the control of tumor growth in patients with metastatic endocrine midgut tumors. Eighty-five treatment-naïve patients were randomly assigned to receive 30 mg monthly octreotide LAR or placebo. Median time to tumor progression in the octreotide LAR and placebo groups was 14.3 and 6 months, respectively (hazard ratio = 0.34; 95% CI, 0.20–0.59; $P = 0.000072$). After 6 months of treatment, stable disease

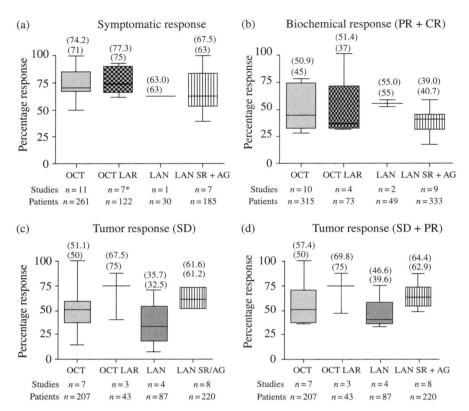

FIGURE 5.3.2 A compilation of the efficacy of different somatostatin analogues and different formulations. (a) Symptomatic response; (b) biochemical response; (c, d) tumor response. PR, partial response; CR, complete response; SD, stable disease. Mean (top) and median in brackets. From Modlin et al. [1]. By permission of Elsevier, Trends in Endocrinology & Metabolism, 2010.

was observed in 67% of patients in the octreotide LAR group and 37% of patients in the placebo group, a highly significant difference. Functionally active and inactive tumors responded similarly. The most favorable effect was observed in patients with low (<10%) hepatic tumor load (median time to progression 29.4 vs. 6.1 months, respectively) and resected primary tumor (median time to progression 29.4 vs. 5.9 months, respectively). In conclusion, octreotide LAR significantly lengthens time to tumor progression compared with placebo in patients with functionally active and inactive metastatic midgut endocrine tumors (Fig. 5.3.3). Because of the low number of observed deaths, survival analysis was not confirmatory.

A randomized phase III trial evaluating the effect of lanreotide versus placebo on progression-free survival in patients with nonfunctioning endocrine tumors is currently ongoing.

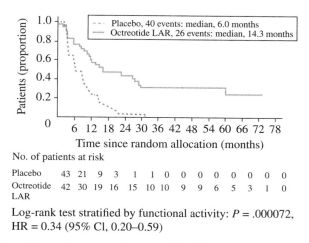

Log-rank test stratified by functional activity: $P = .000072$,
HR = 0.34 (95% CI, 0.20–0.59)

FIGURE 5.3.3   Kaplan–Meier curve demonstrating time to tumor progression in patients treated with octreotide LAR versus placebo in the PROMID phase III trial. From Ref. [11]. © Wiley.

Predictive Factors of Tumor Control under SST Analogues

A small intestinal tumor origin has been suggested as likely to predict a better tumor stabilizing response [12]. In a clinical trial with high doses of lanreotide, higher rates of tumor growth inhibition were observed in midgut tumors in contrast to foregut tumors [13]. Recently, a retrospective study reported that Ki-67 index ≤5%, pretreatment stability, and liver tumor invasion ≤25% were significantly associated with disease stability under lanreotide. Gender, age, origin of the primary tumor, discovery mode, functionality, and presence of extrahepatic metastases were not predictive factors of tumor control by lanreotide [14].

Results of Combination Therapies

The question whether SST analogues and interferon exhibit a synergistic effect in the management of neuroendocrine tumor is controversial. Some studies reported that the biochemical response rate was higher with the association than with single agent, but the combination conferred more adverse events and patients had a significantly poorer quality of life compared to patients who received SST analogues alone. No improved effect on tumor size compared with single agent treatment has been shown in two prospective randomized clinical trials [13, 15].

In a phase II study, the combination of octreotide and bevacizumab (monoclonal antibody against vascular endothelial growth factor) improved progression-free survival and increased the rate of partial remission, in comparison with the combination of octreotide and pegylated interferon [16].

The activity of the oral inhibitor of mammalian target of rapamycin (mTOR) (everolimus) in combination with octreotide LAR was recently reported in a prospective, randomized phase III study including 410 patients with advanced, low-grade, or intermediate-grade pancreatic endocrine tumors [17]. The median progression-free survival was 11.0 months with everolimus as compared with 4.6 months with placebo.

WHAT ABOUT NEW SST ANALOGUES?

New classes of SSTR subtype-selective analogues are being developed and tested. These new analogues may prove valuable for the treatment of tumors that express SSTR subtypes other than SSTR2 and SSTR5.

The multi-SST analogue SOM230 (pasireotide) has high affinity for SSTR1, SSTR2, SSTR3, and SSTR5. In an *in vitro* study evaluating the use of octreotide and pasireotide on HEK293 cells expressing SSTR2 subtype on the cell membrane, treatment with octreotide resulted in an internalization of SSTR2 at 30 min, whereas treatment with pasireotide did not lead to SSTR2 internalization. Such findings may suggest that a persistent and more durable efficacy could be obtained with pasireotide.

Pasireotide is currently under evaluation in phase II and III trials in patients with octreotide-resistant carcinoid tumors. A phase II, open-label, multicenter trial evaluated the efficacy and safety of SOM230 in 44 patients with metastatic digestive endocrine tumors whose symptoms (diarrhea and flushed skin) were resistant to standard treatment with octreotide LAR. At subcutaneous doses up to 900 µg, SOM230 controlled these symptoms in up to 27% of patients [18].

Moreover, SOM230 inhibits IGF-1 action, inhibits VEGF secretion, and induces apoptosis in human cell cultures, supporting the hypothesis that pasireotide may have potential in the treatment of these tumors.

Chimeric molecules that bind SSTR2 and the dopamine receptor D2 (BIM-23A760) are also being developed and are in clinical trials in carcinoid tumors.

CONCLUSIONS

SST analogues are the best therapeutics for reducing symptoms in patients with digestive neuroendocrine tumors [2]. Long-acting formulations of SST analogues have significantly improved the quality of life of these patients. The antiproliferative effects of SST analogues require further investigation, as well as future studies to confirm the findings of the PROMID study. It will also be important to determine whether high-dose therapy with SST analogues might increase the antiproliferative effects in tumor cells in clinical and in preclinical trials. The effects of combination of SST analogues with drugs such as interferon, mTOR inhibitors, or vascular endothelial growth factor inhibitors must also be studied in prospective randomized clinical trials. Studies of pan-receptor analogues, such as pasireotide, might provide further insight into the antiproliferative effects of SST analogues and provide better control of patients' symptoms. Clinical trials of chimeric molecules that bind SSTR2 as well as D2 will provide information about the crosstalk between different G-protein-coupled receptors in the development of neuroendocrine tumors. Further development of these types of drugs has the potential to improve the efficacy of neuroendocrine therapies and clarify the pathogenesis of these tumor types.

REFERENCES

[1] Modlin, I. M.; Pavel, M.; Kidd, M.; et al. Alimentary Pharmacology and Therapeutics 2010, 31, 169–188.

[2] Oberg, K. E.; Reubi, J. C.; Kwekkeboom, D. J.; et al. Gastroenterology 2010, 139, 742–753.

[3] Ruszniewski, P.; Ducreux, M.; Chayvialle, J. A.; et al. Gut 1996, 39, 279–283.

[4] Wymenga, A. N.; Eriksson, B.; Salmela, P. I.; et al. Journal of Clinical Oncology 1999, 17, 1111–1117.

[5] O'Toole, D.; Ducreux, M.; Bommelaer, G.; et al. Cancer 2000, 88, 770–776.

[6] Kvols, L. K.; Moertel, C. G.; O'Connell, M. J.; et al. New England Journal of Medicine 1986, 315, 663–666.

[7] Kulke, M. H.; Anthony, L. B.; Bushnell, D. L.; et al. Pancreas 2010, 39, 735–752.

[8] Grozinsky-Glasberg, S.; Shimon, I.; Korbonits, M.; et al. Endocrine Related Cancer 2008, 15, 701–720.

[9] Strosberg, J.; Kvols, L. World Journal of Gastroenterology 2010, 16, 2963–2970.

[10] Kulke, M. H.; Siu, L. L.; Tepper, J. E.; et al. Journal of Clinical Oncology 2011, 29, 934–943.

[11] Rinke, A.; Müller, H. H.; Schade-Brittinger, C.; et al. Journal of Clinical Oncology 2009, 27, 4656–4663.

[12] Aparicio, T.; Ducreux, M.; Baudin, E.; et al. European Journal of Cancer 2001, 37, 1014–1019.

[13] Faiss, S.; Pape, U. F.; Bohmig, M.; et al. Journal of Clinical Oncology 2003, 21, 2689–2696.

[14] Palazzo, M.; Lombard-Bohas, C.; Cadiot, G.; et al. Gastroenterology 2011, 140, S-873 (abstract).

[15] Arnold, R.; Rinke, A.; Klose, K. J.; et al. Clinical Gastroenterology and Hepatology 2005, 3, 761–771.

[16] Yao, J. C.; Phan, A.; Hoff, P. M.; et al. Journal of Clinical Oncology 2008, 26, 1316–1323.

[17] Yao, J. C.; Shah, M. H. ; Ito, T.; et al. New England Journal of Medicine 2011, 364, 514–523.

[18] Kvols, L.; Wiedenmann, B.; Oberg, K.; et al. Journal of Clinical Oncology 2006, 24, 185 s (abstract 4082).

5.4

SOMATOSTATIN ANALOGUES USE IN OTHER THAN ENDOCRINE TUMOR INDICATIONS

ALEKSANDRA GILIS-JANUSZEWSKA[1], MALGORZATA TROFIMIUK-MÜLDNER[1], AGATA JABROCKA-HYBEL[1,2], AND DOROTA PACH[1,2]

[1]*Department of Endocrinology with Nuclear Medicine Unit, Medical College, Jagiellonian University, Krakow, Poland*
[2]*Department of Endocrinology, University Hospital in Krakow, Krakow, Poland*

ABBREVIATIONS

CAS	clinical activity score
CH	congenital hyperinsulinism
CI	confidence interval
ERCP	endoscopic retrograde cholangiopancreatography
GO	Graves' ophthalmopathy
HCC	hepatocellular carcinoma
NSCLC	non-small cell lung cancer
SSA	somatostatin analogues
SSTRs	somatostatin receptors

Somatostatin Analogues: From Research to Clinical Practice, First Edition. Edited by
Alicja Hubalewska-Dydejczyk, Alberto Signore, Marion de Jong, Rudi A. Dierckx,
John Buscombe, and Christophe Van de Wiele.

INTRODUCTION

Somatostatin analogues (SSA) are actually widely used in the treatment of pituitary and neuroendocrine tumors to control clinical symptoms and tumor growth. Due to somatostatin properties, like decreasing hormone and enzyme secretion, inhibiting cell proliferation, reducing splanchnic pressure, and modulating immune response, SSA have also been used in other, both endocrine-related and non-endocrine-related, diseases, usually off-label or in experimental setting.

Recent investigations show increase in therapeutic applications of SSA. Development of new SSA with high affinity to somatostatin receptor (SSTR) sub-types other than type 2 (SSTR type 2) might bring new therapeutic targets, as some tumors present the high expression of SSTR1 or SSTR5. Other than generally approved applications of SSA are discussed in the following.

SSA USE IN GASTROENTEROLOGY

The most vast evidence for nonconventional use of somatostatin and its analogues exists for gastric tract disease treatment. Somatostatin properties to inhibit secretion of many hormones, such as gastrin, cholecystokinin, glucagon, growth hormone, insulin, secretin, pancreatic polypeptide, TSH, and vasoactive intestinal peptide, have been widely investigated. However, somatostatin also reduces secretion of fluids in the intestine and pancreas, reduces gastrointestinal motility, and inhibits contraction of the gallbladder. Somatostatin and its analogues can cause vasoconstriction and reduce pressure in portal circulation.

Variceal and Other Gastric Tract Bleedings

The first attempts to use somatostatin and its analogues to control variceal bleeding date to the early 1980s. Variceal bleeding is an important life-threatening consequence of cirrhotic liver disease related to high mortality rates. Endoscopic treatment is the first-line therapy; however, several drugs have been tested to assist the endoscopic interventions, including SSA. One of the mechanisms of action of somatostatin is through reducing portal pressure. Somatostatin and all commercially marketed analogues (octreotide, lanreotide, and vapreotide) have been tested for that indication. The early results show their efficacy to control acute variceal bleeding as well as to reduce the risk of recurrent bleeding and death, especially when started prior to endoscopy. Somatostatin and SSA are easy to administer and have few minor side effects.

Meta-analysis published in 2003, in which 15 trials were included, comparing emergency sclerotherapy with vasoactive drugs for variceal bleeding, demonstrated that sclerotherapy was not superior to terlipressin, somatostatin, or octreotide for mortality, blood transfusions, use of balloon tamponade, initial hemostasis, and rebleeding. Sclerotherapy was associated with significantly more adverse events than somatostatin [1].

However, later data have not supported the routine use of SSA in that setting. The Cochrane Database systematic review from 2008 analyzed somatostatin and its analogues as the agents to reduce the need for blood transfusions and to improve the survival rate in patients with bleeding esophageal varices. 21 trials were included (2588 patients). The drugs did not reduce mortality significantly (relative risk (RR) 0.97 and 95% confidence interval (CI) 0.75–1.25 for the trials with a low risk of bias and RR 0.80 and 95% CI 0.63–1.01 for the other trials). Units of blood transfused were 0.7 (0.2–1.1) less with drugs in the trials with a low risk of bias and 1.5 (0.9–2.0) less in the other trials. The number of patients failing initial hemostasis was reduced, RR 0.68 (0.54–0.87). The number of patients with rebleeding was not significantly reduced for the trials with a low risk of bias, RR 0.84 (0.52–1.37), while it was substantially reduced in the other trials, RR 0.36 (0.19–0.68). The author concluded that slight reduction in the volume of the transfused blood did not support the regular SSA therapy in the acute variceal bleeding [2].

Recently published meta-analysis compared the efficacy of vasoactive medications (terlipressin, vapreotide, and octreotide) in patients having acute variceal bleeding. The search identified 30 trials with a total of 3111 patients. The use of vasoactive agents was associated with a significantly lower risk of acute all-cause mortality and transfusion requirements and improved control of bleeding and shorter hospital stay. Studies comparing different vasoactive medications failed to demonstrate a difference in efficacy [3].

Considering the differences in the studies and meta-analyses outcomes, the current evidence to routinely apply SSA in acute bleeding from esophageal varices is too limited. However, it seems worthwhile to conduct the large-scale, randomized trial, focused also on long-term mortality.

Somatostatin and SSA have also been tested in gastric bleeding of nonvariceal etiology. The experience is limited to a few small trials in which somatostatin seemed to effectively control gastric bleeding and performed better than, for example, ranitidine [4].

Pancreatic Surgery and Pancreatitis

Somatostatin and SSA have been also used to prevent postoperative complications in patients undergoing major pancreatic surgery. Continuous pancreatic juice secretion and soft structure of the pancreatic parenchyma are major risk factors of serious postoperative complications like fistula, fluid collection, and leakage from the anastomosis. Several studies suggested that SSA given preoperatively through inhibition of perioperative pancreatic secretion reduces serious postoperative complications [5].

Two meta-analyses assessing the effectiveness of SSA for pancreatic surgery have been published recently. In a 2010 paper, seventeen trials involving 2143 patients were identified. The overall number of patients with postoperative complications was lower in the SSA group (RR, 0.71; 95% CI, 0.62–0.82) with no difference between the groups in perioperative mortality (RR, 1.04; 95% CI, 0.68–1.59), reoperation rate (RR, 1.15; 95% CI, 0.56–2.36), and hospital stay (mean difference (MD), −1.04 days; 95% CI, −2.54 to 0.46). The incidence of pancreatic fistula was lower in the SSA group (RR, 0.64; 95% CI, 0.53–0.78). Analysis of results of trials that clearly distinguished clinically significant fistulas revealed no difference between the two groups (RR, 0.69; 95% CI, 0.34–1.41). Subgroup analysis revealed a shorter hospital stay in the SSA group than among controls for patients with malignant etiology (MD, −7.57 days; 95% CI, −11.29 to −3.84) [6].

The most recent 2012 meta-analysis was aimed to determine whether prophylactic SSA should be used routinely in pancreatic surgery. Nineteen trials were identified (17 trials of high risk of bias) involving 2245 patients. There was no significant difference in the perioperative mortality (RR, 0.80; 95% CI, 0.56–1.16) or the number of patients with drug-related adverse effects between the two groups (RR, 2.09; 95% CI, 0.83–5.24). The overall number of patients with postoperative complications was significantly lower in the SSA group (RR, 0.69; 95% CI, 0.60–0.79), but there was no significant difference in the reoperation rate (RR, 1.26; 95% CI, 0.58–2.70) or hospital stay (MD, −1.04 days; 95% CI, −2.54 to 0.46). On inclusion of trials that clearly distinguished clinically significant fistulas, there was no significant difference between the two groups (RR, 0.69; 95% CI, 0.34–1.41; $N=247$). The authors concluded that SSA may reduce perioperative complications but do not reduce perioperative mortality. The authors stressed the need of further adequately powered trials with low risk of bias [7].

The use of octreotide has been also investigated in several clinical studies in patients with moderate to severe acute pancreatitis, particularly ERCP associated. In the meta-analysis published in 2010, which included 3818 participants, efficacy of somatostatin or octreotide for this ERCP complication prevention was assessed. Authors concluded that somatostatin and high-dose octreotide may prevent post-ERCP pancreatitis, particularly if administered to the pancreatic duct, or if biliary sphincterotomy is performed, or high-dose drug is administered over 12 h [8].

Similarly favorable outcomes in terms of reducing post-ERCP pancreatitis and hyperamylasemia rates by intravenous bolus injection of somatostatin were reported in the meta-analysis published in 2007 [9].

Abdominal pain, which is the most important clinical symptom of chronic pancreatitis, possibly results from an increased intraductal pressure during secretion. The inhibiting effect of octreotide treatment on the pain has been also investigated. In the study in which octreotide was used for 3 weeks, significantly less pain and analgesic use were recorded during octreotide treatment than during placebo. It was also shown that combined therapy with diclofenac and somatostatin can reduce the frequency and severity of postendoscopic retrograde cholangiopancreatography (ERCP) pancreatitis [10].

A scarce evidence exists that SSA may be useful in the prevention of the exacerbation of pancreatitis. Recently published results of a small, multicenter, randomized controlled trial showed that prolonged intravenous administration of octreotide hours in the early stage of acute pancreatitis could prevent the development of severe form of the disease in the obese patients [11].

However, it should be remembered that SSA treatment is the risk factor for cholelithiasis and choledocholithiasis, and the acute pancreatitis cases complicating the long-term treatment with SSA have been also reported.

Postsurgical Pain Relief

The effect of intravenous infusion of octreotide has also been tested on postoperative pain relief in patients after major abdominal surgery. It was demonstrated that perioperative octreotide infusion could be an adjuvant to opioid analgesia due to a piritramide opioid-sparing effect. Octreotide infusion was suggested as alternative treatment to patients who cannot tolerate the adverse effect of opioids [12].

Refractory Diarrhea

Refractory diarrhea, lasting more than 2 weeks and nonresponding to the standard treatment, is one of the conditions, in which SSA have been tested off-label. They have been tried in diarrhea of different etiologies: cancer-related diarrheas, including related to chemotherapy and radiotherapy, and also noncancer related, including short bowel syndrome, Crohn's disease, ileo- and jejunostomy dumping syndrome, graft-versus-host disease, and AIDS-related diarrhea. Most of the experiences come however from case reports and uncontrolled small-scale trials [13].

In 2001, a systematic review of SSA use in refractory diarrhea of different etiology was published. The response percentage was 73% overall in case series and 64% in randomized controlled trials (not significant). In the most homogenous group of AIDS patients, somatostatin and octreotide were less effective, whereas in highly heterogenous postchemotherapy group, both drugs were highly effective [14].

SSA USE IN OPHTHALMOLOGY

Retinopathy

In retinopathies, especially proliferative diabetic retinopathy, the rationale for somatostatin as a therapeutic agent for retinal neovascularization is under discussion. Retinal cells express SSTR2 and SSTR3. They are the most important receptor subtypes mediating growth hormone secretion and endothelial cell cycle arrest, retinal endothelial cell apoptosis and release of insulin. SSA have shown to be an effective treatment for severe proliferative diabetic retinopathy through blocking the local and systemic production of growth hormone and insulin-like growth factor type 1 associated with angiogenesis and endothelial cell proliferation. Long-acting SSA use in patients who fail panretinal photocoagulation is of particular interest. There are studies on the possibility of the use of SSA in other retinal vascular proliferative diseases such as retinopathy of prematurity and age-related macular degeneration [15, 16].

SSA has been also used in the treatment of cystoid macular edema in diabetic patients. It is believed that in this pathology, SSA, through direct antiproliferative effect on the retinal endothelium, may help to restore the inner blood–retinal barrier and intensify the active transport of fluid out of the retina toward the choroids [17].

In the studies on nonfunctioning pituitary adenomas, long-term high-dose treatment with octreotide caused rapid improvement in visual impairment with no change of the tumor size that could be explained more likely by the direct effect of SSA on the retina and the optic nerve [18].

The novel SSA pasireotide is also of interest as the retina-protective agents. In animal studies, it showed neuroprotective properties at the concentration or dose of 100 nM regardless of the model of retinal ischemia [19].

Graves' Ophthalmopathy

Graves' ophthalmopathy (GO) is an autoimmune condition affecting orbital tissues. Due to immunomodulatory effects of somatostatins, they can be used in some GO cases not responding to standard treatment. Somatostatin receptors have been identified on orbital fibroblasts and activated lymphocytes. In primary cultured retro-orbital fibroblasts of patients with GO expression of SSTR genes, all the SSTR subtypes (SSTR2, SSTR3, SSTR5) that are required for the negative cell growth signal and that bind SSA were found. It was also demonstrated that octreotide inhibits *in vivo* growth and activity of retro-orbital fibroblasts from GO patients. Activity of the GO has been shown to correlate with the uptake of somatostatin analogues in orbital scintigrams.

Somatostatin inhibits lymphocyte proliferation and production of factors involved in the orbital autoimmune process: colony-stimulating factor, inflammatory cytokines, and immunoglobulins [20]. Results of small pilot studies suggested benefit from SSA in GO; however, recently published randomized clinical trials with the use of lanreotide or octreotide have shown limited improvements. In a meta-analysis of randomized, controlled trials of treatment modalities for GO with the clinical activity score (CAS) as the primary outcome, four studies comparing long-acting release formulations of SSA (octreotide and lanreotide) with placebo were included. The treatment duration ranged from 3 to 8 months, with follow-up periods of 3–14 months. The combined results at the end of follow-up showed a minor but statistically significant lower CAS for patients treated with SSA over placebo (MD, −0.63; 95% CI, −0.98 to −0.28). There was no advantage for somatostatin analogues in other outcomes, including diplopia, proptosis, and lid aperture. Patients in the SSA group had significantly more adverse events (OR, 2.57; 95% CI, 1.09–6.05), mostly gastrointestinal (in 60% of patients), that did not result in the discontinuation of the drug. The use of novel SSA with the broader binding affinity in GO needs to be tested [21].

SSA USE IN THE TREATMENT OF HYPOGLYCEMIA

The other applications of SSA have included the treatment of hypoglycemia of different etiology. In animal model, selective antagonism of somatostatin receptor type 2 (SSTR2) normalizes glucagon and corticosterone responses to hypoglycemia and largely ameliorates insulin-induced hypoglycemia. In the same model, catecholamine responses were not affected by somatostatin [22].

SSA have been used in toxicology in the treatment of prolonged recurrent hypoglycemia after sulfonylurea overdose. The available data suggest that octreotide could be considered first-line therapy in both pediatric and adult sulfonylurea poisoning with clinical and laboratory evidence of hypoglycemia. Maintenance doses of octreotide may be required to prevent recurrent hypoglycemia [23].

Congenital hyperinsulinism (CH) is a very rare disease referring to a variety of inborn disorders in which hypoglycemia is caused by excessive insulin secretion.

Conservative management option for CH are SSA added to frequent feeding, diazoxide, and, in some cases, glucagon. Such treatment can be effective in terms of achieving normoglycemia, normal growth, and neurodevelopment. Typical side effects of SSA in the treatment of CH were in most cases transient. Octreotide was administered subcutaneously either by multiple injections (every 6–8 h) or by continuous infusion via a pump. The results of studies on the effective use of long-acting somatostatin analogues in these disorders showed improved life quality with long-acting somatostatin analogue formulation treatment when compared to continuous infusion of the drug via a pump [24].

ONCOLOGY

The frequent expression of SSTRs in various malignancies has given the promise of the new effective treatment modalities. However, with some exceptions, SSA in oncology outside of the neuroendocrine tumors have been tested in *in vitro* or animal models or very small human series. The published results are somewhat conflicting and no clinical recommendations have been issued in this field.

One of the most examined neoplasms in which those agents have been tried is hepatocellular carcinoma (HCC). Meta-analysis of nine randomized controlled in HCC showed that the 6-month and 12-month survival rates in the octreotide group were significantly higher than those of the control group (6 months, RR, 1.41; 95% CI, 1.12–1.77; $P=0.003$; 12 months, RR, 2.66; 95% CI, 1.30–5.44; $P=0.008$). When including the studies using no treatment as control or being performed in China, including more than 50 patients and with follow-up longer than 2 years, the sensitivity analyses tended to confirm the primary meta-analysis. Whereas, when including the studies using placebo as control or being performed in Western countries, the difference was not significant [25].

There were also studies on the use of SSA in breast cancer; however, octreotide depot failed to show any clinical benefit, when added to the tamoxifen in breast cancer patients [26].

The treatment of the postsurgical lymphorrhea complications after axillary node dissection in breast cancer with lanreotide was also ineffective [27]. However, the new cytotoxic SSA with SSTR2 affinity (among others, AN-162) are being tested in *in vitro* models with promising result [28].

A meta-analysis of uncontrolled studies on SSA plus dexamethasone as the treatment for castration-resistant prostate cancers revealed that such drug combination resulted in partial response in about 60% of patients, with median progression-free survival of 7 months and median overall survival of 16 months.. Other available data address rather biochemical markers of prostate cancer, with no sustained decline in PSA levels after octreotide administration [29, 30].

New cytotoxic SSA AN-162 has been tried in studies on non-small cell lung cancer cell (NSCLC) lines. AN-162 induced apoptosis-related genes expression in NSCLC cells, and it should be considered for further therapy [31, 32].

Substantial expression of SSTR in meningiomas has induced the concept of the therapeutic use of SSA in inoperable/recurrent/progressive tumors. Observational studies have indicated that although SSA are not able to induce the tumor regression, they may halt the meningioma growth and arrest tumor progression [33].

In 2011, results of phase II study of subcutaneous octreotide in patients with progressive or recurrent meningioma and meningeal hemangiopericytoma have been published. Twelve patients were treated with subcutaneous short-acting octreotide; the primary measure was radiographic response rate. Eleven of the twelve patients experienced progression, with a median time to progression of 17 weeks. Two patients experienced long progression-free intervals (30 months and ≥18 years). Octreotide failed to produce objective tumor response [34].

INFLAMMATORY DISEASES

Somatostatin is known for having immunomodulating properties. Experimental models of arthritis revealed that antinociceptive and anti-inflammatory actions of octreotide and pasireotide are largely mediated via the SSTR2 receptor [35].

However, the studies on SSA in inflammatory or autoimmune diseases are scarce. Long-acting octreotide in refractory rheumatoid arthritis was tested in a small, open-labeled pilot study. Although the results were encouraging and treatment led to clinical improvement in a subset of patients with active refractory arthritis, the placebo-controlled trials were not conducted or their results not published. In contrast, a few cases of arthritis related to the SSA treatment have been reported [36].

REFERENCES

[1] D'Amico, G.; Pietrosi, G.; Tarantino, I.; et al. Gastroenterology 2003, 124, 1277–1291.

[2] Gøtzsche, P. C.; Hrobjartsso, A. Cochrane Database of Systematic Reviews 2008, 16, CD000193.

[3] Wells, M.; Chande, N.; Adams, P.; et al. Alimentary Pharmacology and Therapeutics 2012, 35, 1267–1278.

[4] Saruç, M.; Can, M.; Küçükmetin, N.; Tuzcuoglu, I.; et al. Medical Science Monitor 2003, 9, Pl84-87.

[5] Berberat, P. O.; Friess, H.; Uhl, W. et al. Digestive Surgery 1999, 60, 15–22.

[6] Koti, R. S.; Gurusamy, K. S.; Fusai, G.; et al. HBP (Oxford) 2010, 12, 155–165.

[7] Gurusamy, K. S.; Koti, R.; Fusai, G.; et al. Cochrane Database of Systematic Reviews 2012, 13, 6.

[8] Omata, F.; Deshpande, G.; Tokuda, Y.; et al. Gastroenterology 2010, 45, 885–895.

[9] Rudin, D.; Kiss, A.; Wetz, R. V.; et al. Journal of Gastroenterology and Hepatology 2007, 22, 977–983.

[10] Katsinelos, P.; Fasoulas, K.; Paroutoglou, G.; et al. Endoscopy 2012, 44, 53–59.

[11] Yang, F.; Wu, H.; Li, Y.; et al. Pancreas 2012, 41, 1206–1212.

[12] Dahaba, A. A.; Mueller, G.; Mattiassich, G.; et al. European Journal of Pain 2009, 13, 861–864.

[13] Peeters, M.; Van den Brande, J.; Francque, S. Acta Gastro-Enterologica Belgica 2010, 73, 25–36.

[14] Szilagyi, A.; Shrier, I. Alimentary Pharmacology & Therapeutics 2001, 15, 1889–1897.

[15] Grant, M. B.; Caballero, S. Drugs of Today (Barcelona) 2002, 38, 783–791.

[16] Davis, M. I.; Wilson, S. H.; Grant, M. B. Hormone and Metabolic Research 2001, 33, 295–299.

[17] Hernaez-Ortega, M. C.; Soto-Pedre, E.; Martin, J. J. Diabetes Research and Clinical Practice 2004, 64, 71–72.

[18] Lamberts, S. W. J.; de Herder, W. W.; van der Lely, A. J.; Hofland, L. J. Endocrinologist 1995, 5, 448–451.

[19] Kokona, D.; Mastrodimou, N.; Pediaditakis, I.; et al. Experimental Eye Research 2012, 103, 90–98.

[20] Pasquali, D.; Vassallo, P.; Esposito, D.; et al. Journal of Molecular Endocrinology 2000, 25, 63–71.

[21] Stiebel-Kalish, H.; Robenshtok, E.; Hasanreisoglu, M.; et al. The Journal of Clinical Endocrinology and Metabolism 2009, 94, 2708–2716.

[22] Yue, J. T.; Riddell, M. C.; Burdett, E.; et al. Diabetes 2013, 62, 2215–2222.

[23] Glatstein, M.; Scolnik, D.; Bentur, Y. Clinical Toxicology 2012, 50, 795–804.

[24] Modan-Moses, D.; Koren, I.; Mazor-Aronovitch, K.; et al. The Journal of Clinical Endocrinology and Metabolism 2011, 96, 2312–2317.

[25] Ji, X. Q.; Ruan, X. J.; Chen, H.; et al. Medical Science Monitor 2011, 17, 169–176.

[26] Pritchard, K. I.; Shepherd, L. E.; Chapman, J. A.; et al. Journal of Clinical Oncology 2011, 10, 3869–3876.

[27] Gauthier, T.; Garuchet-Bigot, A.; Marin, B.; et al. European Journal of Surgical Oncology 2012 38, 902–909.

[28] Seitz, S.; Buchholz, S.; Schally, A. V.; et al. Anticancer Drugs 2013, 24, 150–157.

[29] Friedlander, T. W.; Weinberg, V. K.; Small, E. J.; et al. Urologic Oncology 2012, 30, 408–414.

[30] Toulis, K. A.; Goulis, D. G.; Msaouel, P.; et al. Anticancer Research 2012, 32, 3283–3289.

[31] Treszl, A.; Schally, A. V.; Seitz, S.; et al. Peptides 2009, 30, 1643–1650.

[32] Shen, H.; Hu, D.; Du, J.; et al. European Journal of Pharmacology 2008, 28, 23–29.

[33] Schulz, C.; Mathieu, R.; Kunz, U.; et al. Neurosurgical Focus 2011, 30, E11.

[34] Johnson, D. R.; Kimmel, D. W.; Burch, P. A.; et al. Neuro-Oncology 2011, 13, 530–535.

[35] Imhof, A. K.; Gluck, L.; Gajda, M.; et al. Arthritis and Rheumatism 2011, 63, 2352–2362.

[36] Paran, D.; Elkayam, O.; Mayo, A.; et al. Annals of the Rheumatic Diseases 2001, 60, 888–891.

6

PEPTIDE RECEPTOR RADIONUCLIDE THERAPY USING RADIOLABELED SOMATOSTATIN ANALOGUES: AN INTRODUCTION

JOHN BUSCOMBE

Addenbrookes Hospital, Cambridge, UK

Radionuclide therapy, sometimes called molecular radiotherapy, remains something of a minority interest in nuclear medicine. Many nuclear medicine departments do not offer radionuclide therapy or limit its use to I-131. However, the use of radiolabeled somatostatin analogues provides the possibility of a truly holistic service from imaging to treatment and reassessment. The basic principle is that if you can see it, you can treat it. Though there has been a desire to pursue such a policy with a wide range of solid tumors, good clinical results have remained elusive. There appear to be two main reasons for this. To be effective, there needs to be a high tumor-to-background ratio, which may be difficult to achieve with antibody-based products but seems to be possible with radiolabeled somatostatin analogues. The second factor is the tumor itself. The most common forms of tumor treated are represented by the neuroendocrine family of cancers. The most common of these is the carcinoid tumor, but there are many others, some producing excretory peptides and others appearing to be "silent." Most of these tumors arise from the midgut. They tend to share some characteristics. They all overexpress somatostatin receptor subtype 2, which can be visualized with a range of products including In-111 pentetreotide, and they are often

Somatostatin Analogues: From Research to Clinical Practice, First Edition. Edited by
Alicja Hubalewska-Dydejczyk, Alberto Signore, Marion de Jong, Rudi A. Dierckx,
John Buscombe, and Christophe Van de Wiele.
© 2015 John Wiley & Sons, Inc. Published 2015 by John Wiley & Sons, Inc.

slow growing. This second factor is important as it means that most chemotherapy drugs that target cell division have little impact on neuroendocrine tumors (NETs) and there is time to give multiple cycles of treatment and assess results.

THE SOIL

Neuroendocrine tumors (NETs) represent a family of tumors, which can be characterized either in terms of morphology, special immunostaining, or functionally by their expression of somatostatin receptors. NETs are associated with the gut and often have parallel names, which reflect this origin including gastroenteropancreatic (GEP) tumors. However this approach excludes some other tumor types that could also be included as NETs such as lung, thymic, renal, and ovarian NETs those these are much rarer as well as medullary cell thyroid cancer (MTC), paraganglioma, and pheochromocytoma. A further division can be related to the part of the embryonic gut from which the cells in the tumor originate. Foregut tumors, which include those of the lung and pancreas, tend to be more aggressive and have a better response to chemotherapy agents [1, 2]. The pancreatic tumors may also produce other hormones such as insulin, gastrin, and even somatostatin. A higher proportion of these pancreatic tumors do not appear to express somatostatin receptor type 2 in high concentration, and radionuclide therapy may be less useful. In addition, these pancreatic tumors can be associated with multiple endocrine neoplasia type 1 of which 50% are familial and due to a defect in a protein called menin [3]. The majority of foregut tumors probably do not excrete an endocrine substance that we can detect.

Midgut carcinoids tend to be much slower growing, resistant to chemotherapy, and often present with liver metastases. The majority are secretory and produce serotonin. Once tumor is in the liver, the patients can present with classical carcinoid syndrome. These midgut-originated tumors tend to have high expression of various somatostatin subtypes, but subtype 2 predominates. The primary method of control is biological through somatostatin analogues [4, 5]. Hindgut NETs are rare normally nonsecretory with little or no expression of somatostatin receptors.

Many other tumors especially with a gut origin can express some neuroendocrine features; however, this does not mean they are true NETs and would not normally be treated as such.

Other types of NETs such as MTC, paragangliomas, and pheochromocytomas fall outside of this classification. Recently, these tumors also have been considered for radiolabeled somatostatin analogue treatment, especially as it has been found that Ga-68-labeled products with PET are finding tumors considered negative on In-111 pentetreotide imaging [6, 7].

THE PEPTIDES

The initial proposal from Krenning's group was to use the Auger electron present in In-111 pentetreotide for therapy. This would require giving very high activities (up to 11 GBq In-111 pentetreotide 3–6 weekly). This is an expensive form of treatment, but in three published trials [8–10], there is evidence for some efficacy and minimal toxicity. The real problem is that it was difficult to give a sufficient radioactive dose to the tumor. This problem is explored in the chapter by Bodei in her explanation of the development of radiolabeled somatostatin therapy. The range of an Auger electron is 1–2 cells so to just irradiate the tumor at least one out of every two its cells has to have uptake of the In-111-pentetreotide. The quantity of this product, which needed to be administered to obtain the therapeutic results, has made it not feasible for treatment purposes.

Much better tumor kill was needed, so the Basle group came up with a DOTA linker that enabled Y-90 to be attached to the somatostatin. Early results were encouraging, but the effect was limited by renal toxicity. This was found to be due to direct cross-reaction with receptors in proximal tubules of the nephrons. It was rapidly proposed and then conformed by good science that the administration of negatively charged amino acids such as lysine and arginine in sufficient quantities was renoprotective.

By the late 1990s, there was some initial interest in a clinical trial from Novartis, which was the legacy holder of the patent for octreotide. An international multicenter phase II trial was performed but not published until 10 years after the trial completion [11].

As part of the trial, some superb dosimetric work was performed by Jamar's group using Y-86 DOTATOC PET imaging [12]. This modeled the radiation dose to the kidneys and showed the reduction that could be obtained using amino acid coinfusion.

Obtaining sufficient peptide for labeling proved problematic for many centers, but Virgolini's group from Vienna and then Innsbruck developed a Y-90 DOTA lanreotide. Lanreotide had a much less precise receptor profile and could be considered a pan-receptor agent. In a multicenter trial of 154 patients, just over half the patients had stable disease or partial response [13]. Renal toxicity was not seen, but there was sufficient bone marrow suppression to limit the activity that could be administered to about 1.3 GBq in three cycles. This appeared to be insufficient for a good therapeutic response. One novel solution was to administer the product intra-arterially; this doubled the partial response rate [14], and the same technique was also applied to treat primary brain tumors [15]. This method is explored in the chapter by Virgolini.

Soon though, additional sources of Y-90 octreotide or DOTATATE became available and have been used by a variety of centers worldwide. The biggest group of just over 1000 patients came from the Basel group that has the longest experience using this agent [16].

Concerns about toxicity especially in the kidneys where between 1 and 4% of patients may end up with significant renal damage lead to the search for an alternative isotope to Y-90 with its energetic beta-emission.

Lu-177-labeled somatostatin analogues appeared to be the answer; its lower penetrating beta and also a gamma for imaging allowed activities to be given up to 7 GBq and posttherapy imaging for dosimetry. This search for the ideal agent is documented in the chapter by van Vliet. Most of the published work originates from Rotterdam with over 500 patients treated so far. It does appear to be well tolerated with renal damage occurring in less than 1% of treated patients. Survival looks to be better than with Y-90, but twice as many patients fail treatment than fail with Y-90 [16, 17]. It is still unclear when Y-90-labeled somatostatin analogues should be given in preference to Lu-177-labeled somatostatin analogues. However, following animal work from De Jong at Rotterdam, sequential Y-90 and Lu-177 DOTATATE have been given, and early results suggest a synergistic effect [18]. Some early work had looked at the coadministration of Y-90- and Lu-177-labeled somatostatin analogues [19], and this is explored in the chapter by Baum.

SUCCESS AND FAILURE

With the reports of over 2000 patients treated with radiolabeled somatostatin analogues and now good results being obtained from the most up-to-date combinations of peptides and radionuclides, this should be a story of unqualified success [20]. This has not been the case. Though there has been commercial interest in both Lu-177 DOTATATE and Y-90 octreotide, there has been little progress toward registration. This has primarily been due to the fact that the research has been led by academia where the questions asked from any study are different from those needed for drug registration.

There have as of the date of publication been any randomized phase III trials comparing treatment with any form of radiolabeled somatostatin analogue with a placebo, alternate therapy, or best supportive care. For most drugs, this is considered necessary for registration purposes. While there has been some phase II trial data published, often, the results are not comparable. Some centers will only treat patients with evidence of tumor progression either radiologically or biochemically. In these cases, even stable disease can be considered a favorable outcome [21]. Other trial centers have also treated patients with stable disease. In this case, it is difficult to determine if the treatment had any efficacy as they may have remained with stable disease even without therapy.

What is needed is a standardized approach as to when treatment should be started, and it would not be too controversial to use the normal oncological standard of commencing treatment only when there is evidence for progression. The special nature of NETs would allow that to be biochemical as well as radiological. An example would be that treatment would be started if the patient's carcinoid syndrome was no longer being controlled by somatostatin analogues.

DOES IT WORK?

The second major issue is how to measure success. Here, there are wide variations in the reporting of different series. Some concentrate on radiological response, and others measure symptom relief. Only a few trials look at measures such as progression-free survival or overall survival, and even fewer use methods such as Kaplan–Meier statistics. This failure to record basic information concerning tumor response and survival makes it very difficult to produce a reasoned meta-analysis and Forest plots, which could be used as a substitute for a randomized controlled trial.

The main standard used in oncological trials to determine response is to use radiological criteria that define a successful treatment as either complete radiological disappearance of the tumor or a reduction of 50% in the tumor mass volume. All the criteria for the WHO to RECIST1.1 are based on this principle. However, again, many authors do not use these formal criteria when measuring response to radiopeptide therapy. Some authors have applied their own criteria such as "minimal response" [17], but without some consensus as to what this means, which is widely accepted, it is of little value. Timing has also been shown to be vital with reports that a year may need to pass between the past treatment and any meaningful tumor size reduction [21]. Therefore, other markers of response are needed. It would appear that the strongest of these is biochemical and symptomatic response, which has been shown to influence survival [21]. Again, measuring symptoms should use a validated system such as the EORTC QLQC30 for NETs [22]. Therefore, for many studies, the only meaningful results that can be compared to alternate treatments such as biological and chemotherapy regimens are median progression-free and overall survival. The provision of this data from different trials has been better but is often incomplete. For example, in one reported series of 500 patients, follow-up data and therefore good quality survival data were only available in 60% of patients [17]. Did the remaining 40% feel so well that they did not come back for follow-up or were they dead? If either was the case, it would have a profound effect on the results as published. What most studies do agree on is that if the NET patient feels worse after treatment or has radiological progression of tumor, then the outcome for them is bleak measured in only a few months of life [17, 21].

DOSIMETRY

Much of the published work on radiolabeled somatostatin analogues focuses on the clinical response. Activities given have been related to pragmatic issues such as cost and availability. It was quickly found that the primary critical organ was the kidney [23]. Understanding of the biodistribution and pharmacokinetics of these radiopeptides meant that it would be possible to reduce the possibility of toxicity with Y-90-labeled products if a renal radiation dose of less than 30 Gy could be achieved by use of amino acid coinfusion and keeping the accumulated activity to less than 15 GBq [24]. Much of this work was performed using Y-86 as a PET tracer [12]. However, the move to treatment with Lu-177 DOTATATE allows for posttherapeutic imaging. This has resulted in regimes that use the dosimetry for one treatment to modify subsequent therapies and attempt to produce a personalized medicine solution to tumor treatment [25].

WAYS FORWARD?

Much work has been done and thousands of patients treated, but radiolabeled somatostatin analogues remain outside of the mainstream oncological treatments. Some of this is structural as oncologists may not wish to refer patients for treatments outside of their control but also some of this is because of the way we have reported our work. While understanding that the reason trials have been reported the way they have may have been for good pragmatic reasons, it is important to understand that if radiolabeled somatostatin analaogues are to have a sustained future in the treatment of patients with disseminated NETs, we must present results in a way that is understandable to oncologists. This means: (a) the necessity of the development of randomized controlled, preferably multicentre, trials, (b) the need to deliver results using the standard terms such as partial and complete response for radiological assessment of tumors, (c) the use of Kaplan–Meier statistics to calculate survivals and also to include data on quality of life measured in a way that has been accepted.

There are other tumors that express somatostatin including medullary cell cancers of the thyroid, paragangliomas, malignant pheochromocytomas, and neuroblastomas [26, 27]. However, they tend to be rarer tumors. The real challenge will be the possibility of application of radiopeptides against the somatostatin receptor in non-small cell lung cancer and primary brain tumors which are much more common.

CONCLUSIONS

Nuclear medicine has come a long way in the development of radiolabeled somatostatin analogues showing that best results come from both optimal combinations of therapeutic isotope such as Lu-177 and Y-90 and optimized peptides such as DOTATATE. Wide acceptance of this work means that investigators must be disciplined in their approach both to methodology and reporting results so that the studies allow easier comparison to other forms of radionuclide therapy.

REFERENCES

[1] Caplin, M. E.; Buscombe, J. R.; Hilson, A. J.; Jones, A. L.; Watkinson, A. F.; Burroughs, A. K. Lancet 1998, 352, 799–780.

[2] Desai, K. K.; Khan, M. S.; Toumpanakis, C.; Caplin, M. E. Minerva Gastroenterologica e Dietologica 2009, 55, 425–443.

[3] Mayr, B.; Apenberg, S.; Rothämel, T.; von zur Mühlen, A.; Brabant, G. European Journal of Endocrinology 1997, 137, 684–687.

[4] Toumpanakis, C.; Garland, J.; Marelli, L.; et al. Alimentary Pharmacology & Therapy 2009, 30, 733–740.

[5] Wymenga, A. N.; Eriksson, B.; Salmela, P. I.; et al. Journal of Clinical Oncology 1999, 17, 1111–1117.

[6] Srirajaskanthan, R.; Kayani, I.; Quigley, A. M.; Soh, J.; Caplin, M. E.; Bomanji, J. Journal of Nuclear Medicine 2010, 51, 875–882.

[7] Conry, B. G.; Papathanasiou, N. D.; Prakash, V.; et al. European Journal of Nuclear Medicine and Molecular Imaging 2010, 37, 49–57.

[8] Valkema, R.; de Jong, M.; Bakker, W. H.; Breeman, W. A. P.; Kooij, P. P. M.; Lugtenburg, P. J. Seminars in Nuclear Medicine 2002, 32, 110–122.

[9] Anthony, L. B.; Woltering, E. A.; Espenan, G. D.; Cronin, M. D.; Maloney, T. J.; McCarthy, K. E. Seminars in Nuclear Medicine 2002, 32, 123–132.

[10] Buscombe, J. R.; Caplin, M. E.; Hilson, A. J. W. Journal of Nuclear Medicine 2003, 44, 1–6.

[11] Bushnell, D. L. Jr.; O'Dorisio, T. M.; O'Dorisio, M. S.; et al. Journal of Clinical Oncology 2010, 28, 1652–1629.

[12] Jamar, F.; Barone, R.; Mathieu, I.; et al. European Journal of Nuclear Medicine and Molecular Imaging 2003, 30, 510–518.

[13] Virgolini, I.; Britton, K.; Buscombe, J.; Moncayo, R.; Paganelli, G.; Riva. Seminars in Nuclear Medicine 2002, 32, 148–155.

[14] McStay, M. K.; Maudgil, D.; Williams, M.; et al. Radiology 2005, 237, 718–722.

[15] Turner, N.; Amlot, P.; Platts, A.; Jones, A.; Buscombe, J. Lancet Oncology 2004, 5, 193–194.

[16] Imhof, A.; Brunner, P.; Marincek, N.; et al. Journal of Clinical Oncology 2011, 29, 2416–2423.

[17] Kwekkeboom, D. J.; de Herder, W. W.; Kam, B. L.; et al. Journal of Clinical Oncology 2008, 26, 2124–2130.

[18] De Jong, M.; Breeman, W. A.; Valkema, R.; Bernard, B. F.; Krenning, E. P. Journal of Nuclear Medicine 2005, 46(Suppl 1), 13S–17S.

[19] Kunikowska, J.; Królicki, L.; Hubalewska-Dydejczyk, A.; Mikołajczak, R.; Sowa-Staszczak, A.; Pawlak, D. European Journal of Nuclear Medicine and Molecular Imaging 2011, 38, 1788–1797.

[20] Buscombe, J.; Navalkissoor, S. Clinical Medicine 2012, 12, 381–386.

[21] Cwikla, J. B.; Sankowski, A.; Seklecka, N.; et al. Annals of Oncology 2010, 21, 787–794.

[22] Scott, N. W.; Fayers, P. M.; Aaronson, N. K.; et al. Quality of Life Research 2009, 18, 381–388.

[23] Bodei, L.; Cremonesi, M.; Zoboli, S.; et al. European Journal of Nuclear Medicine and Molecular Imaging 2003, 30, 207–216.

[24] Rolleman, E. J.; Valkema, R.; de Jong, M.; Kooij, P. P.; Krenning, E. P. European Journal of Nuclear Medicine and Molecular Imaging 2003, 30, 9–15.

[25] Wehrmann, C.; Senftleben, S.; Zachert, C.; Müller, D.; Baum, R. P. Cancer Biotherapy & Radiopharmaceuticals 2007, 22, 406–416.

[26] van Essen, M.; Krenning, E. P.; Kooij, P. P.; et al. Journal of Nuclear Medicine 2006, 47, 1599–1606.

[27] Gains, J. E.; Bomanji, J. B.; Fersht, N. L.; et al. Journal of Nuclear Medicine 2011, 52, 1041–1047.

6.1

SOMATOSTATIN ANALOGUES AND RADIONUCLIDES USED IN THERAPY

ESTHER I. VAN VLIET, BOEN L.R. KAM, JAAP J.M. TEUNISSEN, MARION DE JONG, ERIC P. KRENNING, AND DIK J. KWEKKEBOOM

Department of Nuclear Medicine and Radiology, Erasmus MC, University Medical Center, Rotterdam, The Netherlands

ABBREVIATIONS

cAMP	cyclic adenosine monophosphate
DOTA	1,4,7,10-tetraazacyclododecane-1,4,7,10-tetraacetic acid
DOTANOC	[DOTA^0-1-Nal^3]octreotide
DOTATATE	[DOTA^0,Tyr^3]octreotate
DOTATOC	[DOTA^0,Tyr^3]octreotide
DTPA	diethylene triamine pentaacetic acid
GEPNET	gastroenteropancreatic tumor
PRRT	peptide receptor radionuclide therapy
sst	somatostatin
TETA	1,4,8,11-tetraazacyclotetradecane-N,N',N'',N'''-tetraacetic acid

Somatostatin Analogues: From Research to Clinical Practice, First Edition. Edited by Alicja Hubalewska-Dydejczyk, Alberto Signore, Marion de Jong, Rudi A. Dierckx, John Buscombe, and Christophe Van de Wiele.
© 2015 John Wiley & Sons, Inc. Published 2015 by John Wiley & Sons, Inc.

INTRODUCTION

Gastroenteropancreatic neuroendocrine tumors (GEPNETs), which comprise functioning and nonfunctioning endocrine pancreatic tumors and carcinoids, are usually slow growing and are often metastasized at diagnosis. Treatment with somatostatin analogues such as octreotide and lanreotide can reduce hormonal overproduction and results in symptomatic relief in most patients with metastasized disease. Objective responses however in terms of tumor size reduction are seldom achieved [1–3]. However, it was recently shown that the long-acting somatostatin analogue octreotide LAR (Sandostatin LAR®; Novartis, Basel, Switzerland) significantly lengthens time to tumor progression when compared with placebo in patients with functionally active and inactive metastatic midgut neuroendocrine tumors (NETs) [4].

The majority of GEPNETs abundantly express somatostatin receptors (ssts), and these tumors can be visualized in patients using the radiolabeled somatostatin analogue [¹¹¹Indium-DTPA⁰]octreotide (OctreoScan®; Covidien, Petten, the Netherlands). A logical next step to this tumor visualization *in vivo* was to also try to treat these patients with radiolabeled somatostatin analogues.

SOMATOSTATIN ANALOGUES AND CHELATORS

Radiolabeled somatostatin analogues that are used for therapy consist of three parts: a cyclic octapeptide, a chelator, and a radionuclide. Combinations of different peptides, chelators, and radionuclides have been tested *in vitro* and/or *in vivo* for their use in peptide receptor radionuclide therapy (PRRT).

Peptides that have been used are octreotide, [Tyr³]octreotide, octreotate, [Tyr³]octreotate, lanreotide, and vapreotide. In [Tyr³]octreotide, phenylalanine (Phe) in position 3 in the amino acid sequence is being replaced by tyrosine (Tyr), resulting in a more hydrophilic peptide. In octreotate, threoninol (Thr(ol)) at the C-terminus (as used in octreotide) is replaced by the natural amino acid threonine (Thr).

As chelators, diethylene triamine pentaacetic acid (DTPA), 1,4,7,10-tetraazacyclododecane-1,4,7,10-tetraacetic acid (DOTA), and 1,4,8,11-tetraazacyclotetradecane-*N*,*N'*,*N''*,*N'''*-tetraacetic acid (TETA) can, among others, be used. Figure 6.1.1 shows the chemical structures of these chelators.

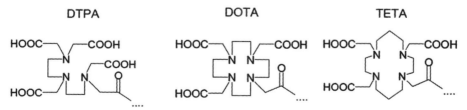

FIGURE 6.1.1 Chemical structures of DTPA, DOTA, and TETA (Note: … is the site for binding a peptide).

The most used combinations of peptide–chelator complexes are [DOTA^0,Tyr^3] octreotide (DOTATOC) and [DOTA^0,Tyr^3]octreotate (DOTATATE). DTPA-octreotide (as used in OctreoScan® when it is coupled to ^{111}Indium (^{111}In) ([^{111}In-DTPA^0] octreotide)) was used when PRRT started in the 1990s. Another peptide–chelator complex used is [DOTA^0-1-NaI^3]octreotide (DOTANOC), in which the third amino acid of octreotide (Phe) is replaced by (1-naphthyl)-alanine. DOTANOC has a good affinity for sst$_2$ (somatostatin receptor subtype 2) but also a high affinity for sst$_3$ and sst$_5$ as opposed to DOTATOC and DOTATATE [5]. This may be important for imaging and treatment of tumors with less pronounced expression of sst$_2$ and more of sst$_3$ and sst$_5$. A preliminary report [6] stated that scintigraphy with ^{111}In-DOTANOC showed higher tumor uptake compared to [^{111}In-DTPA^0]octreotide (^{111}In-octreotide) in most of the patients. Unfortunately, uptake in the background was also higher, which resulted in lower tumor-to-background ratios: in only 2 of 31 patients (1 with follicular thyroid carcinoma and 1 with hepatocellular carcinoma), tumor-to-background ratio was higher with ^{111}In-DOTANOC. Figure 6.1.2 shows the chemical structures of DTPA-octreotide, DOTATOC, DOTATATE, and DOTANOC.

Changing the peptide, chelator, or radionuclide can considerably affect the binding affinities for the somatostatin receptor, as has been shown by Reubi et al. [7]. These authors tested various combinations of peptides, chelators, and radiometals for their binding affinities for the five different somatostatin receptor subtypes. Table 6.1.1 gives an overview of the affinity profiles of various somatostatin analogues. Most importantly, Reubi et al. demonstrated what effect even apparent small changes in, for example, peptide structure have on binding affinities for the various ssts.

The replacement of Phe by the more hydrophilic Tyr at position 3 in [Yttrium-DOTA,Tyr^3]octreotide or in [Gallium-DOTA,Tyr^3]octreotide resulted in an improved binding affinity to the sst$_2$ receptor. The more lipophilic Phe showed improved binding affinity to sst$_3$ when compared to Tyr^3. Addition of a radiometal (Y or Ga) to [DOTA,Tyr^3]octreotide resulted in a better sst$_2$ binding affinity. A possible explanation for this could be the difference in charge of the molecule when adding a radiometal. Furthermore, addition of Ga to [DOTA,Tyr^3]octreotate resulted in an eightfold improvement in sst$_2$ binding affinity when compared with [Y-DOTA,Tyr^3]octreotate. The same held true when these radiometals were labeled to octreotide, with a fivefold improvement in sst$_2$ binding affinity with labeled [DOTA,Tyr^3]octreotide when Y was replaced by Ga. Substitution of a chelator, DTPA, by DOTA, coupled to the octreotide molecule, improved the sst$_3$ binding affinity 14-fold. Furthermore, substitution of DTPA by DOTA when coupled to [Tyr^3]octreotate improved binding to the sst$_2$. [Y-DOTA]lanreotide and [Y-DOTA]vapreotide showed higher binding affinity to the sst$_5$ compared to [Y-DOTA,Tyr^3]octreotide. Since the hydrophilicity of DOTA-vapreotide and DOTA-lanreotide is less than that of [DOTA,Tyr^3]octreotide, it seemed that lower hydrophilicity improves the affinity for sst$_5$. [In-DTPA,Tyr^3] octreotate and [Y-DOTA,Tyr^3]octreotate showed better binding affinity for sst$_2$ than [In-DTPA]octreotide and [Y-DOTA,Tyr^3]octreotide, respectively.

In conclusion, Reubi et al. demonstrated that the affinity for the five different somatostatin receptors is influenced mostly by changes in the peptide and, to a lesser degree, by metal replacement or chelator substitution. These differences in affinity

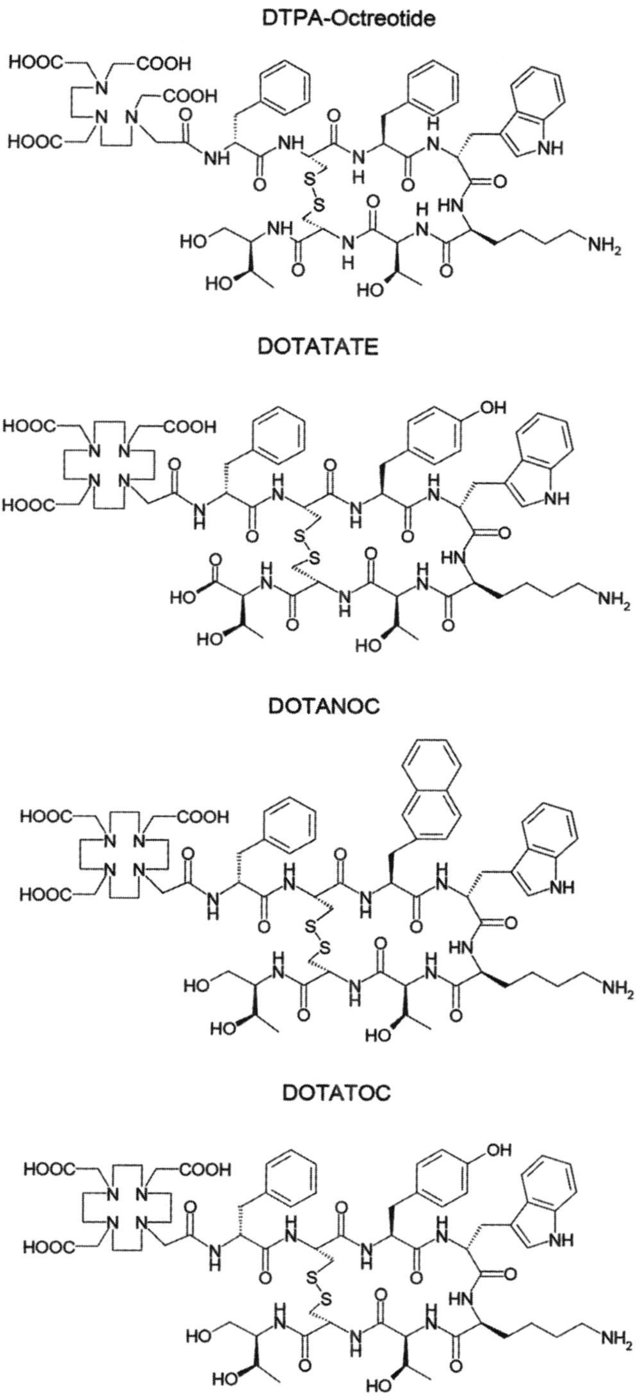

FIGURE 6.1.2 Structures of DTPA-octreotide, DOTATOC, DOTATATE, and DOTANOC.

TABLE 6.1.1   Affinity profiles (IC50) for human sst_1–sst_5 receptors of a series of somatostatin analogues[a]

Peptide	sst_1	sst_2	sst_3	sst_4	sst_5
Somatostatin-28	5.2 (0.3)	2.7 (0.3)	7.7 (0.9)	5.6 (0.4)	4.0 (0.3)
Octreotide	>10,000	2.0 (0.7)	187 (55)	>1,000	22 (6)
DTPA-octreotide	>10,000	12 (2)	376 (84)	>1,000	299 (50)
In-DTPA-octreotide	>10,000	22 (3.6)	182 (13)	>1,000	237 (52)
In-DTPA-[Tyr³]octreotate	>10,000	1.3 (0.2)	>10,000	433 (16)	>1000
DOTA-[Tyr³]octreotide	>10,000	14 (2.6)	880 (324)	>1,000	393 (84)
DOTA-[Tyr³]octreotate	>10,000	1.5 (0.4)	>1000	453 (176)	547 (160)
DOTA-lanreotide	>10,000	26 (3.4)	771 (229)	>10,000	73 (12)
Y-DOTA-[Tyr³]octreotide	>10,000	11 (1.7)	389 (135)	>10,000	114 (29)
Y-DOTA-[Tyr³]octreotate	>10,000	1.6 (0.4)	>1000	523 (239)	187 (50)
Y-DOTA-lanreotide	>10,000	23 (5)	290 (105)	>10,000	16 (3.4)
Y-DOTA-vapreotide	>10,000	12 (2)	102 (25)	778 (225)	20 (2.3)
Ga-DOTA-[Tyr³]octreotide	>10,000	2.5 (0.5)	613 (140)	>1,000	73 (21)
Ga-DOTA-[Tyr³]octreotate	>10,000	0.2 (0.04)	>1000	300 (140)	377 (18)

[a]Adapted from Reubi et al. [7].
All values are half maximal inhibitory concentration (IC50) (SEM) in nM. No data are available for Lu-loaded somatostatin analogues.

for the somatostatin receptor are reflected by the differences in tumor uptake when using either octreotide- or octreotate-based radiolabeled somatostatin analogues for the treatment of somatostatin receptor-positive tumors. Various preclinical and clinical studies demonstrated an advantage of the use of [Tyr³]octreotate in comparison with [Tyr³]octreotide when labeled with various radiometals in the amount of tumor uptake and antitumor activity in animals and also in man [8–10].

Pansomatostatins

Newly developed peptides are the so-called pansomatostatin agonists. These are somatostatin agonists characterized by the ability to bind to all five somatostatin receptors. Reubi et al. [11] described the binding affinity of KE108, a somatostatin agonist that showed high affinity for all five somatostatin receptors. They demonstrated that KE108, as well as SS-28 (the 28-amino-acid form of somatostatin, one of the two natural forms of somatostatin), inhibits forskolin-stimulated cAMP accumulation, hence showing its agonistic properties, in all five receptor subtypes. Another multireceptor ligand somatostatin analogue is pasireotide (SOM230) [12], which has high binding affinities for sst_1, sst_2, sst_3, and sst_5. The use of such analogues in combination with radiometals via a suitable chelator is of interest for both diagnostic and therapeutic purposes. However, these combinations have to be tested to establish their *in vivo* stability, biodistribution, and affinity profiles, as well as their pharmacokinetic and toxicological properties, before clinical trials with these compounds can be initiated. Another study [13] investigated the use of a chelated pansomatostatin analogue, KE88, coupled to, among others, [111]In. It showed a high internalization rate of [^{111}In-DOTA]KE88 in sst$_3$-expressing cells, whereas internalization into sst$_2$- and sst$_5$-expressing cells was low to absent. The same was demonstrated in an animal model in which [^{111}In-DOTA]KE88 had a high and persistent uptake in sst$_3$-expressing tumors but a low uptake in sst$_2$-expressing tumors. Since most NETs overexpress the sst$_2$, the role of KE88 in tumor targeting and treatment of NETs is therefore not promising.

However, it was demonstrated that NETs may express different somatostatin receptors concomitantly [14, 15]; therefore, the use of pansomatostatins may be of importance for those tumors for optimal tumor targeting.

Recently, Cescato et al. [16] demonstrated that binding of KE108 and SOM230 to the sst$_{2a}$ receptor had different effects than native somatostatin (SS-14 (the 14-amino-acid form of somatostatin)) on its postreceptor signaling. Together with SS-14, both analogues were agonists for the inhibition of adenylyl cyclase but antagonized SS-14 actions on intracellular calcium and extracellular signal-regulated kinase (ERK) phosphorylation. Hence, the pansomatostatin agonists KE108 and SOM230 cannot be considered as simple mimics of somatostatin. Since NETs mainly express sst$_2$ [15], these observations are of utmost importance for the possible use of these pansomatostatins in the treatment of NETs. Furthermore, the pansomatostatin agonist SOM230 was not able to trigger sst$_2$ internalization *in vitro* or *in vivo* [17]. Internalization of the receptor–ligand complex is one of the crucial steps for success of tumor cell targeting by radionuclides, as is discussed in the following.

Somatostatin Antagonists

All studies on PRRT in patients so far have been performed with somatostatin receptor agonists, because agonists are internalized in the (tumor) cells and are therefore relatively long retained within the tumor. Somatostatin receptor antagonists however are not internalized in the (tumor) cells and were therefore thought to be inappropriate for PRRT. Recently, however, Ginj et al. [18] demonstrated almost twice as high tumor retention of the radiolabeled sst$_2$ antagonist [^{111}In-DOTA]sst$_2$-ANT during the first 24 h compared to the agonist [^{111}In-DTPA]octreotate and this despite a lower affinity for the sst$_2$ of the antagonist. This is thought to be caused by binding of the antagonist to a larger variety of receptor conformations. These results were unexpected and are very promising in the attempt to increase the radiation dose to tumors. Unfortunately, no data have been published for the amount of activity beyond 24 h after injection, which is a very important factor as well for estimating the total amount of radiation to the tumor. Also, no data have been published yet on toxicity and biodistribution of (radiolabeled) somatostatin antagonists in man. Another study [19] demonstrated that various sst$_2$-selective somatostatin antagonists showed higher binding affinity for the sst$_2$ receptor when compared to sst$_2$-selective somatostatin agonists. All of these somatostatin antagonists were coupled to DOTA as a chelator, which enables their radiolabeling with ^{111}In, ^{90}Yttrium (^{90}Y), or ^{177}Lutetium (^{177}Lu), hence making them potential candidates for tumor targeting.

The use of the [^{64}Cu-CB-TE2A]sst$_2$ antagonist as a somatostatin antagonist for positron emission tomography (PET) imaging was tested by Wadas et al. [20]. This study demonstrated that although the affinity of this antagonist was less than that of the agonist [^{64}Cu-CB-TE2A-Tyr^3]octreotate for the sst$_2$ receptor, the antagonist had superior tumor-to-background ratio, making it an interesting radionuclide for imaging. However, for therapy, the absolute amount of uptake in the tumor is of great importance, and this is not necessarily reflected by a better tumor-to-background ratio.

RADIONUCLIDES USED IN PRRT

Various radionuclides can be used for PRRT. Of course, these radionuclides must be coupled to the peptide in a stable manner, necessitating the use of chelators, because many radionuclides are so-called bone seekers, implying potential grave bone marrow toxicity if the radionuclide is present in an unbound state *in vivo*.

^{111}In, ^{90}Y, and ^{177}Lu are radionuclides that have been used in PRRT.

^{111}In emits Auger electrons with a maximum tissue penetration range of 10 μm. In addition, it has γ-emission at 171 and 245 keV. Its half-life is 2.8 days. Due to its very short tissue penetration range, the presence of Auger electrons close to the DNA is necessary to cause DNA breakage and cellular death, emphasizing the need of sufficient internalization for successful treatment when using ^{111}In.

^{90}Y is a pure, high-energy β-emitter, with a maximum energy of 2.27 MeV. Its maximal tissue penetration is 12 mm and its half-life is 2.7 days. These characteristics make ^{90}Y more suitable for a tumoricidal effect as such. Also, because of the larger tissue penetration range, it may be more suited for larger tumors and for known heterogeneous receptor distribution within the tumor. The latter proposed effect is only possible due to the "cross-fire" effect, that is, the tumoricidal effect of indirect irradiation of tumor cells without any sst expression via radiation originating from receptor-positive cells on neighboring tumor cells. It is very likely that the "cross-fire" effect is becoming more important with increasing tissue penetration range of the used radionuclide. A disadvantage of ^{90}Y, since it is a pure β-emitter, is that no posttreatment imaging and thus no direct dosimetric calculations can be done. For dosimetry, analogues labeled with the positron emitter ^{86}Y are needed. However, this may still be questioned because of the shorter half-life (14.8 h) of ^{86}Y in comparison to ^{90}Y (64 h). Alternatively, the use of [^{111}In-DOTA^0,Tyr^3]octreotide has been advocated as a surrogate for [^{90}Y-DOTA^0,Tyr^3]octreotide for dosimetry. However, as described earlier, Reubi et al. [7] have demonstrated that replacement of the metal element (e.g., In for Y or Ga) influences the receptor affinity of these compounds, even when the peptide–chelator complex is unchanged. Therefore, the assumption that the biodistribution of the In- and Y-labeled peptide analogues is similar or comparable has to be proven *in vivo*, before the use of In-labeled substitutes can be accepted for clinical dosimetry studies.

^{177}Lu is a medium energy β-emitter, with a maximum energy of 0.5 MeV and with a maximal tissue penetration of 2 mm. Its half-life is 6.7 days. ^{177}Lu also emits low-energy γ-rays at 208 and 113 keV with 10 and 6% abundance, respectively, allowing scintigraphy and subsequent dosimetry using the same therapeutic compound. The smaller particle range of ^{177}Lu can in theory be advantageous for the use in small tumors, because less of the radiation is lost to surrounding tissue, if compared with ^{90}Y.

The combination of ^{177}Lu and ^{90}Y for therapy has been tested by de Jong et al. [21]. Rats bearing both small and large tumors were treated with somatostatin analogues labeled with ^{90}Y, ^{177}Lu, or the combination of these two compounds. It was demonstrated that animals treated with the combination of ^{90}Y and ^{177}Lu had a significantly longer median survival compared to rats treated either with ^{90}Y or with ^{177}Lu alone. The authors postulated that ^{177}Lu is more suitable for PRRT in small tumors, whereas PRRT with ^{90}Y is more suitable for larger tumors and for tumors with heterogeneous receptor distribution. However, the best antitumor results were achieved with a combination of ^{90}Y and ^{177}Lu radiolabeled somatostatin analogues. The tumor in this rat model is rapidly growing, which may cause a more heterogeneous receptor distribution because of tumor necrosis associated with rapid tumor growth. In contrast, NETs in man in general have a homogeneous receptor distribution and grow slowly, hence making it difficult to extrapolate these findings directly to NETs in man.

A preliminary report [22] stated that patients treated with the combination of [^{90}Y-DOTA^0,Tyr^3]octreotate and [^{177}Lu-DOTA^0,Tyr^3]octreotate had a longer median time of survival than patients treated with [^{90}Y-DOTA^0,Tyr^3]octreotate only. However, group sizes were small (only 16 patients in each group), so the results are very likely influenced by an inclusion bias and confirmatory results are lacking. Ideally, a trial with patients randomized to one of these compounds or the combination should be performed to establish the best treatment option for GEPNET patients.

That different radionuclides may be more suitable for tumors of different sizes is also supported by a mathematical model evaluating tumor curability in relation to tumor size, using 22 different β-emitting radionuclides [23], where it was shown that the optimal tumor diameter for tumor curability calculated for ^{90}Y was 34 mm, whereas it was 2 mm for ^{177}Lu.

That tumor response is indeed dependent of tumor size has been shown in a rat somatostatin receptor-positive pancreatic model [24]. Animals were treated with 370 MBq [^{90}Y-DOTA^0,Tyr^3]octreotide. In animals bearing medium-sized tumors (3–9 cm^2 (mean, 7.8 cm^2)), 100% complete response was found, whereas in animals bearing small tumors (≤1 cm^2) or large tumors (≥14 cm^2), the amount of complete response was 50 and 0%, respectively. The authors postulated that in tumors ≤1 cm^2, the ^{90}Y radiation energy will not be completely absorbed in the tumor, whereas in tumors ≥14 cm^2, the failure to reach a cure may be explained by the increased number of clonogenic and probably hypoxic tumor cells.

Another radionuclide that is under investigation for its use in PRRT is ^{64}Copper (^{64}Cu). ^{64}Cu has a half-life of 12.8 h. It emits a 0.58 MeV β$^-$ particle (40%), a 0.66 MeV β$^+$ particle (19%), and a γ photon of 1.34 MeV (0.5%), yielding a mean range of penetrating radiation of 1.4 mm in tissue. A preclinical study with [^{64}Cu-TETA-Tyr^3] octreotate showed a favorable high capacity for binding to somatostatin receptors both *in vivo* and *in vitro* [25]. Furthermore, in two animal tumor models used in this study, [^{64}Cu-TETA-Tyr^3]octreotate showed a twofold increase in tumor uptake compared to [^{64}Cu-TETA]octreotide. However, [^{64}Cu-TETA-Tyr^3]octreotate showed a particularly high uptake in the bone. This is less attractive, because the bone marrow is considered to be a dose-limiting organ for PRRT.

Lewis et al. [26] demonstrated that therapy of rats bearing somatostatin receptor-positive CA20948 rat pancreatic tumors with three doses of 20 mCi (740 MBq) of [^{64}Cu-TETA-Tyr^3]octreotate resulted in complete regression of the tumors, with no palpable tumors for 10 days. Furthermore, the mean survival time of these rats was nearly twice that of controls. Toxicity only comprised transient changes in blood and liver chemistry. Absorbed dose estimates showed the kidneys to be the dose-limiting organ.

Also, high linear energy transfer (LET) α-particle emitters have been evaluated for their use in PRRT [27, 28], but the use of these compounds is seriously hampered by major toxicity, especially to the kidneys.

PREREQUISITES FOR THE DESIGN OF RADIOLABELED PEPTIDES

When designing new radiolabeled peptides for therapy, biological *in vivo* stability, strong target (receptor) affinity, target (receptor) specificity, as well as pharmacokinetic characteristics play an important role. For example, the biological stability of native somatostatin (SS-14) is poor due to rapid enzymatic degradation within the range of minutes after intravenous administration. The somatostatin analogue octreotide that contains eight amino acids is more resistant to this enzymatic degradation due to various modifications including the introduction of D-amino acids and shortening of the molecule to the bioactive core sequence.

Internalization

It is important to emphasize that not only binding of the radiolabeled somatostatin analogue to the receptor is of importance for successful targeting of tumor cells by somatostatin receptor-mediated radionuclide therapy but also internalization of the receptor–ligand complex into the cell interior. Furthermore, for successful targeting of tumor cells by somatostatin receptor-mediated radionuclide therapy, the amount of radioactivity concentrated within tumor cells should be sufficient, which, besides high receptor density on the tumor cell membrane, is determined by the rate of internalization of the radioligand and by intracellular retention of the radionuclide. Internalization is especially necessary for radionuclides with a short tissue penetration range like ^{111}In. For radionuclides with a longer tissue penetration range, like ^{90}Y and ^{177}Lu, internalization as such is probably less important for therapy considering the "cross-fire" effect. Furthermore, when tumor uptake is high, effective tumor therapy would be still possible despite lack of internalization, provided that the residence time at the receptor is sufficient, as has been suggested by Ginj et al. in their study using radiolabeled somatostatin receptor antagonists [18].

Other factors with influence on the internalization efficacy were studied by de Jong et al. [29]. In this study, it was demonstrated that the internalization *in vitro* of [^{111}In-DTPA^0]octreotide and of [^{90}Y-DOTA^0,Tyr^3]octreotide in sst$_2$-positive rat pancreatic tumor cells is a highly specific and temperature-dependent process. Internalization was strongly reduced at 6°C when compared to incubation at 37°C. Internalization was also strongly reduced in the presence of 10 μM unlabeled octreotide, indicating that this process is highly receptor specific. Furthermore, the internalized fraction of [^{90}Y-DOTA^0,Tyr^3]octreotide was greater than that of [^{111}In-DTPA^0]octreotide and [^{111}In-DOTA^0,Tyr^3]octreotide.

Lukinius et al. [30] and Janson et al. [31] have shown *in vivo* that, after a diagnostic dose of [^{111}In-DTPA-D-Phe^1]octreotide, the somatostatin receptor–ligand complex is internalized through receptor-mediated endocytosis to endosomes and further into the cell interior. A total of eight patients with midgut carcinoids were injected with 160–200 MBq [^{111}In-DTPA-D-Phe^1]octreotide 2 days before abdominal surgery and during surgery samples of primary and metastatic tumor tissues were collected. The radioactive labeled octreotide was demonstrated by silver precipitation in the cell interior in primary tumors as well as in metastases. In the case report described by Lukinius et al., which comprised one patient, the amount of silver grains (indicating internalization) in the primary tumor was more than that in the mesenteric metastasis, possibly indicative of expression of other subtypes of somatostatin receptors in the primary tumor when compared to metastases, since it is known that the somatostatin receptor is internalized differently according to the sst subtype of the receptor [32], or due to differences in vascularization of the primary tumor compared to the mesenteric metastasis. However, in the seven patients described by Janson et al., similar uptake and intracellular processing of ^{111}In-octreotide in primary tumors and metastases was found.

Cescato et al. [33] demonstrated that various unlabeled somatostatin analogues with a high affinity for sst$_2$ induced internalization of sst$_2$ and the same held true for unlabeled somatostatin analogues with high sst$_3$ affinity. In contrast, not all unlabeled somatostatin agonists with a high affinity for sst$_5$ were able to stimulate sst$_5$ receptor endocytosis. The authors postulate that this could be due to the different cellular distribution characteristics of sst$_5$ in comparison with that of sst$_2$ and sst$_3$. They found that even in cells that are not exposed to agonists, a distinct intracellular perinuclear staining of sst$_5$ in addition to cell-surface sst$_5$ was observed. Another group [34] showed in COS-7 cells exogenously expressing rat sst$_5$ that functional sst$_5$ is maintained at the cell surface even in the presence of somatostatin, both because of the rapid recycling of the internalized receptor to the cell surface and because of massive recruitment of sst$_5$ to the cell surface from an intracellular sst$_5$ reserve pool, emphasizing the different features of sst$_5$ when compared to the other somatostatin receptors.

Specific Activity

When using radiolabeled somatostatin analogues for PRRT, high specific activities are required. At a high specific activity, less unlabeled peptide is present, and therefore, there will be less competition between labeled and unlabeled peptides for the receptor and a higher percentage of the dose of radioactivity can be internalized. It is highly plausible that with a higher specific activity, a higher tumoricidal effect can be achieved, as has been shown *in vitro* by Capello et al. [8]. Another study [35] demonstrated that in rats bearing a pancreatic somatostatin receptor-positive tumor, the uptake in the tumor was highest at 0.5 μg injected peptide (DOTATOC). With increasing amounts of peptide, tumor uptake decreased. This could be the result of a positive effect of increasing amounts of peptide on, for example, receptor clustering and a negative effect of receptor saturation when increasing this dose further.

In conclusion, for the design of radiolabeled peptides for therapy, attention must be paid to biological stability, receptor affinity, receptor specificity, and pharmacokinetic characteristics. Internalization plays a pivotal role for therapy with radiometals with a short tissue penetration range (like ^{111}In), whereas internalization as such is not mandatory for therapy with radiometals with a longer tissue penetration range (like ^{90}Y and ^{177}Lu). However, it remains unclear if compounds that do not internalize have a sufficient residence time on the receptor to ensure a radiation dose to the tumor high enough to be tumoricidal. Furthermore, for PRRT, high specific activities are required. Further animal studies are warranted to elucidate these interesting research questions.

CLINICAL STUDIES

Table 6.1.2 gives an overview of tumor responses in patients with GEPNETs treated with different radiolabeled somatostatin analogues. Chapter 6.3 by L. Bodei et al. gives a more detailed overview about the therapeutic use of radiolabeled somatostatin analogues in patients.

Figure 6.1.3 shows an example of a patient with an insulinoma with liver metastases treated with 29.6 GBq (800 mCi) [^{177}Lu-DOTA[0],Tyr[3]]octreotate with a partial remission (PR) after therapy.

CONCLUSIONS

For PRRT in patients with GEPNETs, various radiolabeled somatostatin analogues are used, which all consist of a cyclic octapeptide, a chelator, and a radionuclide. The affinity of these compounds for the five different somatostatin receptors is mostly influenced by changes in the peptide used and, to a lesser degree, by metal replacement or chelator substitution. Newly developed peptides comprise pansomatostatin agonists, which bind to all five somatostatin receptors. Since some NETs may express different somatostatin receptors concomitantly, the use of pansomatostatins may be of importance for these tumors.

Internalization of the radiolabeled peptide after specific binding to a tumor cell via its receptor is regarded to play a pivotal role for the success of PRRT. Internalization is mandatory for radiometals with a short tissue penetration range like [111]In. In contrast, for radiometals with a longer tissue penetration range like ^{90}Y and ^{177}Lu, internalization as such may not be mandatory for therapy. The same holds true for

TABLE 6.1.2 Tumor responses in patients with GEPNETs, treated with different radiolabeled somatostatin analogues[a]

Center (reference)	Tumor response ligand	Patient number	CR	PR	MR	SD	PD	CR+PR
Rotterdam [37]	[^{111}In-DTPA^0]octreotide	26	0	0	5 (19%)	11 (42%)	10 (38%)	0%
New Orleans [38]	[^{111}In-DTPA^0]octreotide	26	0	2 (8%)	NA	21 (81%)	3 (12%)	8%
Milan [39]	[^{90}Y-DOTA^0,Tyr^3]octreotide	21	0	6 (29%)	NA	11 (52%)	4 (19%)	29%
Basel [40, 41]	[^{90}Y-DOTA^0,Tyr^3]octreotide	74	3 (4%)	15 (20%)	NA	48 (65%)	8 (11%)	24%
Basel [42]	[^{90}Y-DOTA^0,Tyr^3]octreotide	33	2 (6%)	9 (27%)	NA	19 (57%)	3 (9%)	33%
Multicentre [43]	[^{90}Y-DOTA^0,Tyr^3]octreotide	58	0	5 (9%)	7 (12%)	33 (61%)	10 (19%)	9%
Rotterdam [44]	[^{177}Lu-DOTA^0,Tyr^3]octreotate	310	5 (2%)	86 (28%)	51 (16%)	107 (35%)	61 (20%)	29%
Multicentre [45]	[^{90}Y-DOTA^0,Tyr^3]octreotide	78	0	4 (5%)	NA	63 (81%)	11 (14%)	5

[a]Adapted from Kwekkeboom et al. [36].

(a)   (b)

(c)   (d)

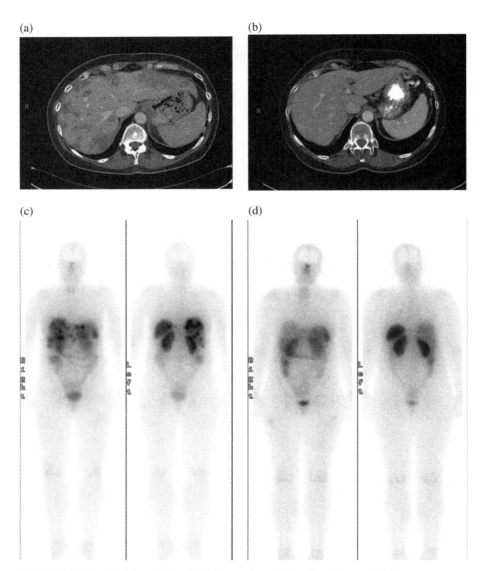

FIGURE 6.1.3   Partial remission (PR) in a patient with an insulinoma with liver metastases treated with 29.6 GBq (800 mCi) [177]Lutetium-octreotate. (a) CT scan showing multiple liver metastases of an insulinoma before treatment with [177]Lutetium-octreotate. (b) CT scan 6 weeks after treatment with [177]Lutetium-octreotate, showing regression of liver metastases, consistent with a PR. (c) Posttherapy scan of the same patient after the first treatment with [177]Lutetium-octreotate, showing intense uptake in multiple liver metastases. (d) Posttherapy scan after the fourth (and last) treatment with [177]Lutetium-octreotate, showing reduced uptake in liver metastases. ANT, anterior; POST, posterior.

somatostatin antagonists, which do not internalize, provided that the residence time of these compounds at the receptor is sufficient.

Radionuclides used for PRRT are ^{111}In, ^{90}Y, and ^{177}Lu. Other radionuclides, as, for example, ^{64}Cu, are under investigation. The use of radiolabeled somatostatin analogues in GEPNETs is a very promising treatment, as is shown by the high rate of remissions achieved after treatment.

ACKNOWLEDGMENTS

We would like to thank E. de Blois for supplying the figures with radiochemical structures of the various chelators and peptide–chelator complexes.

REFERENCES

[1] Arnold, R.; Benning, R.; Neuhaus, C.; et al. Digestion 1993, 54, 72–75.

[2] Janson, E. T.; Oberg, K. Acta Oncologica 1993, 32, 225–229.

[3] Ducreux, M.; Ruszniewski, P.; Chayvialle, J. A.; et al. American Journal of Gastroenterology 2000, 95, 3276–3281.

[4] Rinke, A.; Muller, H. H.; Schade-Brittinger, C.; et al. Journal of Clinical Oncology 2009, 27, 4656–4663.

[5] Wild, D.; Schmitt, J. S.; Ginj, M.; et al. European Journal of Nuclear Medicine and Molecular Imaging 2003, 30, 1338–1347.

[6] Valkema, R.; Froberg, A. C.; Bakker, W. H.; et al. Journal of Nuclear Medicine 2007, 48, 394P (abstract).

[7] Reubi, J. C.; Schar, J. C.; Waser, B.; et al. European Journal of Nuclear Medicine 2000, 27, 273–282.

[8] Capello, A.; Krenning, E. P.; Breeman, W. A.; et al. Cancer Biotherapy and Radiopharmaceuticals 2003, 18, 761–768.

[9] De Jong, M.; Breeman, W. A.; Bakker, W. H.; et al. Cancer Research 1998, 58, 437–441.

[10] Esser, J. P.; Krenning, E. P.; Teunissen, J. J.; et al. European Journal of Nuclear Medicine and Molecular Imaging 2006, 33, 1346–1351.

[11] Reubi, J. C.; Eisenwiener, K. P.; Rink, H.; et al. European Journal of Pharmacology 2002, 456, 45–49.

[12] Schmid, H. A. Molecular and Cellular Endocrinology 2008, 286, 69–74.

[13] Ginj, M.; Zhang, H.; Eisenwiener, K. P.; et al. Clinical Cancer Research 2008, 14, 2019–2027.

[14] Hofland, L. J.; Lamberts, S. W. Endocrine Review 2003, 24, 28–47.

[15] Reubi, J. C.; Waser, B.; Schaer, J. C.; et al. European Journal of Nuclear Medicine 2001, 28, 836–846.

[16] Cescato, R.; Loesch, K. A.; Waser, B.; et al. Molecular Endocrinology 2010, 24, 240–249.

[17] Waser, B.; Cescato, R.; Tamma, M. L.; et al. European Journal of Pharmacology 2010, 644, 257–262.

[18] Ginj, M.; Zhang, H.;, Waser, B.; et al. Proceedings of the National Academy of Sciences of the United States of America 2006, 103, 16436–16441.

[19] Cescato, R.; Erchegyi, J.; Waser, B.; et al. Journal of Medicinal Chemistry 2008, 51, 4030–4037.

[20] Wadas, T. J.; Eiblmaier, M.; Zheleznyak, A.; et al. Journal of Nuclear Medicine 2008, 49, 1819–1827.

[21] De Jong, M.; Breeman, W. A.; Valkema, R.; et al. Journal of Nuclear Medicine 2005, 46, 13S–17S.

[22] Kunikowska, J.; Krolicki, L.; Mikolajczak, R.; et al. Journal of Nuclear Medicine 2009, 50, 106 (abstract).

[23] O'Donoghue, J. A.; Bardies, M.; Wheldon, T. E. Journal of Nuclear Medicine 1995, 36, 1902–1909.

[24] De Jong, M.; Breeman, W. A.; Bernard, B. F.; et al. Journal of Nuclear Medicine 2001, 42, 1841–1846.

[25] Lewis, J. S.; Srinivasan, A.; Schmidt, M. A.; et al. Nuclear Medicine and Biology 1999, 26, 267–273.

[26] Lewis, J. S.; Lewis, M. R.; Cutler, P. D.; et al. Clinical Cancer Research 1999, 5, 3608–3616.

[27] Norenberg, J. P.; Krenning, B. J.; Konings, I. R.; et al. Clinical Cancer Research 2006, 12, 897–903.

[28] Miederer, M.; Henriksen, G.; Alke, A.; et al. Clinical Cancer Research 2008, 14, 3555–3561.

[29] De Jong, M.; Bernard, B. F.; De Bruin, E.; et al. Nuclear Medicine Communications 1998, 19, 283–288.

[30] Lukinius, A.; Ohrvall, U.; Westlin, J. E.; et al. Acta Oncologica 1999, 38, 383–387.

[31] Janson, E. T.; Westlin, J. E.; Ohrvall, U.; et al. Journal of Nuclear Medicine 2000, 41, 1514–1518.

[32] Hukovic, N. ; Panetta, R. ; Kumar, U.; et al. Endocrinology 1996, 137, 4046–4049.

[33] Cescato, R.; Schulz, S.; Waser, B.; et al. Journal of Nuclear Medicine 2006, 47, 502–511.

[34] Stroh, T.; Jackson, A. C.; Sarret, P.; et al. Endocrinology 2000, 141, 354–365.

[35] De Jong, M.; Breeman, W. A.; Bernard, B. F.; et al. European Journal of Nuclear Medicine 1999, 26, 693–698.

[36] Kwekkeboom, D. J.; Mueller-Brand, J.; Paganelli, G.; et al. Journal of Nuclear Medicine 2005, 46, 62S–66S.

[37] Valkema, R.; De Jong, M.; Bakker, W. H.; et al. Seminars in Nuclear Medicine 2002, 32, 110–122.

[38] Anthony, L. B.; Woltering, E. A.; Espenan, G. D.; et al. Seminars in Nuclear Medicine 2002, 32, 123–132.

[39] Bodei, L.; Cremonesi, M.; Zoboli, S.; et al. European Journal of Nuclear Medicine and Molecular Imaging 2003, 30, 207–216.

[40] Waldherr, C.; Pless, M.; Maecke, H. R.; et al. Annals of Oncology 2001, 12, 941–945.

[41] Waldherr, C.; Pless, M.; Maecke, H. R.; et al. Journal of Nuclear Medicine 2002, 43, 610–616.

[42] Waldherr, C.; Schumacher, T.; Maecke, H. R.; et al. European Journal of Nuclear Medicine and Molecular Imaging 2002, 29, 100 (abstract).

[43] Valkema, R.; Pauwels, S.; Kvols, L. K.; et al. Seminars in Nuclear Medicine 2006, 36, 147–156.

[44] Kwekkeboom, D. J.; de Herder, W. W.; Kam, B. L.; et al. Journal of Clinical Oncology 2008, 26, 2124–2130.

[45] Bushnell, Jr., D. L.; O'Dorisio, T. M.; O'Dorisio, M. S.; et al. Journal of Clinical Oncology 2010, 28, 1652–1659.

6.2

PRRT DOSIMETRY

MARK KONIJNENBERG

Erasmus MC, Rotterdam, The Netherlands

ABBREVIATIONS

ABVD	doxorubicin, bleomycin, vinblastine, and dacarbazine
AML	acute myeloblastic leukemia
BED	biologically effective dose
CT	computed tomography
GEPNET	gastroenteropancreatic neuroendocrine tumor
LQ	linear quadratic
MDS	myelodysplastic syndrome
MIRD	Medical Internal Radiation Dose
MOPP	mechlorethamine, vincristine, procarbazine, and prednisone
NCCN	National Comprehensive Cancer Network
PET	positron emission tomography
PRRT	peptide receptor radionuclide therapy
RADAR	Radiation Dose Assessment Resource
RT	radiotherapy
SPECT	single photon emission computed tomography

Somatostatin Analogues: From Research to Clinical Practice, First Edition. Edited by
Alicja Hubalewska-Dydejczyk, Alberto Signore, Marion de Jong, Rudi A. Dierckx,
John Buscombe, and Christophe Van de Wiele.
© 2015 John Wiley & Sons, Inc. Published 2015 by John Wiley & Sons, Inc.

PEPTIDE RECEPTOR RADIONUCLIDE THERAPY

Peptide receptor radionuclide therapy (PRRT) is a low-dose-rate radiotherapy (RT) (as opposed to external beam radiation) developed for treatment of metastasized cancer. The radiolabeled peptide contains a DOTA-chelated radionuclide, Yttrium-90 (^{90}Y) or Lutetium-177 (^{177}Lu). Octreotate binds with high affinity to somatostatin receptors that are overexpressed on the cell surface of most neuroendocrine tumors. The targeting of radiolabeled somatostatin analogues to tumor cells is the basis of the therapeutic effect of these pharmaceuticals.

The radiation toxicity dose estimates used in PRRT are predominately based on the experience and knowledge obtained from external beam radiation. In external beam RT, the recommended radiation toxicity threshold doses are based on empirical data. Over the years, the quality of dosimetry in external beam RT has improved significantly, and the combination of clinical experience and information on normal organ tolerance doses has resulted in consensus guidelines [1]. Although the recommendations by Emami contain several limitations and uncertainties, they address a clinical need in external beam RT, and consequently, the incidence of radiation toxicity to normal organs has been reduced by dosimetry-based treatment planning [2].

In addition to the large base of empirical knowledge that has been obtained from external beam radiation research, there is also more than 60-year experience in radionuclide therapy, in particular with ^{131}I therapy (Iodine-131 sodium iodide) for the treatment of benign and malignant thyroid disorders [3]. The most common radiation-induced toxicity following ^{131}I therapy is bone marrow suppression, mainly thrombocytopenia. Limiting the prescribed cumulative administered radioactivity to a radioactivity level where the bone marrow dose remains below 2 Gy has proven to be a successful guideline for preventing hematological side effects [4, 5].

Dosimetry

Patient-specific dosimetry ideally allows predicting both the risk of radiation-induced organ toxicity and the probability of tumor cure. In PRRT, the emphasis in dosimetry research is aimed at assessing absorbed doses to the kidneys and the bone marrow. Dosimetry-guided treatment planning for PRRT is ultimately the aim for an optimal treatment with a maximal absorbed dose in the target tumor region together with acceptable absorbed doses in normal organs with physiological uptake like kidneys and bone marrow. Several problems seriously hamper the development of this ideal therapy protocol.

The high-quality constraints on the dosimetry are set by the steeply sloped dose–effect limits. Consequentially nonoptimized dosimetry results in the scattered response and unproven dose–effect relations. Hardly any PRRT therapy shows severe acute toxicity, apart from the late occurring renal toxicity observed years after therapy with ^{90}Y-DOTA-octreotide [6]. Dose escalation studies have hardly been performed in PRRT, and when toxicity is reported, the dosimetry often has not been measured or according to a different dosimetry model. By the lack of clear dosimetry-guided trials with toxicity, the dose–effect relations for normal organ toxicity and for tumor cure efficacy are derived from the experience in external beam therapy. Scarcely evidenced radiobiology methods are used to relate the RT absorbed dose limits to equivalent dose limits for PRRT. In RT, absorbed doses are delivered at high dose rates in fractions of 1–2 Gy and conformal to image-derived (CT scan) geometry. The absorbed doses in PRRT, however, are given with low and exponentially decaying dose rates and conformal to physiological, functional, and receptor-mediated uptake of the radiolabeled peptide. It is impossible to define one target volume in PRRT, as the therapy is aimed at treating metastasized disease, of which only the larger lesions (>1 cm size) are visible and thereby evaluable for response. Heterogeneity in the radioactivity uptake by variance in receptor expression or on the functional subunit level is also not visible, but can seriously influence the absorbed dose distribution, especially for low-range particle emitters.

The aim for dosimetry in PRRT should therefore be to develop a robust, high-quality absorbed dose calculation model and to use proven radiobiology models in defining the PRRT-specific dose–effect curves. The high quality is delivered by using enough time points appropriate to determine the normal kinetic distribution pattern and especially identify patient-specific deviations; also, the actual volumes used in the dose calculations should be patient conform. The secondary aim of robustness is necessary when absorbed doses reported by several clinical centers are to be compared and can be filled in by developing an omnipotent dosimetry software code with hardly any user input or preferably a quickly and easily performed dosimetry assessment and calculation model.

Various dosimetry calculation models, practical or more sophisticated, are available and are being used. Dosimetry data needed are radioactivity measurements of blood and urine samples, as well as quantitative scintigraphic images measured at least in three time points after administration. Planar imaging is straightforward and easily performed to determine the kinetics of the radiopharmaceutical, while SPECT and SPECT/CT imaging allows determination of the three-dimensional organ activity distribution with optimal background resolution. Internal dosimetry is usually performed according to the MIRD scheme. Mean absorbed doses in normal organs can be derived with dedicated software (like OLINDA/EXM [7]).

The general scheme for calculating radiation dosimetry with radionuclides has been defined by the MIRD scheme dosimetry formula [8]:

$$D(r_t) = \sum_s \int A_s(t)dt \times S(r_t \leftarrow r_s)$$
$$= \sum_s \tilde{A}_s \times S(r_t \leftarrow r_s) = A_0 \sum_s \tau_s \times S(r_t \leftarrow r_s)$$
$$\Rightarrow D_{\text{kidneys}} \approx A_0 \tau_{\text{kidneys}} \times S(\text{kidneys} \leftarrow \text{kidneys})$$

The dose to the target organ (D; kidney or bone marrow) is calculated by the product of the number of decays in a source organ ($\tilde{A}_s$) and the S-value, which expresses the dose rate per radioactivity for a source (r_s) to target (r_t) combination. More conveniently, the residence time τ is used and is obtained by integrating over the time-activity curve relative to the injected activity A_0 and $\tau = \tilde{A}_s / A_0$. For the beta-particle-emitting radionuclides ^{177}Lu and ^{90}Y, only the self-dose needs to be considered ($r_s = r_t$), because of their short β-particle ranges in tissue. The radioactivity uptake and clearance kinetics in the kidneys of the radiolabeled peptide $A_s(t)$ is needed for calculation of the radiation dose to the kidneys, together with the S-value for the kidney self-dose. These S-values can be adjusted with the actual patient's kidney volume, derived from CT imaging; both the S-values and the mass scaling can be performed within the OLINDA/EXM code [7]:

$$D_{rm} = \tilde{A}_{rm} S(rm \leftarrow rm) + \sum_h \tilde{A}_h S(rm \leftarrow h) + \tilde{A}_{rb} S(rm \leftarrow rb)$$

There are three contributing sources that account for the radiation dose to the bone marrow from radiolabeled peptides: (1) the self-dose by accumulation in the red marrow ($\tilde{A}_{rm}$), either by blood flow or by specific uptake in the bone marrow; (2) cross-dose from activity in major source organs and tumors ($\tilde{A}_h$); and (3) cross-dose from the remainder of the body ($\tilde{A}_{rb}$), assuming homogeneous radioactivity distribution. As with the organ dosimetry, the calculations of the radiation dose to the bone marrow are highly dependent on phantom models, with no possibility for scaling the phantom to patient conformity. In case of heterogeneous uptake in the bone marrow, for example, due to bone metastasis involvement, there is currently no accepted model for the dose calculation.

A direct comparison of the radioactivity concentrations in blood and bone marrow aspirates in 14 patients treated with ^{177}Lu-DOTA[0]-Tyr[3]-octreotate showed that the radioactivity concentrations in blood and bone marrow aspirates were nearly equal and that there is no specific binding of the peptide to bone marrow precursor cells [9]. The contributions to the bone marrow absorbed dose come largely (50% of total) by cross-dose contributions from the remainder of the body, second (34%) by the red marrow itself due to circulating blood, and the least (14%) by other source organs (liver, spleen, kidney) and tumors. The bone marrow absorbed dose has a high inter-patient variation.

Pharmacokinetic compartmental models are helpful to identify the ideal time points in activity quantification and define numerical solutions in calculating the organ cumulated activities (see MIRD pamphlet 16 [10]). The time points depend on the number of compartments in the uptake and clearance pattern that are needed for accurate residence time determination. For peptides, the uptake and initial fast clearance phase are usually neglected due to the longer half-lives used for therapy. For ^{90}Y, the time points used are 4, 24, and 48 h, although when measuring by PET imaging using [86]Y (with $t_{1/2} = 14.74$ h) the sensitivity of the last measurement is much reduced. With ^{177}Lu, the time points are usually 24, 72, and 168 h, with optionally a measurement at 1 h to define the 100% whole body uptake calibration.

As an example, the pharmacokinetic model for the biodistribution of ^{177}Lu-DOTA-octreotate and the resulting distribution curves are shown in Figures 6.2.1, 6.2.2, and 6.2.3.

Absorbed dose calculations are generally performed by using dedicated software codes with residence times for the source organs τ as input (OLINDA/EXM,

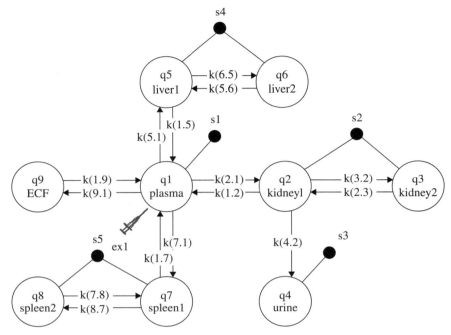

FIGURE 6.2.1 Generalized compartment model for the biodistribution of ^{177}Lu-DOTA[0]-Tyr[3]-octreotate in humans. Abbreviations: ex1, experimental input function; the time of drug administration (defined as T_0); red dots, sample time points for the indicated compartments (organs); ECF, extracellular fraction; q^{1-9}, compartments (or organs) 1–9; k(i,j), kinetic transfer components between the indicated compartments i and j (organs).

RADAR) [7]. These calculations are done for fixed stylized models of the human body (e.g., MIRD phantoms) with the possibility to correct for the actual mass (M_{CT}) of the organ of interest by the correction factor $M_{phantom}/M_{CT}$. More patient conformity can be obtained by calculating the absorbed dose in each voxel, either by assuming local absorption of all beta-energy inside the source voxels, or by using voxel S-values (MIRD pamphlet 17 [11]), or by doing a Monte Carlo calculation of all radiation transport from the source voxels to all surrounding voxels and itself.

Absorbed doses in the tumor lesions are estimated by assuming the lesion to be spherical in shape and, within the activity, uniformly distributed [12]. The pure β-decay nature of ^{90}Y without any γ-emission severely hinders direct dosimetry for ^{90}Y-DOTA-octreotide. Bremsstrahlung images are possible but need extensive reconstruction algorithms and corrections applied [13]. Instead, the imaging can be performed using surrogate radionuclides linked to the peptide, either with ^{111}In or with [86]Y. With its gamma-emission, ^{111}In-DOTA-octreotide is very suitable as a surrogate for ^{90}Y-DOTA-octreotide dosimetry. The positron emitter [86]Y has yielded good quality dosimetry but is very difficult to make as well as demanding to generate good quality images [14, 15]. Quantitative imaging with

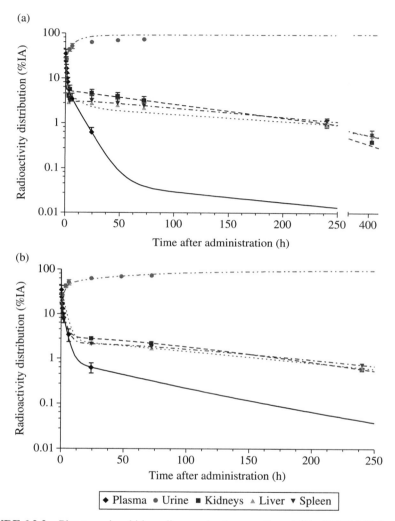

FIGURE 6.2.2 Plasma, urine, kidney, liver, and spleen profiles of ^{177}Lu-DOTA^0-Tyr^3-octreo-tate. Curves fit to mean percentages of infused activity (%IA) data using PK compartment model after treatment with 1.85 GBq (50 mCi) ^{177}Lu-DOTA^0-Try^3-octreotate; (a) without amino acid coinfusion ($N = 5$) and (b) with amino acid coinfusion ($N = 4$) in the same patients.

^{111}In-DTPA-octreotide (OctreoScan) planar or SPECT yields a different biodistri-bution pattern than ^{90}Y-DOTA-octreotide [16]; it did, however, identify the patients with high kidney uptake. PET-based dosimetry with ^{68}Ga-DOTA-octreo-tide is not accurate due to its short physical half-life (68 min). As ^{90}Y also emits a 1.71 MeV positron, it has been feasible to image the ^{90}Y renal uptake distribution by PET imaging [17]. Imaging of ^{177}Lu-DOTA-octreotate with its gamma-ray emissions is much easier to perform and has the advantage of dosimetry

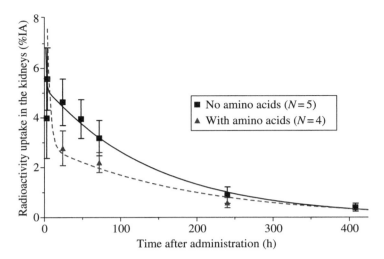

FIGURE 6.2.3 Uptake and clearance of ^{177}Lu-DOTA^0-Tyr^3-octreotate in kidneys of group 1 patients. Data, plotted as the mean percentage of radioactivity (%IA), acquired after administration of 1.85 GBq (50 mCi, dose level 1); (■) without amino acid coinfusion; and (▲) with amino acid coinfusion. Curve fits to data are according to the PK compartmental model presented in Figure 6.2.1.

evaluation during the actual therapy. Usually, dosimetry is performed during the first therapy course with ^{177}Lu-DOTA-octreotate.

PRRT using ^{177}Lu-DOTA^0-Tyr^3-octreotate shows significant similarities to ^{131}I therapy. The data in Table 6.2.1 shows that the physical characteristics of ^{131}I and ^{177}Lu are similar, except for the high-energy γ-rays of ^{131}I, which cause a higher radiation dose to nontarget regions. Biodistribution and radiation dosimetry studies have shown that the critical organ in both radionuclide therapies (^{131}I and ^{177}Lu-DOTA^0-Tyr^3-octreotate) is the bone marrow. In both radionuclide therapies, at the doses shown in Table 6.2.2, the absorbed dose in the bone marrow can exceed the toxicity threshold dose established with external beam RT (see data in Table 6.2.2). ^{177}Lu-DOTA^0-Tyr^3-octreotate treatment results in radiation doses to other organs such as the kidney, spleen, and pituitary that may be higher than the doses resulting from ^{131}I therapy. However, based on clinical experience, the spleen and pituitary are not critical organs in terms of radiation-related damage from PRRT. The kidneys are generally considered to be the critical organs for therapy with ^{177}Lu-DOTA^0-Tyr^3-octreotate. Therefore, it is common practice to coadminister amino acids to protect the kidneys [18–35].

BONE MARROW TOXICITY AND DOSIMETRY IN PRRT

Radiation-induced bone marrow toxicity results from damage to the hematopoietic tissue (the red marrow) where red blood cells, platelets, and white blood cells are produced. In PRRT using ^{177}Lu-DOTA^0-Tyr^3-octreotate or RT using ^{131}I, the radiation

TABLE 6.2.1 Physical characteristics of ^{131}I, ^{90}Y, and ^{177}Lu [36]

Characteristic	^{131}I	^{90}Y	^{177}Lu
Decay half-life (day)	8.0207 day	64.1 h	6.647 day
β-particle energy (keV/decay)	181.9 keV	932.9 keV	133.3 keV
Main γ-ray energy (keV) and abundance (%/decay)	364 keV (81.7%/decay) 637 keV (7.2%/decay)	511 keV (0.0064%/decay) Bremsstrahlung	113 keV (6.4%/decay) 208 keV (11.0%/decay)

dose to red bone marrow accumulates from the β-ray irradiation from ^{131}I or ^{177}Lu radioactivity in the blood circulating in the bone and from penetrating γ-ray radiation from activity dispersed throughout the remainder of the body. This "blood circulation" bone marrow dosimetry model has been safely employed for decades with ^{131}I therapy [3]. It can also be used to determine the maximum dose that can be safely adminis-tered with ^{177}Lu-DOTA^0-Tyr^3-octreotate.

KIDNEY TOXICITY IN PRRT

Radiation-induced kidney toxicity caused by PRRT is due to damage to the nephrons, the basic structural and functional units of the kidney. Nephrons consist of glomeruli and tubuli, which are responsible for filtration and reabsorption, respectively. The threshold dose for radiation damage to the kidneys by PRRT depends on the radionu-clide that is used. To date, the radionuclides used for PRRT with somatostatin analogues are Indium-111 (^{111}In [40, 41]), ^{90}Y ([22, 28–30, 42–47]), and ^{177}Lu ([9, 20, 31, 33, 34, 44, 46, 48–56]). Tubular reabsorption of ^{90}Y-labeled peptides leads to a uniform radiation field to the kidney parenchyma that is comparable to that of external beam RT, despite the lower dose rate of PRRT [28]. A correlation was found between the kidney radiation dose and chronic kidney toxicity after treatment with ^{90}Y-DOTA^0-Tyr^3-octreotide, as depicted in Figure 6.2.4. The dose at which 50% of the patients will encounter kidney toxicity was established to be 35 Gy [39], which is 7 Gy higher than the 28 Gy established for external beam RT [57]. Because ^{177}Lu has a shorter-range β-particle than ^{90}Y, there is less damage to nearby nontarget tissue [58, 59]. Moreover, PRRT using a radionuclide with much shorter-range emissions (^{111}In Auger electrons) does not result in kidney damage despite higher cumulative doses to the kidney [44]. Therefore, 23 Gy is considered to be a conservative value for the radiation toxicity threshold dose for PRRT using ^{177}Lu-DOTA^0-Tyr^3-octreotide. A summary of the previously reported kidney and bone marrow dosimetry findings in studies using ^{177}Lu-DOTA^0-Tyr^3-octreotate is provided in Table 6.2.3.

TABLE 6.2.2 Comparison between radiation threshold doses to normal tissue for external beam radiotherapy, ^{131}I therapy, and PRRT with either ^{90}Y-DOTA[0]-Tyr[3]-octreotide or ^{177}Lu-DOTA[0]-Tyr[3]-octreotate

Organ	Damage endpoints	RT threshold dose[a] (Gy)	[131]I therapy 30 GBq[b] (Gy)	^{90}Y-DOTA-octreotide 13.3 GBq[c] (Gy)	^{177}Lu-DOTA-octreotate 29.6 GBq[d] (Gy)
Thyroid	Ablation	80	Target	0.5–2	0.5–4
Bone marrow	Hypoplasia	2	2–4	30 ± 11	23 ± 8
Kidneys	Late nephritis	23	2–3	30 ± 8	17 ± 3
Urinary bladder	Contracture	65	10–15		
Liver	Radiation-induced liver disease	30	1–2	10 ± 5	1–5
Spleen	Increased infection	40	NR[e]	30 ± 15	64 ± 10

[a] Data from Marks L et al. [2] and ICRP 41 [37].
[b] Iodine-131 therapy with a cumulative injected radioactivity of 30 GBq (ICRP 53 [38]).
[c] Peptide receptor radionuclide therapy with 3 times 4.4 GBq (cumulative 13.3 GBq) ^{90}Y-DOTA[0]-Tyr[3]-octreotide.
[d] PRRT with 4 times 7.4 GBq (cumulative 29.6 GBq) ^{177}Lu-DOTA[0]-Tyr[3]-octreotate.
[e] NR, not reported.

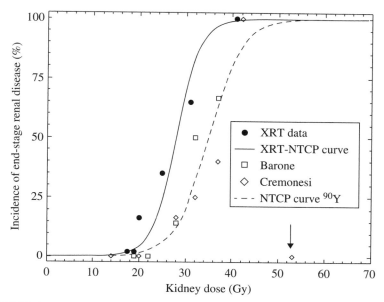

FIGURE 6.2.4   Dose–response curve for late radiation-induced kidney toxicity after external beam radiotherapy (XRT, closed dots and solid curve) and after PRRT with 90Y-DOTA-octreotide (open signs and dashed curve). The normal tissue complication probability curve for the XRT data (26) is fitted with TD50 = 28 Gy and at slope of $m = 0.15$. The NTCP curve was fitted through ^{90}Y data with TD50 = 35 Gy and $m = 0.18$ (Curve reproduced with permission from B. Wessels et al. [39]).

KIDNEY DOSIMETRY IN PRRT

Most dosimetry data analyzed are based on planar measurements. Despite the disadvantages of the use of planar image acquisition, it is a widely used and accepted technique, and it has been widely used in PRRT [31, 46, 54, 55]. The use of planar data based on conjugate view methods is known to be less accurate for determining kidney and other normal organ radiation doses. Studies that directly compare SPECT and planar imaging show that the planar imaging method consistently overestimates organ dosimetry [32–34]. This is primarily due to the occurrence of overlapping radioactivity that appears on planar images such as from the overprojection of the liver with tumor nodules into the right kidney region of interest and also from intestinal radioactivity potentially overlapping the left kidney [31, 55]. For this reason, in the study reported here, the organ uptake (abdominal region) was not differentiated into separate organs and instead was considered to be a single source region. One issue with this approach, however, is that dosimetry models, like the OLINDA/EXM code [7], consider separate source organs only and do not facilitate dose calculations for dispersed radioactivity by widespread receptor-positive metastases. Most dosimetry models are based on rigid stylized phantoms with a possibility for correcting the dose by the actual organ volumes, but do not allow correction for

TABLE 6.2.3 Overview of peer-reviewed publications on dosimetry for peptide Receptor radionuclide therapy with ^{177}Lu-DOTA^0-Tyr^3-octreotate

Study	Method	N^a	IA^b (GBq)	Amino acids	Kidney dose per IA (Gy/GBq)	Bone marrow dose per IA (Gy/GBq)
Kwekkeboom [19]	Planar	5	1.85	Lys/Arg	0.88 ± 0.19	0.070 ± 0.009
Bodei [46]	Planar	5	3.7–5.18	NR	0.9 ± 0.5	NR^c
Forrer [9]	Planar	15	7.47 ± 0.10	Lys/Arg	NR	0.02–0.13
Wehrmann [31]	Planar	69	3–7	Lys/Arg	0.9 ± 0.3	0.05 ± 0.02
Sandström [34]	SPECT/CT	24	7.4	VAMIN-14	0.71 ± 0.24	NR
Sandström [54]	SPECT and planar	200	7.4	VAMIN-14	0.62 (0.49–0.75)	0.016 (0.012–0.022)
Garkavij [33]	SPECT/CT	16	7.4	VAMIN-14	0.90 ± 0.21	0.07 ± 0.02
Swärd [55]	Planar	26	8	NR	0.9 ± 0.4	NR
Claringbold [32]	SPECT/CT	33	7.8	Synthamin	0.31 (0.14–0.46)	NR

^{a}N, number of patients.

^{b}IA, administered radioactivity of ^{177}Lu-DOTA^0-Tyr^3-octreotate.

^{c}NR, not reported.

the actual patient geometry [60]. The technique has limitations for individual patient dosing, because it tends to overestimate organ doses and subsequently leads to conservative limits for the total cumulative dose of administered radioactivity. However, planar imaging-based dosimetry with ^{111}In-DOTA^0-Tyr^3-octreotide has been of value in determining threshold doses in population studies for ^{90}Y-DOTA^0-Tyr^3-octreotide therapy [61].

Studies based on planar measurements for dosimetry calculations were performed by [31, 46, 55]. In each of these studies, patients were treated with cumulative doses of approximately 30 GBq (800 mCi) ^{177}Lu-DOTA^0-Tyr^3-octreotate given in four separate administrations, and their findings are in good agreement with other data. Bodei and coworkers [46] suggest that for patients without risk factors such as hypertension and diabetes, the threshold for kidney damage is set at a biologically effective dose (BED) of 40 Gy, which would correspond to a MIRD-based radiation dose by ^{177}Lu of 36 Gy. In the report by Wehrmann, no individual patient doses are given, but with a mean kidney dose per radioactivity of 0.9 ± 0.3 Gy GBq^{-1}, the cumulative dose to the kidneys will be 27 ± 9 Gy for a cumulative administered radioactivity of 29.6 GBq [31]. There was no indication in their publication of any adjustment of the dosing scheme. In the study by Swärd et al. [55], the patients were scheduled for four administrations of 8 GBq each, unless the total radiation dose to the kidneys was more than 27 Gy. Only 12 out of 26 patients were given the full therapy of four administrations. The mean radiation dose to the kidneys was determined to be 24 ± 6 Gy, but no individual dosing schemes were provided. The reported patient-averaged dose per radioactivity is in the same order of magnitude: 0.9 ± 0.4 Gy GBq^{-1}. Therefore, while these studies show general agreement on the radiation dose per unit of administered radioactivity to the kidneys, the lack of long-term follow-up data prevents any conclusions about long-term renal toxicity.

Studies comparing the accuracy of radioactivity uptake measurements and dosimetry by quantitative SPECT/CT to those obtained with conventional planar imaging in patients treated with ^{177}Lu-DOTA^0-Tyr^3-octreotate were performed by Sandström et al. [34, 54], Garkavij et al. [33], and Claringbold et al. [32]. Sandström and coworkers [34] showed that planar and SPECT-based dosimetry results are comparable in regions with low tumor density but that planar imaging-based dosimetry in highly neoplastic regions overestimates the dose. The individual patient ratios of absorbed dose between the planar and SPECT-derived kidney dose ranged between 0.7 and 5 for the right kidney and between 0.8 and 9 for the left kidney. The mean radiation doses to the kidneys in their report are somewhat lower than these previously published by others. The observed variations and uncertainties were attributed to overlapping radioactivity from organ and tumor lesions. Garkavij and coworkers [33] determined that there was a ~30% overestimation of radiation dose using planar radioactivity data with background correction outside the abdominal cavity in comparison to using SPECT imaging data where all β-ray energy from ^{177}Lu is assumed to be absorbed within each source voxel self (planar, 1.15 ± 0.29 Gy GBq^{-1}; SPECT, 0.90 ± 0.21 Gy GBq^{-1}). In their study, the cumulative administered radioactivity was tailored such that the total absorbed dose to the kidneys would remain below 27 Gy (based on the planar imaging method), usually by reducing the number of treatments. However, 4 of the 12 evaluated patients were subsequently shown to have received planar imaging-based doses exceeding 30 Gy. But only two patients showed radiation doses to the kidneys above 30 Gy (and only slightly) when the SPECT method for dosimetry was used. The other two patients with 30 Gy plus doses (by planar imaging-based dosimetry) were found to have doses of about 19 Gy using the SPECT method. Furthermore, it was observed that there was an almost equal contribution (within 10%) to the total radiation dose to the kidneys from each of the treatment cycles.

In the study by Claringbold et al. [32], 33 patients were treated with four times 7.8 GBq (200 mCi) ^{177}Lu-DOTA^0-Tyr^3-octreotate. PRRT was followed by treatment with capecitabine, which is, given the 14-day time schedule per treatment, considered to have no effect on the pharmacokinetics of the peptide. Quantitative SPECT/CT imaging was used to perform voxel-based dosimetry using a point kernel to model radiation transport. They reported that the cumulative absorbed dose to the kidneys for the completed 4 treatments (31.2 GBq cumulative administered dose) is 10 Gy (range: 4.5–14.6 Gy). This is much lower than the values reported by other groups. It is claimed that this difference is due to the use of patient-specific geometry for the dosimetry calculations instead of the stylized MIRD phantom used in the OLINDA code [7]. Unfortunately, there was no direct comparison between the outcomes of both methods.

LINEAR QUADRATIC MODEL AND BED

Radiation-induced damage (cell death) results from the combination of "direct damage," the induction of double-stranded DNA breaks, and "indirect damage," the induction of multiple single-stranded DNA breaks that combine to yield a lethal double-stranded break. Dividing the radiation dose for external beam therapy into fractions has shown to result in enhanced repair of sublethal damage in normal tissue during the treatment intervals compared to the repair in tumor tissue. The repair of sublethal damage dose rates is described by the linear quadratic (LQ) model, which when stated in terms of a specific biological effect (E) (e.g., toxicity) per fraction (i) of fractional dose (D) can be expressed as

$$E_i = \alpha D_i + \beta D_i^2$$

where D_i is the radiation dose in Gy, α is the biological effect (per Gy) of the initial linear component, and β is the quadratic component of the effect (per Gy^2). In terms of an observed biological effect such as cell survival, α represents direct, irreparable damage (e.g., lethal DNA strand breaks), and β represents lethal combinations of indirect, repairable damage (e.g., multiple single-stranded breaks).

The ratio of both factors (α/β) determines the susceptibility of tissues to different dose rates, achieved either by dose fractionation schemes or by using radionuclides with different half-lives. Normal tissue with a low α/β (2–4 Gy) shows less damage by a fixed dose delivered at low dose rates, whereas tumor tissue with a high α/β

(5–10 Gy) shows relatively constant damage from a fixed dose delivered at varying dose rates.

The LQ model can also be used to compare the biological effect to tissue resulting from different dose rates due to RT dose fractionation schemes (or different radionuclides) by calculating the "BED". BED comparisons are based on the assumption that biological effects at a defined endpoint in specific tissues are equivalent, regardless of the dosing scheme. Studies on the induction of chronic kidney disease induced by external beam radiation and PRRT with ^{90}Y showed equivalent BEDs [39], as shown in Figure 6.2.5. The commonly defined endpoint for chronic kidney disease in these studies is a 20% decline in creatinine clearance per year [28, 44], as the clinical endpoint for creatinine clearance toxicity is not always reached within the follow-up after therapy.

The BED (or E/α) is an approximate quantity by which different RT fractionation regimens may be intercompared. For instance, for an external beam RT regimen employing n equal fractions, the BED will be

$$\mathrm{BED} = \frac{E}{\alpha} = nD\left(1 + \frac{D}{\alpha/\beta}\right)$$

At very low dose rates, as is in the case with PRRT, an additional factor must be added to account for the decreasing biologically effectiveness of the repairable component (β)

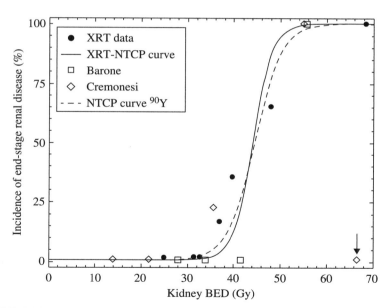

FIGURE 6.2.5 Dose–response curve for correlation between late radiation-induced kidney toxicity and biologically effective dose (BED) to the kidneys after external beam radiotherapy (XRT, closed dots and solid curve) and after PRRT with 90Y-DOTA-octreotide (open signs and dashed curve). The NTCP curve was fitted through ^{90}Y data with TD$_{50}$ = 44 Gy and slope $m = 0.12$ (Figure reproduced with permission from B. Wessels et al. [39]).

of damage. In this case, as shown in the equation below, BED is determined from the cumulative dose (D) that is delivered in i fractions (D_i), the effective half-life (T_{eff}) of the radioactivity in a specific tissue/organ, the repair half-life (T_μ), and the ratio between direct and combinatorial damage (α/β). The latter values were deduced from external RT practice—$T_\mu = 2.8\,h$ and $\alpha/\beta = 2.6\,Gy$ [28, 62]:

$$\mathrm{BED} = \sum_i D_i \left(1 + \frac{T_\mu}{T_{eff} + T_\mu} \frac{D_i}{\alpha / \beta} \right)$$

In external beam RT, kidney radiation doses of 23 and 28 Gy are considered to result in 5 and 50% probabilities, respectively, of developing radiation nephropathy within 5 years after treatment [1]. The 23 Gy dose delivered by fractionated external beam RT corresponds to a BED of 34 Gy. A BED of 34 Gy corresponds to a kidney radiation dose of 30 Gy for ^{177}Lu-DOTA^0-Tyr^3-octreotate treatment with a cumulative amount of radioactivity of 29.6 GBq given in four administrations (fractions). By using radionuclides with long half-life and by fractionation of the therapy, radiation effect in normal tissue will be reduced [39, 63]. Since ^{177}Lu has a longer half-life than ^{90}Y (6.65 days vs. 2.67 days), the threshold radiation dose for kidney damage will be higher. Bodei et al. [46] calculated that in patients with multiple risk factors for renal toxicity, like diabetes and/or high blood pressure, the threshold BED for kidney radiation toxicity with PRRT will be 28 Gy and without risk factors a BED limit of 40 Gy is assumed, thereby substantiating dose limits to the risk factors already identified by Valkema et al. [44].

BONE MARROW DOSIMETRY CONSIDERATIONS

There are three contributing sources that account for the radiation dose to the bone marrow from PRRT: (1) the self-dose by accumulation in the red marrow, either by blood flow or by specific uptake in the bone marrow; (2) cross-dose from major source organs and tumors; and (3) cross-dose from the remainder of the body, assuming homogeneous radioactivity distribution. As with the organ dosimetry, the calculations of the radiation dose to the bone marrow are highly dependent on phantom models. In case of heterogeneous uptake in the bone marrow, for example, due to bone metastasis involvement, there is currently no accepted model for the dose calculation.

A direct comparison of the radioactivity concentrations in blood and bone marrow aspirates in the 14 patients treated with ^{177}Lu-DOTA^0-Tyr^3-octreotate showed that the radioactivity concentrations in blood and bone marrow aspirates were nearly equal and that there is no specific binding of the peptide to bone marrow precursor cells [9]. As discussed by Forrer et al., the contributions to the bone marrow absorbed dose come largely from cross-dose from other source organs (liver, spleen, kidney, and blood), tumors, and the remainder of the body. The bone marrow absorbed dose has a high interpatient variation.

BONE MARROW RADIATION TOXICITY LITERATURE

In the study by Forrer, no correlation was found between the radiation dose to the bone marrow and the decrease in platelets following treatments (as assessed from blood samples taken 6 weeks after the treatment) for patients treated with up to a maximum 29.6 GBq (800 mCi) of ^{177}Lu-DOTA^0-Tyr^3-octreotate [9]. Wehrmann and coworkers [31] and Garkavij and coworkers [33] also found no correlation between the bone marrow dose and blood parameters after therapy. They report mean radiation doses to the bone marrow per administration of 7.4 GBq ^{177}Lu-DOTA^0-Tyr^3-octreotate of 0.4 ± 0.1 Gy and 0.5 ± 0.1 Gy, respectively. Walrand and coworkers [47] reported that ^{86}Y-DOTA^0-Tyr^3-octreotide is taken up in the spine at later time points. It is not clear whether the observed spinal uptake is unconjugated ^{86}Y or the intact radiolabeled peptide. The study of Walrand did show a correlation between the bone marrow dose and reduction in platelets after therapy, only when the absorbed dose was modulated for the patients with low platelet counts before PRRT.

INDUCTION OF MYELODYSPLASTIC SYNDROME AND ACUTE LEUKEMIA

Leukemia and myelodysplastic syndromes (MDS), which have the potential to evolve into leukemia, are prominent late effects of exposure to radiation. In the general population, MDS occurs in 5 per 100,000 people and increases in incidence to about 34 per 100,000 among individuals older than age 70 (NCCN, Practice Guidelines in Oncology v.4.2006; Myelodysplastic Syndromes). The natural incidence of MDS and acute leukemia in nonradiated GEPNET patients is not known—the same holds for GEPNET patients treated with chemotherapy. Chemotherapy (alkylating agents, nitrosoureas, topoisomerase II inhibitors, e.g., teniposide and etoposide), in the presence or absence of external beam radiation, is known to induce toxicity to the bone marrow. Various figures for 15-year cumulative risk of AML after chemotherapy have been reported, as high as 10 percent, specifically in Hodgkin lymphoma patients treated with mechlorethamine, vincristine, procarbazine, and prednisone (MOPP). The 15-year cumulative risk of AML after doxorubicin, bleomycin, vinblastine, and dacarbazine (ABVD) has been estimated to be less than 1 percent. This is reflected in a decrease in leukemia risk among those treated for Hodgkin lymphoma after 1980 when treatment with MOPP/ABVD largely replaced MOPP-like combinations (2.1 vs. 6.4%) (In UpToDate®, P.M. Mauch, Second malignancies after treatment of classical Hodgkin lymphoma).

The *Atlas of Genetics and Cytogenetics in Oncology and Haematology* indicates that secondary MDS/AML after chemotherapy is diagnosed after lymphoma therapy with a percentage of relative risk ranging from 2.2 to 3.3 at 15 years [64]. The incidence rate of leukemia after ^{131}I therapy was reported to be 0.4% and was not dose dependent [65]. Usual treatment doses in those patients are below 18.5 GBq (500 mCi) ^{131}I.

KIDNEY DOSIMETRY CONSIDERATIONS

The determination of the radiation dose to organs, including the kidneys, has traditionally been based on the conjugate view method of planar imaging. The method
produces generally acceptable results, but not in regions that have widespread
tumor involvement. In these regions, planar imaging methods fail because images
have overlapping radioactivity from uptake in the kidney and from metastases and/
or tumor uptake. This leads to an overestimation of the actual kidney radiation
dose. The use of quantitative SPECT to measure the uptake of ^{177}Lu-DOTA^0-Tyr^3-
octreotate is less prone to errors caused by this overlapping radioactivity.
Additionally, CT image can be used to define the actual patient geometry and
kidney volume, which further improves the quality of the dosimetry. All planar
imaging-based dosimetry studies show mean doses per IA of 0.9 Gy GBq^{-1} with a
33% variation between patients because of differences in kidney kinetics (see
Table 6.2.3, planar imaging method). When quantitative SPECT is used, the mean
radiation dose to the kidneys is calculated to be 0.57 Gy GBq^{-1} (weighted mean
from SPECT data in Table 6.2.3) with the same 33% variation induced by difference
in kinetics. From these data, the cumulative administration of 29.6 GBq ^{177}Lu-
DOTA^0-Tyr^3-octreotate will result in a kidney radiation dose of 27 ± 9 Gy according
to the planar imaging method or, more accurately, 16.8 ± 5.7 Gy according to the
quantitative SPECT method.

KIDNEY DOSIMETRY: HETEROGENEITY OF THE DOSE
DELIVERED TO KIDNEYS

With respect to the biological effect of radiation, the distribution of the dose delivered to the kidneys by ^{90}Y or ^{177}Lu is different, due to their different β-energies [59];
^{90}Y emits β-particles with a mean energy of 0.933 MeV, which has a range in tissue
of 3.5 mm, whereas the β-particles from ^{177}Lu with a mean energy of 0.133 MeV
have a range in tissue of only 0.2 mm. The homogeneous radiation exposure of the
kidneys obtained by ^{90}Y is more similar to the exposures resulting from external RT
than obtained by the shorter-range ^{177}Lu. As the radiolabeled peptide is reabsorbed by
the proximal tubuli, ^{177}Lu will irradiate the tubuli, but because of the limited range of
^{177}Lu β-emissions, it will have little effect on the glomeruli. Glomeruli are known to
be more radiation sensitive and show less repair capacity than tubuli. Glomeruli and
tubuli may have different repair kinetics and therefore different radiation responses,
but the actual rates have never been measured. Moreover, in general, dosimetry calculations are based on the MIRD scheme, with the assumption that radioactivity in the
kidney is homogeneously distributed. This assumption was investigated by de Jong
et al. [58] by determining the distribution of radioactivity in the normal human
kidney after i.v. injection of ^{111}In-DTPA^0-octreotide. Three renal cancer patients,
scheduled for nephrectomy, received i.v. injection of ^{111}In-DTPA^0-octreotide, and the
distribution of radioactivity was assessed by SPECT scanning before surgery and by
ex vivo autoradiography of the kidney after surgery (Fig. 6.2.6).

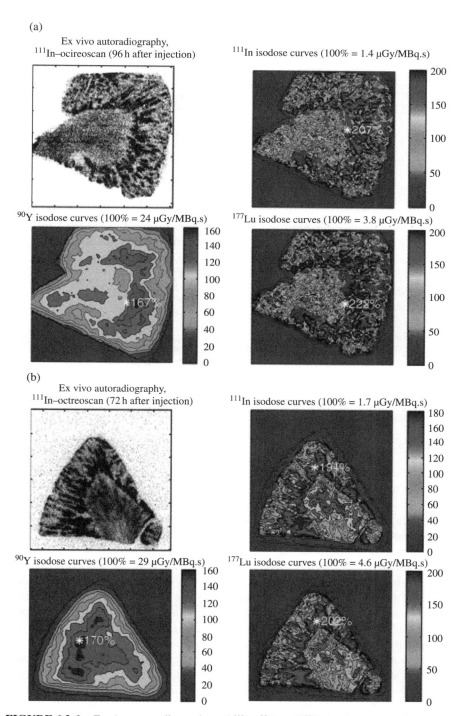

(a)

Ex vivo autoradiography,
^{111}In–ocireoscan (96 h after injection)

^{111}In isodose curves (100% = 1.4 μGy/MBq.s)

^{90}Y isodose curves (100% = 24 μGy/MBq.s)

^{177}Lu isodose curves (100% = 3.8 μGy/MBq.s)

(b)

Ex vivo autoradiography,
^{111}In–octreoscan (72 h after injection)

^{111}In isodose curves (100% = 1.7 μGy/MBq.s)

^{90}Y isodose curves (100% = 29 μGy/MBq.s)

^{177}Lu isodose curves (100% = 4.6 μGy/MBq.s)

FIGURE 6.2.6 *Ex vivo* autoradiographs and ^{111}In, ^{90}Y, and ^{177}Lu isodose curves for kidney sections from 3 patients (a, b, and c) at different time intervals after administration of ^{111}In-DTPA-octreotide (Figure reproduced with permission from M. Konijnenberg et al. [59]). (*See insert for color representation of the figure.*)

(c)

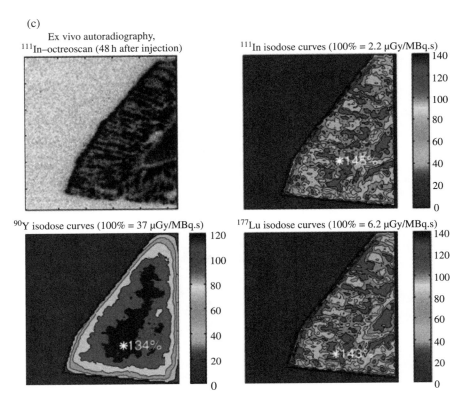

FIGURE 6.2.6 *(Continued)*

Based on autoradiography, radioactivity was found to be localized predominantly in the cortex of the kidney. In the cortex, radioactivity was not distributed homogeneously but formed a striped pattern, with most of the radioactivity centered in the inner cortical zone. These findings showed that average dose calculations using the MIRD scheme, assuming homogeneous renal radioactivity distribution, are inadequate to estimate the radiation dose to various parts of the kidney after radionuclide therapy. Different effects due to inhomogeneity can be expected from radionuclide therapy using radionuclides emitting particles with short particle ranges, for example, Auger electron emitters, alpha-emitters, and low-energy beta-emitters. From the aforementioned discussion, it can be seen that the threshold dose constraints determined from external beam RT are directly applicable to defining the maximum treatment dose in PRRT with ^{90}Y-DOTA^0-Tyr^3-octreotide, but less so with ^{177}Lu-DOTA^0-Tyr^3-octreotate where it most likely overestimates toxicity. This distribution is not observed in SPECT imaging, and when the dose distributions inside the kidney are compared for ^{111}In, ^{90}Y, and ^{177}Lu, based on quantitative SPECT/CT-based dosimetry, no difference between the distribution patterns was observed [66].

REFERENCES

[1] Emami, B.; Lyman, J.; Brown, A.; et al. International Journal of radiation Oncology, Biology, Physics 1991, 21, 109–122.

[2] Marks, L. B.; Yorke, E. D.; Jackson, A.; et al. International Journal of radiation Oncology, Biology, Physics 2010, 76, S10–S19.

[3] Lassmann, M.; Hänscheid, H.; Chiesa, C.; et al. European Journal of Nuclear Medicine and Molecular Imaging 2008, 35, 1405–1412.

[4] Benua, R.; Cicale, N.; Sonenberg, M.; et al. American Journal of Roentgenology 1962, 87, 171–182.

[5] Verburg, F. A.; Hänscheid, H.; Biko, J.; et al. European Journal of Nuclear Medicine and Molecular Imaging 2010, 37, 896–903.

[6] Imhof, A.; Brunner, P.; Marincek, N.; et al. Journal of Clinical Oncology 2011, 29, 2416–2423.

[7] Stabin, M. G.; Sparks, R. B.; Crowe, E. Journal of Nuclear Medicine 2005, 46, 1023–1027.

[8] Sgouros G. Journal of Nuclear Medicine 2005, 46, 18S–27S.

[9] Forrer, F.; Krenning, E. P.; Kooij, P.; et al. European Journal of Nuclear Medicine and Molecular Imaging 2009, 36, 1138–1146.

[10] Siegel, J. A.; Thomas, S. R.; Stubbs, J. B.; et al. Journal of Nuclear Medicine 1999, 40, 37S–61S.

[11] Bolch, W. E.; Bouchet, L. G.; Robertson, J. S.; et al. Journal of Nuclear Medicine 1999, 40, 11S–36S.

[12] Cremonesi, M.; Ferrari, M.; Bodei, L.; et al. Journal of Nuclear Medicine 2006, 47, 1467–1475.

[13] Minarik, D.; Sjögreen-Gleisner, K.; Linden, O.; et al. Journal of Nuclear Medicine 2010, 51, 1974–1978.

[14] Walrand, S.; Jamar, F.; Mathieu, I.; et al. European Journal of Nuclear Medicine and Molecular Imaging 2003, 30, 354–361.

[15] Walrand, S.; Flux, G. D.; Konijnenberg, M. W.; et al. European Journal of Nuclear Medicine and Molecular Imaging 2011, 38, S57–S68.

[16] Helisch, A.; Förster, G. J.; Reber, H.; et al. European Journal of Nuclear Medicine and Molecular Imaging 2004, 31, 1386–1392.

[17] Walrand, S.; Jamar, F.; van Elmbt, L.; et al. Journal of Nuclear Medicine 2010, 51, 1969–1973.

[18] Hammond, P. J.; Wade, A. F.; Gwilliam, M. E.; et al. British Journal of Cancer 1993, 67, 1437–1439.

[19] Kwekkeboom, D. J.; Bakker, W. H.; Kooij, P. P.; et al. European Journal of Nuclear Medicine 2001, 28, 1319–1325.

[20] Kwekkeboom, D. J.; de Herder, W. W.; Kam, B. L.; et al. Journal of Clinical Oncology 2008, 26, 2124–2130.

[21] Kwekkeboom, D. J.; Krenning, E. P.; Lebtahi, R.; et al. Neuroendocrinology 2009, 90, 220–226.

[22] Jamar, F.; Barone, R.; Mathieu, I.; et al. European Journal of Nuclear Medicine and Molecular Imaging 2003, 30, 510–518.

[23] Bodei, L.; Cremonesi, M.; Zoboli, S.; et al. European Journal of Nuclear Medicine 2003, 30, 207–216.

[24] Bodei, L.; Cremonesi, M.; Grana, C.; et al. European Journal of Nuclear Medicine and Molecular Imaging 2004, 31, 1038–1046.

[25] Rolleman, E. J.; Valkema, R.; de Jong, M.; et al. European Journal of Nuclear Medicine and Molecular Imaging 2003, 30, 9–15.

[26] Rolleman, E. J.; Melis, M.; Valkema, R.; et al. European Journal of Nuclear Medicine and Molecular Imaging 2010, 37, 1018–1031.

[27] Barone, R.; Pauwels, S.; De Camps, J.; et al. Nephrology Dialysis Transplantation 2004, 19, 2275–2281.

[28] Barone, R.; Borson-Chazot, F.; Valkema, R.; et al. Journal of Nuclear Medicine 2005, 46, 99S–106S.

[29] Bushnell, D.; O'Dorisio, T. M.; O'Dorisio, M. S.; et al. Cancer Biotherapy and Radiopharmaceuticals 2004, 19, 35–41.

[30] Bushnell, D. L.; O'Dorisio, T. M.; O'Dorisio, M. S.; et al. Journal of Clinical Oncology 2010, 28, 1652–1659.

[31] Wehrmann, C.; Senftleben, S.; Zachert, C.; et al. Cancer Biotherapy and Radiopharmaceuticals 2007, 22, 406–416.

[32] Claringbold, P. G.; Brayshaw, P. A.; Price, R. A.; et al. European Journal of Nuclear Medicine and Molecular Imaging 2011, 38, 302–311.

[33] Garkavij, M.; Nickel, M.; Sjögreen-Gleisner, K.; et al. Cancer 2010, 116, 1084–1092.

[34] Sandström, M.; Garske, U.; Granberg, D.; et al. European Journal of Nuclear Medicine and Molecular Imaging 2010, 37, 212–225.

[35] Grozinsky-Glasberg, S.; Barak, D.; Fraenkel, M.; et al. Cancer 2011, 117, 1377–1385.

[36] Eckerman, K.; Endo, A., ed. *MIRD: Radionuclide data and decay schemes*, 2nd edition, Society of Nuclear Medicine, Reston, 2009.

[37] ICRP publication 41, Non-stochastic effects of ionizing radiation, Ann ICRP, 1984, 14.

[38] ICRP Publication 53 Radiation Dose to Patients from Radiopharmaceuticals, Ann ICRP, 1988, 18.

[39] Wessels, B. W.; Konijnenberg, M. W.; Dale, R. G.; et al. Journal of Nuclear Medicine 2008, 49, 1884–1899.

[40] Valkema, R.; De Jong, M.; Bakker, W. H.; et al. Seminars in Nuclear Medicine 2002, 32, 110–122.

[41] Barone, R.; Walrand, S.; Konijnenberg, M.; et al. Nuclear Medicine Communications 2008, 29, 283–290.

[42] Waldherr, C.; Pless, M.; Maecke, H. R.; et al. Annals of Oncology 2001, 12, 941–945.

[43] Paganelli, G.; Zoboli, S.; Cremonesi, M.; et al. European Journal of Nuclear Medicine 2001, 28, 426–434.

[44] Valkema, R.; Pauwels, S. A.; Kvols, L. K.; et al. Journal of Nuclear Medicine 2005, 46, 83S–91S.

[45] Valkema, R.; Pauwels, S.; Kvols, L. K.; et al. Seminars in Nuclear Medicine 2006, 36, 147–156.

[46] Bodei, L.; Cremonesi, M.; Ferrari, M.; et al. European Journal of Nuclear Medicine and Molecular Imaging 2008, 35, 1847–1856, Erratum in Eur J Nucl Med Mol Imaging 2008, 35, 1928.

[47] Walrand, S.; Barone, R.; Pauwels, S.; et al. European Journal of Nuclear Medicine and Molecular Imaging 2011, 38, 1270–1280.

[48] Kwekkeboom, D. J.; Bakker, W. H.; Kam, B. L.; et al. European Journal of Nuclear Medicine and Molecular Imaging 2003, 30, 417–422.

[49] Kwekkeboom, D. J.; Teunissen, J. J.; Bakker, W. H.; et al. Journal of Clinical Oncology 2005, 23, 2754–2762.

[50] Van Essen, M.; Krenning, E. P.; Bakker, W. H.; et al. European Journal of Nuclear Medicine and Molecular Imaging 2007, 34, 1219–1227.

[51] De Keizer, B.; van Aken, M. O.; Feelders, R. A.; et al. European Journal of Nuclear Medicine and Molecular Imaging 2008, 35, 749–755.

[52] Esser, J. P.; Krenning, E. P.; Teunissen, J. J.; et al. European Journal of Nuclear Medicine and Molecular Imaging 2006, 33, 1346–1351.

[53] Teunissen, J. J.; Krenning, E. P.; de Jong, F. H.; et al. European Journal of Nuclear Medicine and Molecular Imaging 2009, 36, 1758–1766

[54] Sandström, M.; Garske-Román, U.; Granberg, D.; et al. Journal of Nuclear Medicine 2013, 54, 33–41.

[55] Swärd, C.; Bernhardt, P.; Ahlman, H.; et al. World Journal of Surgery 2010, 34, 1368–1372.

[56] Khan, S.; Krenning, E. P.; van Essen, M.; et al. Journal of Nuclear Medicine 2011, 52, 1361–1368.

[57] Dawson, L. A.; Kavanagh, B. D.; Paulino, A. C.; et al. International Journal of Radiation Oncology, Biology, Physics 2010, 76, S108–S115.

[58] De Jong, M.; Valkema, R.; Van Gameren, A.; et al. Journal of Nuclear Medicine 2004, 45, 1168–1171.

[59] Konijnenberg, M.; Melis, M.; Valkema, R.; et al. Journal of Nuclear Medicine 2007, 48, 134–142.

[60] Stabin, M. G. Journal of Nuclear Medicine 2008, 49, 853–860.

[61] Cremonesi, M.; Ferrari, M.; Zoboli, S.; et al. European Journal of Nuclear Medicine 1999, 26, 877–886.

[62] Konijnenberg, M. W. Cancer Biotherapy and Radiopharmaceuticals 2003, 18, 619–625.

[63] Cremonesi, M.; Ferrari, M.; Di Dia, A.; et al. The Quarterly Journal of Nuclear Medicine and Molecular Imaging 2011, 55, 155–167.

[64] Flandrin, G. Classification of myelodysplasic syndromes. Atlas of Genetics and Cytogenetics in Oncology and Haematology. May 2002.

[65] Hall, P.; Boice, J.; Berg, G.; et al. Lancet 1992, 340, 1–4.

[66] Baechler, S.; Hobbs, R. F.; Boubaker, A.; et al. Medical Physics 2012, 39, 6118–6128.

6.3

PEPTIDE RECEPTOR RADIONUCLIDE THERAPY (PRRT): CLINICAL APPLICATION

LISA BODEI AND GIOVANNI PAGANELLI

Division of Nuclear Medicine, European Institute of Oncology, Milan, Italy

ABBREVIATIONS

BED	biological effective dose or bioeffective dose
DLT	dose-limiting toxicity
DOTA	$1,4,7,10$-tetraazacyclododecane-N,N',N'',N'''-tetraacetic acid
GBq	gigabecquerel
PRRT	peptide receptor radionuclide therapy
PET	positron emission tomography
PET/CT	positron emission tomography/computed tomography
QoL	quality of life
Sqm	square meter
SWOG	Southwestern Oncology Group
RECIST	Response Evaluation Criteria In Solid Tumors
WHO	World Health Organization
^{90}Y-DOTA-Tyr^3-octreotide	or ^{90}Y-DOTATOC or ^{90}Y-octreotide
^{177}Lu-DOTA-Tyr^3-octreotate	or ^{177}Lu-DOTATATE or ^{177}Lu-octreotate

Somatostatin Analogues: From Research to Clinical Practice, First Edition. Edited by Alicja Hubalewska-Dydejczyk, Alberto Signore, Marion de Jong, Rudi A. Dierckx, John Buscombe, and Christophe Van de Wiele.
© 2015 John Wiley & Sons, Inc. Published 2015 by John Wiley & Sons, Inc.

INTRODUCTION

Neuroendocrine tumors are generally slow growing and are frequently discovered when metastatic spread has occurred. This leaves room for multiple treatments, individualized through a multidisciplinary approach that considers tumor type, extension, and related symptoms.

The therapeutic choice depends on the aim of the therapy, whether to slow the tumor growth and ameliorate the symptoms or to eradicate the disease. Available options include surgery, interventional radiology techniques, and medical therapies, such as somatostatin analogues, alpha-interferon, chemotherapy, new biomolecular targeted therapies, and peptide receptor radionuclide therapy (PRRT) with radiolabeled somatostatin analogues [1].

Neuroendocrine tumors overexpress somatostatin receptors on their cell membrane, particularly the subtype 2. This constitutes the biomolecular basis for the use of somatostatin analogues in therapy.

"Cold" somatostatin analogues have an important symptomatic effect, and as recently demonstrated, they exert an action on tumor growth by limiting its natural progression [2]. Nevertheless, the escape from the pharmacologic effect of octreotide or lanreotide, the so-called tachyphylaxis, commonly and early occurs [3].

In the past two decades, a new approach to neuroendocrine tumors based on specific receptor targeting with radionuclides was introduced in the clinical practice in many European centers (the Netherlands, Switzerland, Italy, Germany, Poland, etc.). PRRT consists in the systemic administration of a synthetic analogue, radiolabeled with a suitable beta-emitting radionuclide. These compounds are able to irradiate tumors and their metastases via the internalization through a specific receptor subtype, generally overexpressed on the cell membrane. Preclinical studies have indicated many potential receptor candidates for PRRT. To date, the mostly exploited system is the somatostatin receptor. After initial experiences with ^{111}In-pentetreotide, to date, the most used radiopharmaceuticals are ^{90}Y-DOTA-Tyr^3-octreotide or ^{90}Y-DOTATOC or ^{90}Y-octreotide and ^{177}Lu-DOTA-Tyr^3-octreotate or ^{177}Lu-DOTATATE or ^{177}Lu-octreotate.

PRRT with either ^{90}Y-octreotide or ^{177}Lu-octreotate can deliver consistent radiation doses to lesions, adequate to achieve significant tumor responses.

RATIONALE

The biomolecular basis of PRRT consists in the selective irradiation of tumor cells, which derives from the radioactivity transported inside the tumor cell after the internalization of the receptor–radioanalogue complex. Since newly formed blood vessels growing around tumor cells express somatostatin receptors, mainly of subtype 2, PRRT can also generate a parallel antiangiogenic response.

First experiences of PRRT began in the early 1990s, when the high-energy and short-range Auger and conversion emissions of In-111, normally used in diagnostics, began to be exploited for therapeutic purposes. Clinical trials using high activities

(~6.7 GBq) of ^{111}In-pentetreotide in patients affected by neuroendocrine tumors were carried out [4].

Despite these premises, objective responses were rare (2.5% partial responses). The efficiency of Auger electron emitters is known to be dependent on their incorporation in the double strand of DNA. The nuclear uptake of ^{111}In-pentetreotide, in fact, is not significant, and this is possibly the cause of the poor results observed in patients.

The next logical step was to utilize high-energy beta-emitters, such as Yttrium-90. The higher energy (maximum 2.2 MeV) and penetration range (R_{95} 5.7 mm) of electrons from ^{90}Y resulted more advantageous, with a direct killing of somatostatin receptor-positive cells and a cross-fire effect that kills nearby receptor-negative tumor cells.

A new somatostatin analogue, named Tyr[3]-octreotide, with receptor affinity similar to that of octreotide, was developed. The 1,4,7,10-tetraazacyclododecane-N, N',N'',N'''-tetraacetic acid (DOTA) chelated form can be easily labeled with radiometals, such as In-111 and Y-90. The newly formed molecule is [^{90}Y-DOTA][0]-Tyr[3]-octreotide or ^{90}Y-DOTATOC or ^{90}Y-octreotide.

More recently, a new analogue, named octreotate (chemically a Tyr[3],Thr[8]-substituted octreotide), was synthesized [5]. The new compound has a six- to ninefold higher affinity for somatostatin receptor subtype 2, the one that is most frequently overexpressed by neuroendocrine tumors. Its DOTA chelated form, [DOTA][0]-Tyr[3]-octreotate or DOTATATE, has been labeled with the beta–gamma-emitter Lutetium-177, which has a lower energy (maximum 0.49 MeV) and penetration range (maximum 2 mm), a longer half-life (6.7 days), and a gamma photonic emission allowing imaging and dosimetry at the same time [6].

CLINICAL PROTOCOL

Patients candidate to PRRT with radiolabeled somatostatin analogues are those with tumor lesions significantly overexpressing somatostatin receptors. Among the inclusion criteria, the presence of functioning receptors, namely, able to internalize the receptor–analogue complex and retain the radioactivity inside the cell, is considered of crucial importance for an efficient therapy.

In order to be admitted to the therapeutic phase, in fact, patients must be selected according to their OctreoScan scintigraphy (or, more recently, their receptor PET with Gallium-68-labeled octreotide). These images must show an adequate uptake (at least equal to the one of the normal liver), namely, an adequate expression of somatostatin receptors, to the lesions in order to plan a high dose to the tumor and a low one on normal tissues.

Scintigraphic evaluation is to date the most accurate method to check for the overexpression of functioning receptors. Compared to immunohistochemistry, which shows that detail at the time of the biopsy, the scintigraphic method allows the evaluation of the receptor density and the internalization capacity at present time.

When evaluating receptor scintigraphy for PRRT selection, it is important to exclude possible false positives, such as the gallbladder, accessory spleens, recent surgical scars, previous radiotherapy, and any other cause of granulomatous-lymphoid infiltrate that can mimic neuroendocrine tumor tissue.

Potential causes of false negatives must be excluded as well. These are mainly represented by small, subcentimetric lesions, under the resolution limit of gamma-camera (although this limitation is partially overcome by receptor PET/CT), and receptor-negative lesions due to recent chemotherapy or poorly differentiated disease.

PRRT consists in the systemic administration of the radiopeptide. The treatment is fractionated in multiple cycles until the maximum cumulative administrable activity is reached. This amount is able to efficiently irradiate the tumor, without trespassing the conventional 25–27 Gy absorbed dose threshold to the kidneys, which are the dose-limiting organs. Recently, it has been observed that switching from absorbed dose to biological effective dose, or BED, this dose threshold value is slightly higher.

Therapy cycles are given 6–9 weeks apart, in order to cover from possible hematological toxicity.

In order to reduce renal dose, patients are premedicated with an intravenous infusion of positively charged amino acids, starting from the 2–3 preceding up to the 2–3 h following PRRT.

Lysine or arginine, 25 g/day, is coinfused with the aim of conveniently hydrating the patients and reducing the renal dose by competitive inhibition of the proximal tubular reabsorption of the radiopeptide.

Radiopeptide therapy is slowly administered intravenously in about 100 ml of saline in about 20 min.

Possible gastrointestinal symptoms, generally a mild nausea, more rarely vomit, may occur with the amino acid coadministration.

EFFICACY

In almost 15 years of academic phase II trials, despite the lack of homogeneity among studies, PRRT with either ^{90}Y-octreotide or ^{177}Lu-octreotate proved to be efficient, with tumor responses in more than 30% of patients, symptom relief and quality of life (QoL) improvement, biomarker reduction, and, ultimately, an impact on survival.

The most used radiopeptide in the first 8–10 years of experience is ^{90}Y-octreotide. All the published results derive from different phase I–II studies, inhomogeneous as to inclusion criteria and treatment schemes. Hence, a direct comparison is virtually impossible to date. Nevertheless, even with these limitations, objective responses are registered in 10–34% of patients [7–12].

Figure 6.3.1 reports an example of objective response in a patient affected by a well-differentiated neuroendocrine carcinoma from unknown primary with diffuse bone marrow metastases.

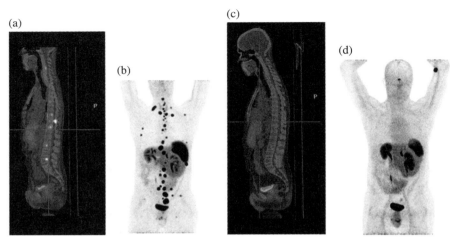

FIGURE 6.3.1   Morphologic response in a patient affected by bone marrow metastases and previous liver metastases from a well-differentiated neuroendocrine carcinoma, Ki-67 10% with an unknown primary. The patient underwent a right liver hepatectomy with a homolateral adrenalectomy. After surgery, the patient was treated with ¹⁷⁷Lu-DOTATATE (24.6 GBq cumulative activity). The evaluation of the status at baseline by means of ⁶⁸Ga-DOTATOC PET/CT (a, fusion sagittal image; b, MIP image) shows diffuse bone marrow involvement, particularly in the axial skeleton. The evaluation performed after the end of PRRT by means of ⁶⁸Ga-DOTATOC PET/CT (c, fusion sagittal image; d, MIP image) shows the almost complete disappearance of the lesions. Please note the hypertrophy of the left liver lobe, as a consequence of the right lobe resection. (*See insert for color representation of the figure.*)

In a study carried out at Basel University, 39 patients with neuroendocrine tumors, mostly of gastroenteropancreatic (GEP) origin, were treated with 4 cycles of ⁹⁰Y-octreotide, with a cumulative activity of 7.4 GBq/sqm. Objective responses, according to WHO criteria, were described in 23%, with a complete remission in 2 patients, partial in 7, and a disease stabilization in 27. Pancreatic neuroendocrine tumors (13 patients) showed a better objective response (38% partial + complete) than the other classes did. A significant amelioration of related symptoms occurred in the majority of patients [10].

In another multicenter phase I study, carried out at Rotterdam, Louvain, and Tampa Universities, 60 patients affected by GEP neuroendocrine tumors were treated with 4 cycles of 0.9, 1.8, 2.8, 3.7, 4.6, and 5.5 GBq/sqm administered 6–9 weeks apart. In a first evaluation of the results, published in 2002 in 32 evaluable patients, objective responses, according to SWOG criteria, consisted in about 9% of partial responses and 9% of minor responses [13]. In a subsequent analysis of the same patient population, published in 2006 on 58 assessable patients who were treated with cumulative activities of 1.7–32.8 GBq, a 57% clinical benefit, including stabilization and minor responses, was observed, according to SWOG criteria. A true objective response was described in 5% of the patients. The most relevant finding of the study was the observed overall survival, with a median value of about 37 months and a median progression-free survival (PFS) of about 29 months. These results compared well with the 12-month overall survival of the historical group treated with ^{111}In-pentetreotide. The median PFS was 29 months [11].

The results of two phase I–II studies and a retrospective evaluation in 141 patients were published by the Milano group in 2004. Patients were affected mainly by neuroendocrine tumors and were treated with a cumulative activity of 7.4–26.4 GBq of ^{90}Y-octreotide, divided into 2–16 cycles, administered 4–6 weeks apart. Objective response rate was 26%, including partial and complete responses, according to SWOG criteria. Disease stabilization was observed in 55% of the patients and disease progression in 18%. The mean duration of response ranged between 2 and 59 months (median 18). The majority of the patients who responded had GEP neuroendocrine tumors. The study showed that, by dividing the objective response according to the basal status, stable patients at baseline had better outcome (partial and complete responses in 32%) than did progressive ones (partial and complete responses in 24%) [9].

Recently, a multicenter study aimed at studying the role of ^{90}Y-octreotide in 90 patients with symptomatic, metastatic "carcinoids," namely, neuroendocrine tumors originating from the small intestines, was published. This study showed that PRRT with ^{90}Y-octreotide was able to induce stabilization or tumor response, according to SWOG criteria, in 74% of patients as well as a durable amelioration of all the symptoms related to the tumor mass and the hypersecretion of bioactive amines. Interestingly, the symptomatic response had an impact on survival, since PFS was significantly longer in those who had durable diarrhea improvement [14].

More recently, the Basel group published the results of their open-label phase II trial in 1109 patients treated with ^{90}Y-octreotide, divided in multiple cycles of 3.7 GBq/m^2 each. Objective morphologic responses, according to RECIST criteria, were observed in 378 (34.1%), biochemical response in 172 (15.5%), and symptomatic in 329 (29.7%). Longer survival was correlated with tumor and symptomatic response. The best predictor of survival was the tumor uptake at baseline [12].

Since its introduction in clinical trials in 2000, the new radiopeptide DOTATATE labeled with Lutetium-177, ^{177}Lu-DOTATATE or ^{177}Lu-octreotate, has gained a great popularity, due to its higher affinity for sst$_2$, its easier manageability, and the possibility of imaging. Nowadays, it is the mostly used radiopeptide.

The first clinical trials were carried out at Rotterdam University. In a preliminary report, 35 patients with GEP neuroendocrine tumors were treated with 3.7, 5.6, or 7.4 GBq of ^{177}Lu-octreotate, up to a final cumulative dose of 22.2–29.6 GBq, with complete and partial responses in 38%, according to WHO criteria. No serious side effects were observed [15].

In a subsequent series enlargement, 131 patients with somatostatin receptor-positive GEP neuroendocrine tumors were treated with cumulative activities of ^{177}Lu-octreotate ranging from 22.2 to 29.6 GBq. In the 125 evaluated patients, according to SWOG criteria, complete remissions were observed in 3 patients (2%), partial remission in 32 (26%), minor response in 24 (19%), and stable disease in 44 patients (35%), while 22 patients (18%) progressed. Better responses were more frequent in case of a high uptake on baseline octreotide scintigraphy and in case of limited liver involvement. On the other hand, progression was significantly more frequent in patients with a low performance score and extensive disease at enrolment. Median time to progression (TTP) was more than 36 months, comparing favorably to chemotherapy [16].

In addition, ^{177}Lu-octreotate significantly improved the global health/QoL and various symptom scales, particularly fatigue, insomnia, pain, as well as role, emotional, and social functions, in patients with metastatic GEP tumors. The effect was more frequent in patients obtaining tumor regression, but surprisingly, it was also observed in progressing patients [17].

Recently, an evaluation of the enlarged series of 504 patients treated with ^{177}Lu-octreotate, 310 of which evaluated for efficacy, confirmed the occurrence of complete and partial remissions in 2 and 28% of cases, with minor responses in 16% and stability in 35%, according to SWOG criteria. However, the most precious information of this study was the impact of PRRT on survival, with a median overall survival >48 months and a median PFS of 33 months. A direct comparison with literature data obtained from similar patients showed an important 40–72-month survival benefit. Although these data are not derived from proper randomized phase III trials, the huge difference in survival is most likely to be caused by a real impact of PRRT. These data compare favorably to other treatments, such as chemotherapy, from the cost/benefit and tolerability point of view [18].

A categorization of objective response showed once more that pancreatic tumors tended to respond better than other GEP tumors, although pancreatic gastrinomas tended to relapse in a shorter interval (median TTP 20 months vs. >36 in the remaining GEP tumors) [16].

A number of phase II studies, mostly retrospective, oriented at defining the objective response in specific classes of diseases, were published in the past years.

A phase II study included traditionally considered "poor responding" tumors, such as bronchial and gastric neuroendocrine carcinomas. Patients were treated with standard 22.2–29.6 GBq activities. Despite the limited number of patients studied, the observed objective response, according to SWOG criteria, was comparable to the one observed in GEP tumors: in bronchial 5/9 partial responses, 1 minor and 2 stabilizations, and in gastric tumors 1/5 complete response, 1 minor and 2 stabilizations. In thymic tumors, the patient series was too small to draw any conclusions [19].

Another phase II study with ^{177}Lu-octreotate administered at standard activities in a small series of patients with "nontypical" neuroendocrine tumors or simply sst$_2$-positive neoplasms, such as paragangliomas (12 patients), meningiomas (5 patients), small cell lung carcinomas (3 patients), and eye melanomas (2 patients), demonstrated that PRRT is able to yield tumor responses, evaluated according to SWOG criteria, in paragangliomas and meningiomas, although with lower rates compared to "typical" neuroendocrine tumors. In small cell lung carcinomas and in melanomas, PRRT with ^{177}Lu-octreotate did not produce any antitumor effect. The limited number of patients may have hampered the study results, particularly for meningiomas, and the need of further studies is suggested by the authors [20].

As regards meningiomas, a dedicated study of PRRT with ^{90}Y-octreotide on a series of 29 patients, 14 with benign forms, 9 atypical, and 6 malignant, was published in 2009 by the Milano group. Patients received cumulative activities of 5–15 GBq divided in 2–6 cycles. According to SWOG criteria, disease stabilization occurred in 19/29 patients (66%), while progression was observed in the other 10 (34%). Better results occurred in patients with low-grade meningiomas, with a median PFS of 61 months, compared to atypical and malignant forms, with a median PFS of 13 months. These data point out the need of treating patients with an adjuvant approach, shortly after surgery, particularly in high-grade meningiomas, where lesions tend to be large and multiple [21].

[177]Lu-octreotate has also been tested in a small group of 5 patients with iodo-negative or resistant differentiated thyroid carcinomas, among which 3 patients with Hürthle cell carcinoma. According to SWOG criteria, 1 patient with Hürthle cell carcinoma had a partial response, 1 patient with Hürthle cell carcinoma had a minor response, and 2 patients, one with a Hürthle and the other with a papillary form, had disease stabilization. These data outline the importance of PRRT with ^{177}Lu-octreotate in a category of endocrine tumors, especially Hürthle cell carcinomas, that has no valuable therapeutic options [22].

Finally, PRRT has also been administered in children and young adults with somatostatin receptor-positive tumors. A phase I study involved 17 subjects (2–24 years old) affected by neuroblastomas, MEN2b, gliomas, and neuroendocrine tumors. Patients were treated with ^{90}Y-octreotide with activities ranging from 1.1 to 1.85 GBq/m^2/cycle for a total of 3 cycles, administered 6 weeks apart. According to the Pediatric Oncology Group criteria, 12% of partial responses and 29% of minor responses were observed. Improved QoL during the treatment was a major advantage of the therapy. Therefore, PRRT proved to be efficient and tolerated also in young patients [23].

Recently, the Rotterdam group published the results of the salvage protocol with ^{177}Lu-octreotate. Patients were enrolled in progression after an initial response to PRRT with ^{177}Lu-octreotate, administered with standard cumulative activities (22.2–29.6 GBq). As a consequence, 32 patients with bronchial or GEP neuroendocrine tumors received 2 additional cycles of ^{177}Lu-octreotate, with a cumulative activity of 15 GBq. A new objective response was obtained in 8 patients (2 partial and 6 minor responses), while stabilization was registered in another 8. Median TTP was 17 months. Both response rate and duration over time appear, therefore, lower than the one obtained with the primary treatment. Nevertheless, salvage therapy was well tolerated by the majority of patients and is, therefore, a valuable option in this category [24].

Presently, following recent tendencies in oncology, PRRT experiences are directing toward combination therapies. Studies of ^{177}Lu-octreotate combined with the radiosensitizer chemotherapy agent capecitabine have been published in the past years. The first study showing encouraging results was carried out in a small group of 7 patients with progressing GEP neuroendocrine tumors. Patients were treated with 4 cycles of standard activities of ^{177}Lu-octreotate plus capecitabine (1650 mg/sqm) for the following 2 weeks. The evaluation showed the absence of severe toxicity, particularly hand–foot syndrome or hematological or renal toxicity. Objective responses were indeed observed [25].

A recent phase II study was completed in 35 patients with progressing neuroendo-crine tumors. Patients were treated with 4 cycles of 7.8 GBq of ^{177}Lu-octreotate plus capecitabine, 1650 mg/m^2, for the following 2 weeks. According to RECIST criteria, a 24% objective response was observed, with a 70% stable disease and 6% progres-sion, without adjunctive toxicity [26].

TOLERABILITY

After 15 years of experience, we can state that, from the safety point of view, PRRT is generally well tolerated. Acute side effects are usually mild: some of them are related to the coadministration of amino acids, such as nausea (or more rarely vomit). Others are related to the radiopeptide, such as fatigue (commonly), a modest alopecia (possible after ^{177}Lu-octreotate), or the exacerbation of an endocrine syndrome, which rarely occurs in functioning tumors. Chronic and permanent effects on target organs, particularly the kidneys and the bone marrow, are generally mild if the necessary precautions are taken.

Dosimetric studies indicated that it is possible to deliver elevated absorbed doses to the tumor, with a relative sparing of normal organs, especially kidneys and bone marrow.

While hematological toxicity mainly occurs immediately after PRRT and is usually mild and transient, permanent and severe bone marrow toxicity is a rare event after PRRT, being the resulting bone marrow absorbed doses usually well below the threshold of toxicity [27, 28]. In contrast, delayed renal toxicity that may occur if the dose threshold is exceeded is permanent. The kidney is, in fact, the critical organ at the doses that are normally reached with PRRT. Commonly accepted data, from external beam radiotherapy experience, indicate threshold values for toxicity of about 25–27 Gy. Recently, radiobiology concepts applied to nuclear medicine therapy indicated that bioeffective doses (BED) are more accurate in predicting toxicity and 40 Gy was indicated as a more likely threshold for renal toxicity [29, 30]. Nephrotoxicity derives from the proximal tubular reabsorption of the radiopeptide. The strategy of coinfusing positively charged amino acids, such as lysine or arginine, competitively inhibiting the radiopeptide reabsorption, leads to a 9–53% reduction of the renal doses [31].

Despite renal protection, a loss of renal function of variable degree (generally mild) may appear months after the end of PRRT. Cases of severe, end-stage renal damage are nowadays rare and mainly occurred at the beginning of PRRT experience, when renal protection was not in use and very high activities were administered [32].

Nevertheless, independently of the development of a clinically evident renal damage, renal function is "consumed" to some extent after PRRT, and a decline in creatinine clearance occurs with time. This loss is more pronounced for ^{90}Y-octreotide (median 7.3% per year) than for ^{177}Lu-octreotate (median 3.8% per year) [33].

Studies demonstrated that a higher and more persistent decline in creatinine clearance and the subsequent development of renal toxicity occurred in those patients with risk factors for delayed renal toxicity, particularly long-standing and poorly controlled diabetes and hypertension. In these patients, the renal BED threshold results about 28 Gy, lower than the one of patients without risk factors. Therefore, a preliminary screening of the candidate patients before PRRT appears essential [34].

From a hematological point of view, PRRT is generally well tolerated. Severe, WHO grade 3 or 4, toxicity does not occur in more than 13% of cases after ^{90}Y-octreotide and 2–3% after ^{177}Lu-octreotate [8, 12].

Previous dose-finding phase I studies demonstrated that maximum cumulative administrable activity per cycle of ^{90}Y-octreotide, with renal protection, is 5.18 GBq [35]. No dose-finding studies have been conducted for ^{177}Lu-octreotate, and studies employing the dose-limiting toxicity (DLT) method were abandoned, since literature data showed the possibility to safely use 7.4 GBq as a maximum activity per cycle and dosimetric studies showed the advantage of hyperfractionation in lowering the renal and bone marrow dose (L. Bodei, EANM'10 Annual Congress of the European Association of Nuclear Medicine, Vienna, Personal communication, 2010; [28, 16, 34]).

^{177}Lu-octreotate demonstrated a higher tolerability compared to ^{90}Y-octreotide, largely due to the physical characteristics of the two radioisotopes.

Phase I–II studies have demonstrated that absorbed and bioeffective doses delivered to kidneys and bone marrow remain within the thresholds (L. Bodei, EANM'10 Annual Congress of the European Association of Nuclear Medicine, Vienna, Personal communication, 2010).

Finally, the possibility of exacerbation of endocrine syndromes, such as carcinoid, hypoglycemic, or Zollinger–Ellison syndrome, must be taken into account, although quite rare. This appears to be related to the massive cell lysis and hormonal stimulation occurring after PRRT that must be prevented and treated accordingly [36, 37].

CONCLUSIONS

PRRT, either with ^{90}Y-octreotide or ^{177}Lu-octreotate, demonstrated to be efficient and relatively safe up to the known thresholds of absorbed and bioeffective dose. Toxicity profile is acceptable from a renal and hematological point of view, if necessary protective measures are taken.

PRRT proved to induce a significant improvement of the QoL and of all the symptoms related to the disease in more than 70% of treated patients. Differently from other treatments applied in neuroendocrine tumors, PRRT with either ^{90}Y-octreotide or ^{177}Lu-octreotate showed a median TTP of at least 30 months. Moreover, patients responding to PRRT with a stabilization or a tumor reduction (accounting for about 75% of treated population) obtain a significant impact on survival (40–72 months from diagnosis). For these reasons, despite the absence of randomized controlled trials, PRRT is considered as a standard of care in the treatment of neuroendocrine tumors, and it is rightfully inserted in many therapeutic algorithms and considered by scientific societies dedicated to neuroendocrine tumors [38–40].

REFERENCES

[1] Modlin, I. M.; Oberg, K.; Chung, D. C.; et al. Lancet Oncology 2008, 9, 61–72.

[2] Rinke, A.; Müller, H. H.; Schade-Brittinger, C.; et al. Journal of Clinical Oncology 2009, 27, 4656–4663.

[3] Hofland, L. J.; Lamberts, S. W. Endocrine Reviews 2003, 24, 28–47.

[4] Valkema, R.; de Jong, M.; Bakker, W. H.; et al. Seminars in Nuclear Medicine 2002, 32, 110–122.

[5] Heppeler, A.; Froidevaux, S.; Maecke, H. R.; et al. European Journal of Chemistry 1999, 7, 1974–1981.

[6] de Jong, M.; Breeman, W. A.; Bernard, B. F.; et al. International Journal of Cancer 2001, 92, 628–633.

[7] Paganelli, G.; Zoboli, S.; Cremonesi, M.; et al. European Journal of Nuclear Medicine 2001, 28, 426–434.

[8] Kwekkeboom, D. J.; Mueller-Brand, J.; Paganelli, G.; et al. Journal of Nuclear Medicine 2005, 46, 62S–66S.

[9] Bodei, L.; Cremonesi, M.; Grana, C.; et al. European Journal of Nuclear Medicine and Molecular Imaging 2004, 31, 1038–1046.

[10] Waldherr, C.; Pless, M.; Maecke, H. R.; et al. Journal of Nuclear Medicine 2002, 43, 610–616.

[11] Valkema, R.; Pauwels, S.; Kvols, L. K.; et al. Seminars in Nuclear Medicine 2006, 36, 147–156.

[12] Imhof, A.; Brunner, P.; Marincek, N.; et al. Journal of Clinical Oncology 2011, 29, 2416–2423.

[13] de Jong, M.; Valkema, R.; Jamar, F.; et al. Seminars in Nuclear Medicine 2002, 32, 133–140.

[14] Bushnell, D. L.; O'Dorisio, T. M.; O'Dorisio, M. S.; et al. Journal of Clinical Oncology 2010, 28, 1652–1659.

[15] Kwekkeboom, D. J.; Bakker, W. H.; Kam, B. L.; et al. European Journal of Nuclear Medicine and Molecular Imaging 2003, 30, 417–422.

[16] Kwekkeboom, D. J.; Teunissen, J. J.; Bakker, W. H.; et al. Journal of Clinical Oncology 2005, 23, 2754–2762.

[17] Teunissen, J. J.; Kwekkeboom, D. J.; Krenning, E. P. Journal of Clinical Oncology 2004, 22, 2724–2729.

[18] Kwekkeboom, D. J.; de Herder, W. W.; Kam, B. L.; et al. Journal of Clinical Oncology 2008, 26, 2124–2130.

[19] van Essen, M.; Krenning, E. P.; Bakker, W. H.; et al. European Journal of Nuclear Medicine and Molecular Imaging 2007, 34, 1219–1227.

[20] van Essen, M.; Krenning, E. P.; Kooij, P. P.; et al. Journal of Nuclear Medicine 2006, 47, 1599–1606.

[21] Bartolomei, M.; Bodei, L.; De Cicco, C.; et al. European Journal of Nuclear Medicine and Molecular Imaging 2009, 36, 1407–1416.

[22] Teunissen, J. J.; Kwekkeboom, D. J.; Kooij, P. P.; et al. Journal of Nuclear Medicine 2005, 46, 107S–114S.

[23] Menda, Y.; O'Dorisio, M. S.; Kao, S.; et al. Journal of Nuclear Medicine 2010, 51, 1524–1531.

[24] van Essen, M.; Krenning, E. P.; Kam, B. L.; et al. Journal of Nuclear Medicine 2010, 51, 383–390.

[25] van Essen, M.; Krenning, E. P.; Kam, B. L.; et al. European Journal of Nuclear Medicine and Molecular Imaging 2008, 35, 743–748.

[26] Claringbold, P. G.; Brayshaw, P. A.; Price, R. A.; et al. European Journal of Nuclear Medicine and Molecular Imaging 2011, 38, 302–311.

[27] Cremonesi, M.; Ferrari, M.; Bodei, L.; et al. Journal of Nuclear Medicine 2006, 47, 1467–1475.

[28] Cremonesi, M.; Botta, F.; Di Dia, A.; et al. The Quarterly Journal of Nuclear Medicine and Molecular Imaging 2010, 54, 37–51.

[29] Dale, R. Cancer Biotherapy and Radiopharmaceuticals 2004, 19, 363–370.

[30] Barone, R.; Borson-Chazot, F.; Valkema, R.; et al. Journal of Nuclear Medicine 2005, 46, 99S–106S.

[31] de Jong, M.; Krenning, E. P. Journal of Nuclear Medicine 2002, 43, 617–620.

[32] Cybulla, M.; Weiner, S. M.; Otte, A. European Journal of Nuclear Medicine 2001, 28, 1552–1554.

[33] Valkema, R.; Pauwels, S. A.; Kvols, L. K.; et al. Journal of Nuclear Medicine 2005, 46, 83S–91S.

[34] Bodei, L.; Cremonesi, M.; Ferrari, M.; et al. European Journal of Nuclear Medicine and Molecular Imaging 2008, 35, 1847–1856.

[35] Bodei, L.; Cremonesi, M.; Zoboli, S.; et al. European Journal of Nuclear Medicine and Molecular Imaging 2003, 30, 207–216.

[36] de Keizer, B.; van Aken, M. O.; Feelders, R. A.; et al. European Journal of Nuclear Medicine and Molecular Imaging 2008, 35, 749–755.

[37] Davì, M. V.; Bodei, L.; Francia, G.; et al. Journal of Endocrinological Investigation 2006, 29, 563–567.

[38] Kwekkeboom, D. J.; Kam, B. L.; van Essen, M.; et al. Endocrine-Related Cancer 2010, 17, R53–R73.

[39] Boudreaux, J. P.; Klimstra, D. S.; Hassan, M. M.; et al. Pancreas 2010, 39, 753–766.

[40] Kwekkeboom, D. J.; Krenning, E. P.; Lebtahi, R.; et al. Neuroendocrinology 2009, 90, 220–226.

6.4

DUO-PRRT OF NEUROENDOCRINE TUMORS USING CONCURRENT AND SEQUENTIAL ADMINISTRATION OF Y-90- AND LU-177-LABELED SOMATOSTATIN ANALOGUES

RICHARD P. BAUM AND HARSHAD R. KULKARNI

THERANOSTICS Center for Molecular Radiotherapy and Molecular Imaging, ENETS Center of Excellence, Zentralklinik Bad Berka, Germany

ABBREVIATIONS

BBNETC	the Bad Berka neuroendocrine tumor center
BUN	blood urea nitrogen
CgA	chromogranin A
CR	complete remission
CT	computed tomography
GFR	glomerular filtration rate
FDG	fluorodeoxyglucose
MDS	myelodysplastic syndrome
MIRD	Medical Internal Radiation Dose
MR	minor response

Somatostatin Analogues: From Research to Clinical Practice, First Edition. Edited by
Alicja Hubalewska-Dydejczyk, Alberto Signore, Marion de Jong, Rudi A. Dierckx,
John Buscombe, and Christophe Van de Wiele.
© 2015 John Wiley & Sons, Inc. Published 2015 by John Wiley & Sons, Inc.

MRI   magnetic resonance imaging
NET   neuroendocrine tumors
PET   positron emission tomography
pNET   pancreatic NET
PR   partial remission
PRRT   peptide receptor radionuclide therapy
SMS   somatostatin
SSTR   somatostatin receptors
SUV   standardized uptake value
TER   tubular extraction rate

INTRODUCTION

The heterogeneous nature of neuroendocrine tumors (NETs) and their indolent course together with lack of optimal conventional therapy for the management of advanced cases pose a challenge for the oncologists. The overexpression of somatostatin receptors (SSTRs) on their cell surface enables the effective treatment of NETs with peptide receptor radionuclide therapy (PRRT) using radiolabeled somatostatin (SMS) analogues. The previous authors have already discussed the concepts and clinical application of PRRT. In this chapter, we would focus on Duo-PRRT.

RADIONUCLIDES

The most frequently used radionuclides in PRRT are Yttrium-90 (^{90}Y) and Lutetium-177 (^{177}Lu). These radionuclides have different physical characteristics, for example, beta-particles are emitted at different energies, resulting in various tissue penetration ranges. A pronounced "cross-fire effect" is found when using ^{90}Y, since the beta-particles are emitted with a relatively high energy, resulting in a tissue penetration range of up to 12 mm, which may be preferable for larger tumors. While this cross-fire effect is beneficial to overcome the inhomogeneity of receptor expression in cancer cells, allowing irradiation of tumor cells that are not directly targeted by the radiopharmaceutical, the long range of the ^{90}Y beta-particles appears to increase the potential for kidney toxicity [1]. ^{177}Lu, on the other hand, emits intermediate-energy beta-particles, resulting in a tissue penetration range of up to 2 mm, which may be preferable for smaller tumors. In addition, ^{177}Lu has two gamma peaks at 113 and 208 keV, which makes it suitable for imaging with a gamma-camera for posttherapeutic dosimetry. Another difference between these radionuclides is their physical half-life. Although the influence of the physical half-life is not yet fully understood, it is very likely that it influences the therapeutic as well as the secondary effects. For ^{90}Y, it is 2.7 days, whereas the physical half-life of ^{177}Lu is more than double (6.7 days). These differences in characteristics in turn influence the decision on the choice of radionuclide (Table 6.4.1) to be used for PRRT.

TABLE 6.4.1 **Factors taken into consideration for deciding upon the PRRT dose/schedule and regime ^{177}Lu versus ^{90}Y**

Tumor related	Non-tumor related
SSTR expression on the tumor cells	Age
Size of the metastatic lesions	Long-standing diabetes mellitus
Number of lesions	Uncontrolled arterial hypertension
Presence of necrosis in the metastatic lesions	Nephrotoxic chemotherapy
Proliferation Rate (Ki-67/MiB1)	Radiation therapy with kidney in the field of radiation
Neoadjuvant setting primary as in inoperable NET	Single kidney
Possibility/necessity of combined internal radionuclide therapy with external radiation therapy	Preexisting myelosuppression
	Tumor-related cachexia (BMI <18.5)
	Carcinoid heart disease (Hedinger's syndrome)

CONCEPT OF COMBINED PRRT

Pragmatically, during PRRT, the presence of different sizes of tumors with inhomogeneous distribution of SSTR in a particular NET patient should be taken into consideration to ensure that most of the beta-energy from the radioisotope is absorbed and the radiation dose to the tumor is optimized [2]. A conceivable solution to take this heterogeneity into account is the use of a combination of the radionuclides ^{177}Lu and ^{90}Y, with considerably different energies and tissue penetration ranges of the emitted beta-particles. Sequential administration of ^{90}Y- and ^{177}Lu-labeled analogues also is helpful for the treatment of larger tumors, followed by treatment of smaller metastases respectively in further treatment cycles. This concept of Duo-PRRT refers to the use of ^{90}Y- and ^{177}Lu-labeled SMS analogues (DOTATATE or DOTATOC) in sequence, that is, in two different settings, 3–6 months apart from each other. Tandem-PRRT, on the other hand, specifically refers to the concurrent use of these radioisotopes, that is, in the same setting. There has been some difference in opinion about the nomenclature. Where some groups have used the term Tandem-PRRT, in our opinion, it was rather Duo-PRRT that was being investigated [3]. In a study by de Jong et al., favorable results were achieved using a combination of ^{177}Lu- and ^{90}Y-labeled SMS analogues when compared with either ^{90}Y- or ^{177}Lu-labeled analogue alone, in animals bearing tumors of various sizes [4]. The results of a recent study by Kunikowska et al. indicated that Tandem-PRRT (with ^{90}Y/^{177}Lu-DOTATATE) provided longer overall survival than with a single radioisotope (^{90}Y-DOTATATE) and the safety of both methods was comparable [5].

INDIVIDUALIZED PRRT

The PRRT regimen should take into account all clinical aspects and molecular features of NET in a particular patient [6]. Many groups have resorted to schematic treatment with fixed schedules and doses. The administration of PRRT in over 1300 patients to date at the Bad Berka neuroendocrine tumor center (BBNETC) has been individualized and based on the evaluation of Bad Berka score based on specific selection criteria (Table 6.4.2). Apart from the size of the tumor, an important consideration on the choice of radionuclide for PRRT is the localization of metastases of NET. Bone metastases have been demonstrated to be effectively controlled by PRRT using ^{177}Lu with long progression-free and overall survival [7].

The BBNETC group in 2003 since the availability of ^{177}Lu pioneered the systematic use of Duo- as well as Tandem-PRRT in a large patient group of progressive NETs, nonresponsive to octreotide/interferon treatment or chemotherapy (Figs. 6.4.1, 6.4.2, and 6.4.3). More than 450 patients have been treated with Duo-PRRT. Frequent therapy cycles (4–6 and up to 10) applying low or intermediate doses of radioactivity are suitable for these relatively slow-growing tumors (long-term low-dose, not short-term high-dose concept). Despite the documentation of clinical value of PRRT in NETs, the potential renal toxicity must be kept in mind.

Due to the small size, radiopeptides are filtered through the glomerular capillaries in the kidneys and subsequently reabsorbed by and retained in the proximal tubular cells. Hence, kidneys are the dose-limiting organs [1, 8]. Given the high kidney retention of radiopeptides, positively charged molecules, such as L-lysine and/or L-arginine, are used to competitively inhibit the proximal tubular reabsorption of the radiopeptide [8]. In addition, the patients should be well hydrated. Based on the

TABLE 6.4.2   Bad berka score of patient selection for PRRT is based on specific molecular and clinical features of NET

Molecular features	Clinical features
SUV on receptor PET/CT (for referrals: Krenning's score on OctreoScan) for determining SMS receptor density	Renal function (GFR measured by Tc-99m DTPA and TER measured by Tc-99m MAG3 as well as serum creatinine and BUN) and elimination kinetics
Ki-67 index/tumor grade	Hematological status (blood counts)
FDG status (glucose hypermetabolism of tumors/metastases)	Liver involvement
Tumor dynamics (doubling time, new lesions)	Extrahepatic tumor burden
Functional activity of tumor	Karnofsky's performance score (KPS) or ECOG scale
	Weight loss
	Time since first diagnosis
	Previous therapies

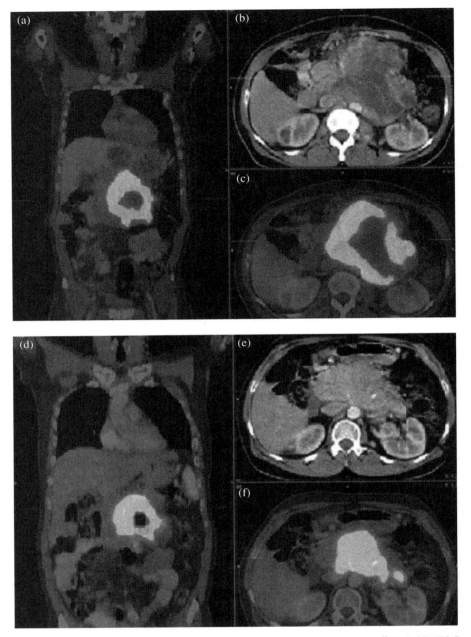

FIGURE 6.4.1 Partial remission after Tandem-PRRT demonstrated by ^{68}Ga-DOTATOC PET/CT (a, b, c, before; d, e, f, after Tandem-PRRT using 4000 MBq ^{177}Lu-DOTATOC and 4500 MBq ^{90}Y-DOTATOC) of a very large, partially necrotic nonfunctional well-differentiated neuroendocrine neoplasm in the pancreas tail and body, with intense somatostatin receptor expression. The tumor was infiltrating the neighboring structures and caused life-threatening recurrent bleedings which stopped after PRRT. SUV_{max} dropped from 39.9 to 35.3, molecular tumor volume dropped from 542 to 214 ml, and the size overall diminished (a and d, coronal fused PET/CT images; b and e, CT images in transverse view; c and f, fused PET/CT images in transverse view). (*See insert for color representation of the figure.*)

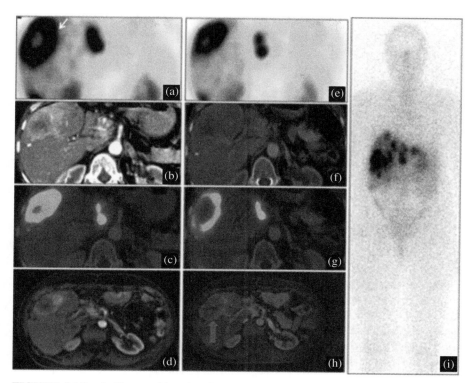

FIGURE 6.4.2 A 52-year-old patient had nonfunctioning, well-differentiated neuroendo-crine tumor of the pancreatic tail with local extension into the stomach and neighboring organs, lymph node metastases, as well as extensive bilobar liver metastases and few bone metastases and underwent Tandem-PRRT with 5.5 GBq of ^{177}Lu and 3.5 GBq of ^{90}Y. ^{68}Ga-DOTATOC PET showed good molecular response in the lesion in segment S8 of the liver (white arrow). There was 47% fall in the SUV from 70.8 to 37.7. T1-weighted MRI with contrast showed posttherapy increase in the size of the lesion but with progressive necrosis (long arrow) and hypervascularity in the rim (short bold arrow) indicating a favorable therapeutic response. The left panel (a–d) shows pretherapy and middle panel (e–h) shows posttherapy images in transverse view (a and e, PET; b and f, CT; c and G-PET/CT fused; d and h, MRI). The whole body planar scan (i) acquired 44 h posttherapy (^{177}Lu gamma-energy window) showed good uptake and long retention in the metastases. (*See insert for color representation of the figure.*)

animal experiments of the groups in Rotterdam and Amsterdam, our group has pio-neered the clinical use of gelofusine in addition to amino acids for kidney protection. Cumulative and per-cycle renal absorbed dose, age, hypertension, and diabetes are considered as contributing factors to the decline of renal function after PRRT [1]. At the BBNETC, renal function is serially determined by ^{99m}Tc-MAG3/tubular extrac-tion rate (TER) and by ^{99m}Tc-DTPA/GFR measurements. All data is entered in a pro-spective structured database (programmed in ACCESS and containing 284 items per patient).

(a)   (b)

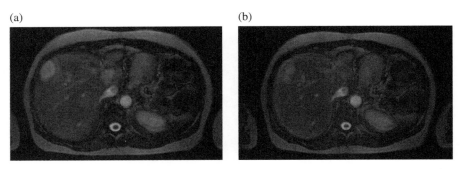

FIGURE 6.4.3 Fused PET/MRI images in transverse view in the same patient (a, using ^{68}Ga-DOTATOC; b, using ^{18}F-FDG) show mismatch of the uptake of the 2 tracers (^{68}Ga-DOTATOC $>^{18}$F-FDG) in the lesion in segment S8 of the liver, demonstrating good differentiation and high somatostatin receptor density. (*See insert for color representation of the figure.*)

Before each new treatment cycle, restaging should be performed by morphologic (CT/MRI) and molecular imaging (^{68}Ga-DOTA-SMS PET/CT; in selected cases, ^{18}F-FDG or ^{18}F-fluoride PET/CT studies should be additionally performed), blood chemistry, and tumor markers (CgA, serotonin, specific hormones). Another very important aspect of personalized medicine and THERANOSTICS is dosimetry. Estimation of tumor and normal organ doses (MIRD/OLINDA) performed after PRRT (using Lu-177-labeled SMS analogues DOTATATE or DOTATOC) is important to ensure that maximum dose is delivered to the tumors and therefore optimizing an individualized treatment protocol [9].

Besides renal toxicity, bone marrow involvement must be considered although it appears not to be a principal dose-limiting factor. Acute hematological toxicity is not uncommon, especially after ^{90}Y-labeled peptide therapy, and the possibility of a mild but progressive impoverishment in bone marrow reserves has to be considered in repeated cycles. In addition, myelodysplastic syndrome (MDS) or overt leukemia may develop in patients receiving high bone marrow doses, especially in those previously treated with alkylating agents [10].

^{177}Lu-DOTATATE was predominantly used for small metastases or in patients with impaired renal or hematological function. A recent analysis of 416 patients (all NET subtypes) treated at the BBNETC showed a median overall survival from the time of first diagnosis of 210 months and a median survival after first PRRT of 59 months (Fig. 6.4.4). This experience confirms an earlier report about overall survival benefit [11]. Long-term follow-up of up to 7 years after Duo-PRRT showed no significant grade 3 or grade 4 nephrotoxicity attributed to concurrent or sequential Duo-PRRT. The median fall in TER was lesser in patients undergoing Duo-PRRT than in those undergoing PRRT with ^{177}Lu or ^{90}Y alone.

In patients with progressive disease (PD) with NET of nonpancreatic origin and pancreatic NET (pNET), tumor response after a mean follow-up of 2 years was as follows: after 3 PRRT cycles, complete remission (CR), partial remission (PR), and

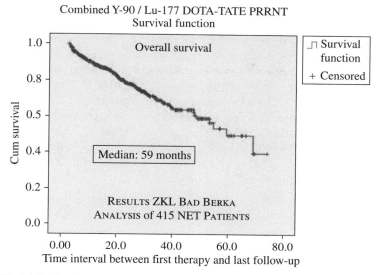

FIGURE 6.4.4 Survival curve depicting the overall survival of patients undergoing Duo-PRRT at the BBNETC.

minor responses (MR) were seen in 52% of patients with pNET (and in 48% in non-pNET); disease was stabilized in 39% with pNET and in 45% of patients with non-pNET. Thirty-six patients with advanced disease died of PD. Objective tumor responses (including improvement of clinical symptoms) were seen in 93% (91% pNET) of the patients. Significant hematological toxicity (mainly erythrocytopenia, rarely neutropenia, and thrombocytopenia) occurred in less than 15% of all patients. MDS developed in five patients (all of whom had also received chemotherapy before). End-stage renal insufficiency was not observed in any of the patients with normal kidney function before PRRT. In most patients receiving ^{177}Lu-DOTATATE alone ($n = 417$ cycles), serum creatinine and TER/GFR did not change significantly. Therefore, the probability and magnitude of renal toxicity can be significantly reduced (or completely avoided) when PRRT is administered in fractionated doses in patients without any preexisting risk factors and under appropriate nephroprotection [12]. Chemotherapy, diabetes mellitus, hypertension, Hedinger's syndrome, and cachexia were identified as the risk factors for nephrotoxicity after PRRT.

We also have treated patients with progressive metastases of NET and with a single functional kidney (24 patients). None of these patients showed grade 3 or 4 nephrotoxicity. PRRT resulted in PR in 36% and stable disease in 36% of the pts; 28% had PD. Fourteen had grade 1 erythrocytopenia, three had grade 1 leukocytopenia, and three had grade 1 thrombocytopenia. In 2009, we have given fractionated low-dose PRRT to two patients on hemodialysis due to end-stage renal insufficiency (to the best of our knowledge, this was the first ever worldwide experience with such a condition). No significant hematotoxicity was observed in the two patients on dialyses, and both showed a good clinical and objective therapy response [13].

SUMMARY

Duo-PRRT using concurrent and sequential administration of [90]Y- and ^{177}Lu-labeled SMS analogues is highly effective in the management of NETs, even in advanced cases, and may be more effective in progressive NETs than using either radionuclide alone. Apart from a benefit in overall survival from time of diagnosis of several years, significant improvement in clinical symptoms and excellent palliation can be achieved. In patients with progressive NET, fractionated, personalized PRRT with lower doses of radioactivity given over a longer period of time results in good therapeutic responses. By this concept, severe hematological and/or renal toxicity can be avoided, and the quality of life can be improved. PRRT should only be performed at specialized centers for individualized interdisciplinary treatment and long-term care.

REFERENCES

[1] Valkema, R.; Pauwels, S. A.; Kvols, L. K.; et al. Journal of Nuclear Medicine 2005, 46, 83S–91S.

[2] O'Donoghue, J. A.; Bardies, M.; Wheldon, T. E. Journal of Nuclear Medicine 1995, 36, 1902–1909.

[3] Seregni, E.; Maccauro, M.; Coliva, A.; et al. The Quarterly Journal of Nuclear Medicine and Molecular Imaging 2010, 54, 84–91.

[4] De Jong, M.; Breeman, W. A.; Valkema, R.; et al. Journal of Nuclear Medicine 2005, 46, 13S–17S.

[5] Kunikowska, J.; Królicki, L.; Hubalewska-Dydejczyk, A.; et al. European Journal of Nuclear Medicine and Molecular Imaging 2011, 38, 1788–1797.

[6] Baum, R. P.; Kulkarni, H. R.; Carreras, C. Seminars in Nuclear Medicine 2012, 42, 190–207.

[7] Ezziddin, S.; Sabet, A.; Heinemann, F.; et al. Journal of Nuclear Medicine 2011, 52, 1197–1203.

[8] Bodei, L.; Cremonesi, M.; Grana, C.; et al. European Journal of Nuclear Medicine and Molecular Imaging 2004, 31, 1038–1046.

[9] Wehrmann, C.; Senftleben, S.; Zachert, C.; et al. Cancer Biotherapy and Radio-pharmaceuticals 2007, 22, 406–416.

[10] Kwekkeboom, D. J.; Mueller-Brand, J.; Paganelli, G.; et al. Journal of Nuclear Medicine 2005, 46, 62S–66S.

[11] Kwekkeboom, D. J.; de Herder, W. W.; Kam, B. L.; et al. Journal of Clinical Oncology 2008, 26, 2124–2130.

[12] Prasad, V.; Hoersch, D.; Zachert, C.; et al. Journal of Nuclear Medicine 2011, 52, 299.

[13] Prasad, V.; Kulkarni, H.; Zachert, C.; et al. Journal of Nuclear Medicine 2011, 52, 1741.

6.5

NONSYSTEMIC TREATMENT OF LIVER METASTASES FROM NEUROENDOCRINE TUMOR

DANIEL PUTZER[1], GERLIG WIDMANN[1], DIETMAR WAITZ[2], WERNER JASCHKE[1], AND IRENE J. VIRGOLINI[2]

[1]Department of Radiology, Innsbruck Medical University, Innsbruck, Austria
[2]Department of Nuclear Medicine, Innsbruck Medical University, Innsbruck, Austria

ABBREVIATIONS

90Y	90Yttrium
177Lu	177Lutetium
68Ga	68Gallium
99mTc-MAA	Technetium-99m albumin aggregated
ALT	alanine transaminase
CT	computed tomography
DOTATOC	DOTA-(0)-Phe(1)-Tyr(3)-octreotide
ECOG	Eastern Cooperative Oncology Group
Gy	Gray
i.a.	intra-arterial
NET	neuroendocrine tumor
MBq	megabecquerel
MRI	magnetic resonance imaging
PET	positron emission tomography
PRRT	peptide receptor radionuclide therapy

Somatostatin Analogues: From Research to Clinical Practice, First Edition. Edited by Alicja Hubalewska-Dydejczyk, Alberto Signore, Marion de Jong, Rudi A. Dierckx, John Buscombe, and Christophe Van de Wiele.
© 2015 John Wiley & Sons, Inc. Published 2015 by John Wiley & Sons, Inc.

RFA   radiofrequency ablation
RILD   radiation-induced liver disease
SIRT   selective internal radiation therapy
SPECT   single photon emission computed tomography
TACE   transarterial chemoembolization
US   ultrasound
srfa   stereotactic radiofrequency ablation
TAME   transarterial mechanical embolizations

TRANSARTERIAL PEPTIDE RECEPTOR RADIONUCLIDE THERAPY

This approach is based on the idea of intra-arterial (i.a.) administration of DOTA-(0)-Phe(1)-Tyr(3)-octreotide (DOTATOC), resulting in regionally intensified peptide receptor radionuclide therapy (PRRT) [1]. McStay et al. showed that hepatic i.a. injection of 90Y-DOTA-lanreotide is a safe and effective palliative treatment for patients with progressive large-volume somatostatin receptor-positive liver metastases from neuroendocrine tumors (NETs) [2].

The selective binding process of SST analogues to the SST receptor on the NET cell surface is the underlying principle for this new treatment, which offers the possibility of combining the effect of a transarterial therapy with selective internal radiation [3]. This makes transarterial PRRT an attractive approach for patients with disseminated liver metastases and patients who are not eligible to radiofrequency ablation (RFA) or TACE.

Requirements can be evaluated before treatment planning and include a normal renal function, which can be confirmed by renal scintigraphy, as well as knowledge on the arterial perfusion of the liver metastases to be treated. However, patients with high tumor burden to the liver should be given special attention, as these patients have an increased risk of developing radiation-induced liver disease (RILD) [4].

DOTATOC is radiolabeled using 68Ga to evaluate the tumor uptake *in vivo*, performing positron emission tomography/computed tomography (PET/CT) imaging. High uptake values have been proven for radiopharmaceuticals applied intra-arterially, leading to lower radiation exposure in healthy tissue. First studies report on differences in uptake of DOTATOC when comparing intravenous (i.v.) injection to i.a. injection [3].

Angiographic intervention forms the basis in this interdisciplinary treatment approach and requires the access to the celiac axis via a transfemoral access. A microcatheter is then inserted coaxially and advanced to the common hepatic artery, proper hepatic artery, taking into consideration possible anatomic variations.

Due to the fact that tissue distribution 72 h after i.a. application equals distribution second to i.v. application, isotopes with shorter half-life periods are of interest for this procedure. 177Lu has some disadvantages when applied intra-arterially in comparison to 90Y, due to a lower binding affinity to the peptide and a longer half-life period. Using less peptide during synthesis is supposed to lead to improved results of i.a. therapy [5].

National radiation protection regulations are of high relevance when considering i.a. application of PRRT, taking into consideration the risk of contamination during the application process as well as the risk of postinterventional complications such as bleeding. Furthermore, the transport of the radioactive substance from the radiopharmaceutical laboratory to the angiographic intervention center as well as the application has to be done according to national regulations [6]. Available treatment results up to now are comparable to results from i.v. PRRT [5, 7].

RFA

In percutaneous image-guided RFA, probes are inserted into the tumor using ultrasound (US), CT, or magnetic resonance imaging (MRI) [8]. The tumor is subsequently devitalized by thermal ablation avoiding surgical resection. Microwave ablation and electrovaporization are less frequently applied ablation technologies of local tumor destruction. The basic principle of electrovaporization is destruction of cell membranes by an electrical current, not by heat.

When comparing surgical resection to RFA, surgery is associated with a morbidity of 15–45% and mortality of 1–5% [9, 10]. RFA represents a low-risk procedure [11]. Major complications requiring intervention such as intraperitoneal bleeding, liver abscess, intestinal perforation, pneumothorax and hemothorax, or bile duct injury are in the range of 2–3% and may often be treated by minor radiological interventions such as drainage or coil embolization [12, 13]. The procedure-related mortality rate is below 1%. Minor complications amount to 5–9% and include the postablation syndrome (fever up to 38.5°C, weakness, fatigue, and leukocytosis). Blood loss, prolonged postoperative intensive care stay, and abdominal wall insufficiency are not associated with RFA. A large prospective study by Mazzaglia et al. of RFA with US guidance included 63 patients with unresectable hepatic metastases from NETs. The median survival after the first ablation was 3.9 years. A significant difference was noted between patients whose largest metastasis exceeded 3 cm in size (median survival <3 years) and those whose dominant lesion was smaller than 3 cm (median survival not reached by study closure). In over 90%, symptomatic improvement was reported posttherapeutically, and the median duration of symptom control was 11 months [14].

Gillams et al. treated 19 patients with 36 percutaneous RFA procedures. During a median follow-up of 21 months, the authors found a complete response (CR) in 6 patients, a partial response (PR) in 7, and stable disease (SD) in 1. Symptomatic disease was reduced in nine (69%) of the 14 patients suffering from hormone overproduction. One patient passed away due to carcinoid crisis. The median postablation survival was 29 months [15].

The feasibility and success of RFA depend on the size and location of the liver lesion. Lesions sized larger than 1 cm usually require more than one probe or several probe positions in order to treat the tumor with overlapping ellipsoidal (single straight electrodes) or spherical (single expandable electrodes) ablation zones, respectively [16]. Incomplete RFA and as a consequence local recurrence are generally related to large size of tumors (>3–5 cm), poor tumor visibility and unfavorable distribution of probes, as well as imprecise probe positioning and cooling effects by larger vessels. The technical limitations of conventional single-needle in-plane techniques using US or CT have been largely overcome by recent multineedle approaches using CT-based 3D treatment planning and stereotactic needle guidance (SRFA) [17]. SRFA allows effective and safe treatment of large-volume disease [18]. In analogy to surgical R0 resection, A0 ablation including a 3D safety margin of at least 5 mm can be objectively verified and documented, by fusion of postablation and preablation images. The use of multimodal fused images from PET/CT during SRFA may permit selective treatment of active metastasis as determined by 68Ga-DOTATOC uptake.

In selected patients, optimized RFA technique can provide comparable oncologic results to resection. Therefore, feasibility of RFA should be discussed in interdisciplinary oncologic boards.

SELECTIVE INTERNAL RADIATION THERAPY

Selective internal radiation therapy (SIRT) is a promising new treatment approach for tumors of various entities, such as hepatocellular carcinoma, colorectal cancer, and NETs [19] (Fig. 6.5.1 and Table 6.5.1).

Compared with transarterial mechanical embolizations (TAME) and chemoembolization (TACE), SIRT combines the therapeutic principles of embolization and internal radiation. Selective treatment of unresectable liver metastases is extending the treatment options. High local radiation doses can be selectively targeted to the tumor due to the fact that metastatic hepatic tumors derive 80–100% of their blood supply from the arterial rather than the portal hepatic circulation, while the normal liver tissue is supplied by the portal vein (60–70%).

Patients considered for SIRT are characterized by unresectable hepatic primary or metastatic cancer, liver-dominant tumor burden, and a life expectancy of at least 3 months [20]. Before patients are selected for SIRT, the arterial blood supply has to be investigated; patency of the portal vein is not mandatory for treatment. However, intrahepatic arteriovenous shunts need to be excluded.

90Y is a β-emitter, resulting from neutron bombardment of 89Y in a reactor, with a tissue penetration of mean, 2.5 mm, and max, 11 mm, and short half-life (64 h), which makes it attractive for i.a. treatment. The emitter is loaded to microspheres in the range of 20–40 μm, enabling delivery to the tumors via the hepatic artery. TheraSphere consists of insoluble glass microspheres where 90Y is an integral constituent of the glass. On the other hand, resin 90Y microspheres have received FDA premarket approval for hepatic metastases from colorectal cancer, concurrent with fluorodeoxyuridine. The microspheres are trapped within the tumor vessels, whereas the lower size limit prevents the microspheres from proceeding into the venous circulation. The mean tissue penetration of 2.5 mm spares normal liver parenchyma. Therefore, SIRT can provide high local tumor doses ranging from 50 to 150 Gy, in contrast to the traditional whole-liver external beam radiation where radiation doses have to be limited to 30 Gy [20].

(a) (b) (c) (d)

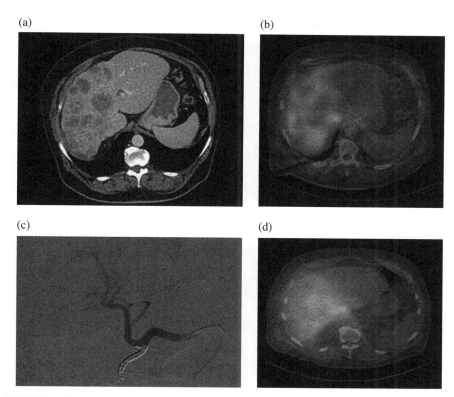

FIGURE 6.5.1 A 67-year-old male patient suffering from an adenocarcinoma of the colon sigmoideum, stage pT3 N1b M1 R0 (K-ras mutation not present), and UICC stage 4. The patient had undergone resection of the primary in 10/2010. He then underwent multiple palliative chemotherapeutic treatments until February 2013, when CT showed progressive liver metastases to the right liver lobe (a). Pulmonary shunt scintigraphy using 150 MBq Tc-99m-MAA on 8/4/2013 revealed faint uptake over both lungs, reaching a pulmonary shunt rate of 14% in the semiquantitative analysis. The scan showed increased liver uptake in the right liver lobe (b). Embolization of the gastroduodenal and right gastric artery was performed using coils (c). Application of 1400 MBq Y-90-SIR-spheres was performed on 18/4/2013. 0.3 GBq was applied in segment IV and 1.11 GBq was applied in the right liver lobe, selectively. Posttherapeutic bremsstrahlung scan revealed increased uptake in the right liver lobe on SPECT/CT imaging, showing no extrahepatic uptake (d). (*See insert for color representation of the figure.*)

The combination of morphological imaging (CT and MRI) with functional imaging, by using combined imaging modalities (PET/CT, SPECT/CT), provides the most reliable information for determining the region of interest. According to the tumor spread in the liver, selective treatment of one or both liver lobes is performed. Sequential treatments may be safer than a whole-liver treatment in 1 session. Sequential treatments allow for a safer approach, maintaining a 30- to 45-day interval between the therapies [19, 21].

Selective application of 99mTc-MAA is done in the right and left hepatic arteries. Planar scintigraphic images are performed for dosimetric measurements. Direct

TABLE 6.5.1 Results from SIRT as treatment from liver metastases from NET

Author	N	Treatment	ORR %	SD %	Symptomatic response	Median TTP	Median survival
Kennedy[a]	148	SIR-spheres	63.2	22.7	n.r.	n.r.	70 months
King[a]	34	SIR-spheres	50	14.7	55	n.r.	35.2
Meranze[a]	10	SIR-spheres	40	60	n.r.	n.r.	28
Wagner[a]	9	SIR-spheres	78	22	n.r.	12 months	36
Kennedy[a]	18	SIR-spheres	89	n.r.	n.r.	n.r.	27
Murthy[b]	8	SIR-spheres	12.5	50	n.r.	n.r.	14 months

n.r., not reported; ORR, objective response rate; SD, stable disease; TTP, time to progression.

[a]SIRT as first-line treatment in therapy-refractory NET.

[b]SIRT as salvage therapy.

tracking of microsphere distribution during application is not feasible and usually not required while using glass microspheres, but is mandatory for resin microspheres because of the relatively high embolic tendency.

Bremsstrahlung scintigraphy up to 24 h after application of the microspheres assists in documenting the distribution of microspheres within the liver.

Extrahepatic distribution of microspheres to visceral sites such as the stomach, duodenum, gallbladder, or pancreas may be associated with the risk of severe radiation damage leading to pain, ulceration, and possibly perforation [22–24]. This complication is to be avoided by obligatory embolization of the relevant sidebranches of the hepatic artery prior to SIRT. Successful embolization of dangerous sidebranches can be assessed by the pretreatment 99mTc-MAA scan. Absolute contraindications for SIRT are excessive shunting to the lungs as quantified by the 99mTc-MAA scan that would result in 30 Gy lung dose on a single administration, a high tumor burden with limited hepatic reserve or signs of reduced liver function as potentially indicated by elevated levels of bilirubin, highly elevated liver enzymes (aspartate transaminase or alanine transaminase (ALT), significantly altered international normalized ratio or partial thromboplastin time, or reduced serum albumin.

Patients who have undergone prior radiotherapy involving the liver should be evaluated on a case-by-case basis. Low ECOG performance score, history of infections to the bile system, and portal hypertension are risk factors.

Posttherapeutic complications have been described, including liver failure. Postinterventional pain has to be considered. Postembolization syndrome includes fatigue, nausea, and abdominal pain. On a regular basis, patients receive a gastrointestinal ulcer prophylaxis and antinausea prophylaxis.

Radiation-induced pneumonitis and elevation of liver function tests, RILD and development of anicteric ascites and increased abdominal girth, as well as rapid weight gain with hypoalbuminemia have been observed posttherapeutically. Bile duct complications include radiation-induced necrosis. First imaging as routine follow-up consisting of abdominal MRI and a metabolic imaging, normally PET/CT, is performed 4 weeks posttherapeutically. The next series of imaging are performed 3, 6, 9, and 12 months after treatment.

CLINICAL RESULTS OF THE SIRT IN PRIMARY AND SECONDARY HEPATIC MALIGNANCIES

Gray et al. [25] reported on 29 consecutive patients with advanced liver metastases from colorectal cancer, showing response rate of 89%. Mean and median survival for patients with metastases confined to the liver was 14.5 and 13.5 months.

In a randomized trial, Van Hazel et al. [26] compared the response rate and time to progression disease in a regimen of systemic fluorouracil/leucovorin chemotherapy versus the same chemotherapy along with a single administration of SIR-spheres in patients with advanced colorectal liver metastases. The time to progressive disease and median survival were significantly longer for patients receiving the combination treatment.

Rhee et al. in a study with 42 patients who underwent SIRT using glass or resin microspheres demonstrated that 92% of glass and 94% of resin patients were classified as PR or SD at 6 months after treatment [27].

King et al. treated 34 patients who had unresectable NET [28]. Symptomatic responses were observed in 18 of 33 patients (55%) at 3 months and in 16 of 32 patients (50%) at 6 months. Radiological liver responses were observed in 50% of patients and included 6 (18%) CR and 11 (32%) PR, and both had a mean overall survival of 29.4 ± 3.4 months. In patients who had evaluable CgA marker levels, there was a decrease in CgA marker levels after RE.

CONCLUSION

In confined liver metastasis from NET, RFA and more recently SRFA are powerful local curative treatment options. Selective i.a. administration of DOTATOC and SIRT are promising new approaches towards a regionally intensified treatment regimen in large bilobar and disseminated liver disease [29–32]. Whether i.a. administration of PRRT results in a higher tumor uptake or intensified therapeutic efficacy in comparison to i.v. PRRT still has to be investigated. However, SIRT has already proven to be a feasible therapeutic option. All available therapeutic approaches have to be inserted in an inter- and multidisciplinary treatment approach for metastatic NET patients.

REFERENCES

[1] Ruszniewski, P.; O'Toole, D. Neuroendocrinology 2004, 80, 74–78.

[2] McStay, M. K.; Maudgil, D.; Williams, M.; et al. Radiology 2005, 237, 718–726.

[3] Kratochwill, C.; Giesel, F. L.; Lòpez-Benitez, R.; et al. Clinical Cancer Research 2010, 16, 2899–2905.

[4] Limouris, G. S.; Chatziioannou, A.; Kontogeorgakus, D.; et al. European Journal of Nuclear Medicine and Molecular Imaging 2008, 35, 1827–1837.

[5] Kratochwil, C.; Lòpez-Benitez, R.; Mier, W.; et al. Endocrine-Related Cancer 2011, 18, 595–602.

[6] Nazario, J.; Gupta, S. Seminars in Oncology 2010, 37, 118–126.

[7] Kwekkeboom, D. J.; Mueller-Brand, J.; Paganelli, G.; et al. Journal of Nuclear Medicine 2005,46, 62S–66S.

[8] Goldberg, S. N.; Gazelle, G. S.; Mueller, P. R. Americal Journal of Roentgenology 2000, 174, 323–331.

[9] Choti, M. A.; Sitzmann, J. V.; Tiburi, M. F.; et al. Annals of Surgery 2002, 235, 759–66.

[10] Virani, S.; Michaelson, J. S.; Hutter, M. M.; et al. Journal of the American College of Surgeons 2007, 204, 1284–1292.

[11] Mulier, S.; Mulier, P.; Ni, Y.; et al. The British Journal of Surgery 2002, 89, 1206–1222.

[12] Livraghi, T.; Solbiati, L.; Meloni, MF.; et al. Radiology 2003, 226, 441–451.

[13] Curley, S. A.; Marra, P.; Beaty, K.; et al. Annals of Surgery 2004, 239, 450–458.

[14] Mazzaglia, P. J.; Berber, E.; Milas, M.; et al. Surgery 2007, 142, 10–19.

[15] Gillams, A.; Cassoni, A.; Conway, G.; et al. Abdominal Imaging 2005, 30, 435–441.

[16] Widmann, G.; Bodner, G.; Bale, R. Cancer Imaging 2009, 9, S63–S67.

[17] Bale, R.; Widmann, G.; Stoffner, D. I. European Journal of Radiology 2010, 75, 32–36.

[18] Widmann, G.; Schullian, P.; Haidu, M.; et al. Cardiovascular and Interventional Radiology 2012, 35, 570–580.

[19] Ahmadzadehfar, H.; Biersack, H. J.; Ezziddin, S. Seminars in Nuclear Medicine 2010, 40, 105–121.

[20] Memon, K.; Lewandowski, R. J.; Mulcahy, M. F.; et al. International Journal of Radiation Oncology Biology Physics 2012, 83, 887–894.

[21] Paprottka, P. M.; Hoffmann, R. T.; Haug, A.; et al. Cardiovascular and Interventional Radiology 2012, 35, 334–342.

[22] Silvanto, A.; Novelli, M.; Lovat, L. Histopathology 2009, 55, 114–115.

[23] Konda, A.; Savin, M. A.; Cappell, M. S.; et al. Gastrointestinal Endoscopy 2009, 70, 561–567.

[24] Crowder, C. D.; Grabowski, C.; Inampudi, S.; et al. The Americal Journal of Surgical Pathology 2009, 33, 93–975.

[25] Gray, B. N.; Anderson, J. E.; Burton, M. A. Australian and New Zealand Journal of Surgery 1992, 62, 105–110.

[26] Van Hazel, G.; Blackwell, A.; Anderson, J.; et al. Journal of Surgical Oncology 2004, 88, 78–85.

[27] Rhee, T. K.; Lewandowski, R. J.; Liu, D. M.; et al. Annals of Surgery 2008, 247, 1029–1035.

[28] King, J.; Quinn, R.; Glenn, D. M.; et al. Cancer 2008, 113, 921–929.

[29] Kulke, M. H.; Scherübl, H. Gastrointestinal Cancer Research 2009, 3, 62–66.

[30] Auernhammer, C. J.; Jauch, K. W.; Hoffmann, J. N. Zentralblatt für Chirurgie 2009, 134, 410–417.

[31] Hoffmann, R.T.; Jakobs, T. F.; Kubisch, C. H.; et al. European Journal of Radiology 2010, 74, 199–205.

[32] Kalinowski, M.; Dressler, M.; König, A.; et al. Digestion 2009, 79, 137–142.

6.6

PEPTIDE RECEPTOR RADIONUCLIDE THERAPY: OTHER INDICATIONS

Agnieszka Stefańska[1], Alicja Hubalewska-Dydejczyk[3], Agata Jabrocka-Hybel[1,2], and Anna Sowa-Staszczak[1]

[1] Department of Endocrinology, University Hospital in Krakow, Krakow, Poland
[2] Department of Endocrinology, Medical College, Jagiellonian University, Krakow, Poland
[3] Department of Endocrinology with Nuclear Medicine Unit, Medical College, Jagiellonian University, Krakow, Poland

ABBREVIATIONS

CCK-2	cholecystokinin 2
EBRT	external beam radiation therapy
GEP-NETs	gastroenteropancreatic neuroendocrine tumors
GH	growth hormone
GLP-1	glucagon-like peptide 1
MIBG	metaiodobenzylguanidine
MTC	medullary thyroid cancer
NET	neuroendocrine tumor
OS	overall survival
PFS	progression-free survival
PRRT	peptide receptor radionuclide therapy
SCLC	small cell lung cancers
SSTR	somatostatin receptors
SRS	somatostatin receptor scintigraphy

Somatostatin Analogues: From Research to Clinical Practice, First Edition. Edited by Alicja Hubalewska-Dydejczyk, Alberto Signore, Marion de Jong, Rudi A. Dierckx, John Buscombe, and Christophe Van de Wiele.

The high expression of somatostatin receptors (SSTRs) is the molecular basis for imaging and radionuclide therapy of neuroendocrine neoplasms. The receptor density and homogeneity are important factors for diagnostic and therapeutic effectiveness of the radiolabeled somatostatin analogues [1, 2]. Gastroenteropancreatic neuroendocrine tumors (GEP-NETs) are not the only neoplasms with high density of SSTRs. Similar density and incidence of SSTRs have been identified in other neuroendocrine tumors such as pheochromocytomas, paragangliomas, and pituitary adenomas but also in tumors of the nervous system such as neuroblastomas, medulloblastomas, and meningiomas [1–12]. Data on the expression of SSTR on glial tumors are conflicting [8, 12–14]. Detection of glial tumors by somatostatin receptor scintigraphy (SRS) depends not only on the expression of SSTR but also on the integrity of the blood–brain barrier [8, 15]. There is limited information on the expression of SSTR in ependymomas—the level of expression is variable and generally lower than that seen in embryonal tumors [8, 16]. The expression of SSTRs, but at lower incidence and density, is also observed on the plasma membrane of the cells of medullary thyroid cancer, small cell lung cancer (SCLC), and melanoma [1, 2, 17, 18]. However, as opposed to neuroendocrine tumors with the homogeneous distribution of the SSTRs, the SSTR expression in those tumors might be heterogeneous [1, 2].

The abovementioned facts are important for the use of radiolabeled somatostatin analogues as the therapeutic agents in case of neoplasms other than gastroenteropancreatic tumors. However, the use of peptide receptor radionuclide therapy (PRRT) in tumors other than GEP-NETs is limited. Most of presented beneath results of clinical trials are promising, but to prove the efficacy and safety of PRRT, more numerous groups of patients should be included.

Paragangliomas are neuroendocrine tumors derived from extra-adrenal autonomic parasympathetic ganglia [19]. Visualization of paragangliomas is possible with the use of metaiodobenzylguanidine (MIBG), which is structurally similar to noradrenaline and is transported into the chromaffin cells and subsequently stored in the secretory vesicles [19]. This allows imaging with [123]I- or ^{131}I-metaiodobenzylguanidine (^{131}I-MIBG), and ^{131}I-MIBG can be used as a therapy [19–21]. Most paragangliomas express also SSTRs [1, 2, 17, 19]. SRS detects more than 90% of known lesions [19, 22]. SRS can be useful also in staging patients with metastatic pheochromocytoma because imaging with labeled MIBG is less sensitive in this group [19, 23]. But in case of primary pheochromocytomas, SRS is less sensitive than ^{123}I-MIBG, partially because of interference from the high physiologic uptake of labeled somatostatin analogue by the kidneys nearby [19]. Radiolabeled somatostatin analogues might be considered as an alternate therapeutic option in paragangliomas and also malignant disseminated pheochromocytomas [19, 24]. Van Essen et al. studied twelve patients with paragangliomas with positive SRS treated with ^{177}Lu-octreotate [19]. Tumor regression was observed in two and disease stabilization in one of four patients with progressive paragangliomas [19]. The authors used also PRRT in five patients with meningioma (results in this group will be discussed in the following), in three patients with SCLC, and two patients with melanoma [19]. The authors concluded that ^{177}Lu-octreotate can have therapeutic effects in paraganglioma and may play a role in the management of this disease, particularly in case of progressive disease with [131]I- or

^{123}I-MIBG-negative lesions, and confirmed expression of the SSTRs [19]. ^{177}Lu-octreotate does not seem to have clinical effects in SCLC and melanoma [19].

In case of the pituitary tumors, treatment depends on the size and hormonal activity of the lesions. Except the prolactin-secreting tumors, for which dopamine agonists are the first-line therapy, the treatment of choice is usually surgery [25–28]. However, prior to the invasive intervention, patients with functioning pituitary adenomas need to be usually prepared with medical therapy. *In vitro* and *in vivo* studies confirmed expression of SSTRs in pituitary tumors [1, 2, 25, 29–31]. Long-acting somatostatin analogues such as octreotide and lanreotide have an established place in the treatment of growth hormone (GH) secreting pituitary adenomas, both in pre- and postsurgical management [32]. There is also novel promising multireceptor ligand somatostatin analogue with affinity for SSTR subtypes 1–3 and 5 used in case of pituitary adenomas causing acromegaly [33]. Somatostatin analogues in combination with dopamine agonists might be used also in case of ACTH-secreting pituitary adenomas [34, 35]. Also, in case of TSH-secreting tumors, medical treatment includes dopaminergic agonists and/or somatostatin analogues [36]. In case of prolactin-secreting tumors resistant to medical treatment and/or with tumor growth-induced visual field impairment, surgery is indicated [25, 28]. Radiotherapy might be also considered; however, the efficacy is delayed and treatment of huge masses causes extensive irradiation of the brain [25, 28]. Similarly, as in case of pheochromocytomas and paragangliomas, there are pituitary tumors resistant to established therapeutic schemes. Therefore, there is a need to search for novel therapies. Baldari et al. reported the effectiveness of radiolabeled somatostatin analogues in case of giant cabergoline- and octreotide-resistant prolactinoma, which caused neurological symptoms [25]. The administration of PRRT led to remarkable tumor shrinkage and a significant improvement in clinical condition [25].

Neuroblastomas derive from the neural crest and synthesize neurotransmitters, including somatostatin [6]. SSTRs (particularly subtype 2) are expressed on the surface of some neuroblastomas, which enables the use of PRRT [4–6]. Neuroblastoma is a typical example of an embryonal tumor of childhood [6, 7, 37, 38]. Neuroblastomas show heterogeneity of behavior ranging from spontaneous regression and differentiation into benign neoplasms in infants to potentially aggressive dissemination in older children [6, 8]. Current treatment for high-risk disease involves an intensive regime with induction chemotherapy, surgery, myeloablative chemotherapy with stem cell rescue, radiotherapy, and continuing therapy with 13-*cis*-retinoic acid and immunotherapy [6, 39–41]. Despite advances in the management of neuroblastoma, the long-term survival rate remains under 40% [6, 39–41]. For patients with relapsed or refractory neuroblastoma, treatment options include further chemotherapy or therapy with ^{131}I-MIBG [6]. The use of radiolabeled somatostatin analogues in neuroblastoma involves a distinct and separate cell-surface molecular target different from the norepinephrine transporter molecule that takes up ^{131}I-MIBG [6]. Menda et al. have conducted phase I trial with the use of ^{90}Y-DOTATOC in children and young adults with refractory solid tumors (neuroblastomas but also paragangliomas and other embryonal and astrocytic brain tumors) expressing SSTRs [7]. PRRT demonstrated a favorable safety profile with no dose-limiting toxicities observed [7].

Gains et al. using data on six children with neuroblastoma who were imaged with ^{68}Ga-DOTATATE PET/CT and subsequently received molecular radiotherapy using ^{177}Lu-DOTATATE provided proof of principle that children with relapsed or primary refractory high-risk neuroblastoma who have SSTR-positive disease can be imaged and treated with SSTR-targeted agents [6].

Medulloblastoma is the most frequent embryonal tumor of the central nervous system [42, 43]. The therapeutic approach for primary medulloblastoma consists of surgery, chemotherapy, and radiotherapy of the craniospinal axis [42]. The recurrence rate for medulloblastoma is observed in more than one third of cases, and recurrences are located either on the primary tumor site or along the cerebrospinal fluid pathways [42]. Based on the fact that medulloblastomas express high levels of SSTR2, Beutler et al. have used targeted radiotherapy (intrathecal administration of ^{90}Y-DOTATOC) to treat tumor remnants, which persisted despite conventional and high-dose chemotherapy and intercurrent resection of the lesion [3, 42]. According to the authors, targeted radiopeptide brachytherapy may represent a promising additional treatment modality for recurrent medulloblastoma when combined with surgery, conventional chemotherapy, and high-dose chemotherapy [42].

Meningiomas arise from cap cells of the arachnoid membrane [44]. In contrast to embryonal tumors such as neuroblastoma and medulloblastoma, the incidence of meningiomas increases with age, but in children, those tumors are more often malignant [44]. Meningiomas are usually benign tumors (WHO grade I), but 6% are atypical (grade II) and 2% malignant (grade III) [44, 45]. Surgery is the treatment of choice and is usually curative [44, 46]. However, recurrence is not uncommon specially for grades II and III [44, 46, 47]. External beam radiation therapy (EBRT) is usually given as a second-line treatment in case of recurrent or inoperable meningiomas [44, 48–53]. But there are still patients left for whom both surgery and radiotherapy are not effective [44]. Therefore, there is a need to search for other new therapeutic modalities [44]. As it was mentioned earlier, meningioma cells express SSTRs usually at high density, which enables both visualization of those tumors by the use of SRS and also is the basis for targeted radionuclide therapy [1, 2, 17, 44, 54, 55]. Bartolomei et al. submit promising results of the therapeutic use of ^{90}Y-DOTATOC in case of 29 patients with recurrent or progressive meningioma after surgery, radiotherapy, and chemotherapy [44]. As the authors concluded, the adjuvant role of radioisotope therapy, soon after surgery, particularly in atypical and malignant meningiomas deserves further investigation [44]. Van Essen et al. confirmed also the efficacy of ^{177}Lu-DOTATATE in case of five patients with meningioma tumors, among them three very large and exophytic tumors; however, response rates in this group of patients were lower than observed in patients with GEP-NETs [19]. Kreissl et al. have also presented results of therapy of 10 patients with unresectable meningioma treated with PRRT (^{177}Lu-DOTATATE or ^{177}Lu-DOTATOC) followed by EBRT [56]. According to the authors, PRRT of meningiomas may be safely used in combination with EBRT [56]. This approach represents an attractive strategy for the treatment of unresectable locally recurring, progressive, or symptomatic meningioma in order to either increase the dose delivered to the tumor or to reduce the dose for organs at risk [56].

SCLC express neuroendocrine markers, among them SSTRs [1, 2, 17, 18]. Therefore, radiolabeled somatostatin analogues can be used for diagnostic scintigraphy and might be considered for therapeutic use [18]. Pless et al. presented report of somatostatin receptor-targeted radiotherapy for SCLC [18]. The authors have treated six patients with SCLC and positive SRS [18]. Each patient received 60mCi/m^2 ^{90}Y-DOTATOC intravenously every 3 weeks, for a total of three cycles [18]. All six patients had tumor progression, median progression-free survival (PFS) was 37.5 days (28–52), and median overall survival (OS) was 103.5 days (28–269) [18]. The authors concluded that in contrast to well-differentiated neuroendocrine tumors, ^{90}Y-DOTATOC seems to be inactive in SCLC [18]. As it was presented earlier, van Essen et al. have used ^{177}Lu-octreotate in case of three patients with SCLC and also concluded that this radiopeptide does not seem to have clinical effects in SCLC [19]. Edelman et al. presented results of a phase I trial, in which patients with metastatic or recurrent SCLC that could not be treated surgically and that were demonstrated to have SSTRs by positive ^{99m}Tc-P2045 (11-amino acid somatostatin analogue) imaging results were treated with escalating doses of ^{188}Re-P2045 [57, 58]. There were no objective responses; however, five of eight patients with progressive disease at baseline remained stable at 8 weeks after therapy [57, 58]. The authors concluded that although responses were not seen, survival for these heavily pretreated patients is interesting and merits further research [57]. Presented results demonstrate that despite the proven expression of SSTRs at the cells of SCLC, PRRT does not seem to be effective in this type of tumor.

Medullary thyroid cancer (MTC) is neuroendocrine neoplasm derived from the parafollicular cells or C cells of the thyroid and accounts for nearly 5–10% of thyroid malignancies [59–61]. Overall, the prognosis for patients with MTC is good. The 10-year survival rate for patients with MTC is 75–85% [62]. Approximately half of the patients with MTC present with disease limited to the thyroid gland [62]. One third of patients present with locally invasive tumors or clinically apparent spread to the regional lymph nodes [62]. Patients with regional disease have a 5-year OS rate of 75.5% [62]. Distant metastases portend a poor prognosis, with a 10-year survival rate of only 40% [62]. Recurrent disease develops in approximately 50% of patients with MTC [62]. The primary treatment for MTC is extensive and meticulous surgical resection. Total thyroidectomy with complete resection of lymph nodes in the central neck, paratracheal, and upper mediastinal region is frequently needed. At present, surgical excision is the only effective treatment for MTC. Patients who have clinically evident disease are best treated with a minimum of total thyroidectomy and bilateral central neck dissection [63, 64]. EBRT does not currently play a significant role in the treatment of patients with MTC [63, 64]. However, radiation therapy has been applied to help palliate local disease when surgery is not a feasible option [62, 64]. Follow-up should start 2–3 months postoperatively by obtaining new baseline calcitonin levels [59–61]. An undetectable basal serum calcitonin level is a strong predictor of complete remission [59–61]. Complete remission may be further confirmed after a provocative test [59–61]. Serum calcitonin should be repeated every 6 months for the first 2–3 years and annually thereafter [59–61]. When the postoperative calcitonin level is elevated, a careful metastatic evaluation must be performed prior to proceeding with operative exploration [59–61]. In some MTC patients despite of the elevated postoperative calcitonin levels and/or abnormal results of the pentagastrin test, there is no evidence of the disease in available imaging techniques. Moreover, these patients are left with few therapeutic choices. Conventional chemotherapy has shown limited efficacy in patients with MTC [65]. Complete responses are very rare and partial responses have been seen in less than one third of patients [65]. Based on the expression of the SSTRs on the surface of the medullary thyroid cancer cells, there is a possibility of the use of the targeted radionuclide therapy [1, 2, 66–69]. Patients with metastatic disease can have significant symptoms from calcitonin excess and those may benefit from treatment with somatostatin analogues [66]. So far, treatment of MTC patients with labeled somatostatin analogues revealed moderate responses, which might be connected with heterogeneous expression of SSTR in these neoplasms [1, 2, 67–69]. It is worth to mention here that medullary thyroid cancer cells overexpress also gastrin and cholecystokinin 2 (CCK-2) receptors [1, 2, 70, 71]. The ability of the medullary thyroid cancers to take up the labeled analogues of gastrin has been demonstrated in the clinical studies [70, 71]. The gastrin and CCK-2 receptors are potential and promising targets for the radionuclide therapies, as their ligands have been successfully labeled with β-emitting radioisotopes [70, 71].

There are other peptide receptors, such as the glucagon-like peptide 1 (GLP-1) receptor in insulinoma, gastrinoma, pheochromocytoma, paraganglioma, and medullary thyroid cancer, being taken into consideration as the promising targets for diagnostic approach and in the future probably also for radionuclide therapy [1, 2, 17, 71].

Presented examples prove that the use of radiolabeled somatostatin analogues is not limited to gastroenteropancreatic neoplasms. PRRT might be an alternate therapeutic option for patients with other tumors expressing SSTRs, particularly in case of recurrent neoplasms or disease resistant to standard therapeutic procedures. However, treatment with radiolabeled somatostatin analogues in those other neoplasms is usually not as efficient as in case of GEP-NETs. Therefore, there is still a need to search for other more specific peptide receptors, which might become targets for novel therapeutic strategies.

REFERENCES

[1] Reubi, JC. Endocrine Review 2003, 24, 389–427.

[2] Kornberg, R. D.; McConnell, H. M. Biochemistry 1971, 10, 1111–1120.

[3] Fruhwald, M. C.; O'Dorisio, M. S.; Pietsch, T.; et al. Pediatric Research 1999, 45, 697–708.

[4] Albers, A. R.; O'Dorisio, M. S.; Balster, D. A.; et al. Regulatory Peptides 2000, 88, 61–73.

[5] O'Dorisio, M. S.; Chen, F.; O'Dorisio, T. M.; et al. Cell Growth Differentiation 1994, 5, 1–8.

[6] Gains, J. E.; Bomanji, J. B.; Fersht, N. L.; et al. Journal of Nuclear Medicine 2011, 52, 1041–1047.

[7] Menda, Y.; O'Dorisio, M. S.; Kao S.; et al. Journal of Nuclear Medicine 2010, 51, 1524–1531.

[8] Khanna, G.; Bushnell, D.; O'Dorisio, M. S. Oncologist 2008, 13, 382–389.

[9] Scheidhauer, K.; Hildebrandt, G.; Luyken, C.; et al. Hormone and Metabolic Research 1993, 27, 59–62.

[10] Schmidt, M.; Scheidhauer, K.; Luyken, C.; et al. European Journal of Nuclear Medicine 1998, 25, 675–686.

[11] Lee, J. D.; Kim, D. I.; Lee, J. T.; et al. Journal of Nuclear Medicine 1995, 36, 537–541.

[12] Reubi, J. C.; Lang, W.; Maurer, R.; et al. Cancer Research 1987, 47, 5758–5764.

[13] Luyken, C.; Hildebrandt, G.; Scheidhauer, K.; et al. Acta Neurochirurgica 1994, 127, 60–64.

[14] Mawrin, C.; Schulz, S.; Pauli, S. U.; et al. Journal of Neuropathology and Experimental Neurology 2004, 63, 13–19.

[15] Haldemann, A. R.; Rösler, H.; Barth, A.; et al. Journal of Nuclear Medicine 1995, 36, 403–410.

[16] Guyotat, J.; Champier, J.; Jouvet, A.; et al. International Journal of Cancer 2001, 95, 144–151.

[17] Fani, M.; Maecke, H. R. European Journal of Nuclear Medicine and Molecular Imaging 2012, 39, 11–30.

[18] Pless, M.; Waldherr, C.; Maecke, H.; et al. Lung Cancer 2004, 45, 365–371.

[19] van Essen, M.; Krenning, E. P.; Kooij, P. P.; et al. Journal of Nuclear Medicine 2006, 47, 1599–1606.

[20] Loh, K. C.; Fitzgerald, P. A.; Matthay, K. K.; et al. Journal of Endocrinological Investigation 1997, 20, 648–658.

[21] Rose, B.; Matthay, K. K.; Price, D.; et al. Cancer 2003, 98, 239–248.

[22] Kwekkeboom, D. J.; van Urk, H.; Pauw, B. K.; et al. Journal of Nuclear Medicine 1993, 34, 873–878.

[23] Van der Harst, E.; De Herder, W. W.; Bruining, H. A.; et al. Journal of Clinical Endocrinology and Metabolism 2001, 86, 685–693.

[24] Gulenchyn, K. Y.; Yao, X.; Asa, S. L.; et al. Clinical Oncology 2012, 24, 294–308.

[25] Baldari, S.; Ferrau, F.; Alafaci, C.; et al. Pituitary 2012, 15, 57–60.

[26] Colao, A.; Di Sarno, A.; Pivonello, R.; et al. Experts Opinion on Investigational Drugs 2002, 11, 787–800.

[27] Colao, A. Best Practice and Research Clinical Endocrinology and Metabolism 2009, 23, 575–596.

[28] Gillam, M. P.; Molitch, M. E.; Lombardi, G.; et al. Endocrine Review 2006, 27, 485–534.

[29] de Jong, M.; Breeman, W. A.; Bakker, W. H.; et al. Cancer Research 1998, 58, 437–441.

[30] Acosta-Gómez, M. J.; Muros, M. A.; Llamas-Elvira, J. M.; et al. British Journal of Radiology 2005, 78, 110–115.

[31] Jaquet, P.; Ouafik, L.; Saveanu, A.; et al. Journal of Clinical Endocrinology and Metabolism 1999, 84, 3268–3276.

[32] Melmed, S.; Colao, A.; Barkan, A.; et al. Journal of Clinical Endocrinology and Metabolism 2009, 94, 1509–1517.

[33] Petersenn, S.; Schopoh, J.; Barkan, A.; et al. Journal of Clinical Endocrinology and Metabolism 2010, 95, 2781–2789.

[34] Feelders, R. A.; Hofland, L. J. Journal of Clinical Endocrinology and Metabolism 2013, 98, 425–438.

[35] Colao, A.; Petersenn, S.; Newell-Price, J.; et al. New England Journal of Medicine 2012, 366, 914–924.

[36] Kienitz, T.; Quinkler, M.; Strasburger, C. J.; et al. European Journal of Endocrinology 2007, 157, 39–46.

[37] Stiller, C. A.; Parkin, D. M. International Journal of Cancer 1992, 52, 538–543.

[38] Spix, C.; Pastore, G.; Sankila, R.; et al. European Journal of Cancer 2006, 42, 2081–2091.

[39] Pearson, A. D.; Pinkerton, C. R.; Lewis, I. J.; et al. Lancet Oncology 2008, 9, 247–256.

[40] Matthay, K. K.; Reynolds, C. P.; Seeger, R. C.; et al. Journal of Clinical Oncology 2009, 27, 1007–1013.

[41] Haupt, R.; Garaventa, A.; Gambini, C.; et al. Journal of Clinical Oncology 2010, 28, 2331–2338.

[42] Beutler, D.; Avoledo, P.; Reubi, J. C.; et al. Cancer 2005, 103, 869–873.

[43] Bouffet, E. European Journal of Cancer 2002, 38, 1112–1120.

[44] Bartolomei, M.; Bodei, L.; De Cicco, C.; et al. European Journal of Nuclear Medicine and Molecular Imaging 2009, 36, 1407–1416.

[45] Kleihues, P.; Louis, D. N.; Scheithauer, B. W.; et al. Journal of Neuropathology and Experimental Neurology 2002, 61, 215–229.

[46] Simpson D. Journal of Neurology Neurosurgery and Psychiatry 1957, 20, 22–39.

[47] Marimanoff, R. O.; Dosoretz, D. E.; Linggood, R. M.; et al. Journal of Neurosurgery 1985, 62, 18–24.

[48] Jaaskelainen, J.; Haltia, M.; Servo, A. Surgical Neurology 1986, 25, 233–242.

[49] Taylor Jr., B. W.; Marcus Jr., R.B.; Friedman, W. A.; et al. International Journal of Radiation Oncology Biology Physics 1988, 15, 299–304.

[50] Condra, K. S.; Buatti, J. M.; Mendenhall, W. M.; et al. International Journal of Radiation Oncology Biology Physics 1997, 39, 427–436.

[51] Mesic, J. B.; Hanks, G. E.; Doggett, R. L. American Journal of Clinical Oncology 1986, 9, 337–340.

[52] Dziuk, T. W.; Woo, S.; Butler, E. B.; et al. Journal of Neuro-Oncology 1998, 37, 177–188.

[53] Goldsmith, B. J.; Wara, W. M.; Wilson, C. B.; et al. Journal of Neurosurgery 1994, 80, 195–[201] Erratum in: Journal of Neurosurgery 1994, 80, 777.

[54] De Jong, M.; Valkema, R.; Jamar, F.; et al. Seminars in Nuclear Medicine 2002, 32, 133–140.

[55] Reubi, J. C.; Maurer, R.; Klijn, J. G.; et al. Journal of Clinical Endocrinology and Metabolism 1986, 63, 433–438.

[56] Kreissl, M. C.; Hänscheid, H.; Löhr, M.; et al. Radiation Oncology 2012, 7, 99.

[57] Edelman, M. J.; Clamon, G.; Kahn, D.; et al. Journal of Thoracic Oncology 2009, 4, 1550–1554.

[58] Graham, M. M.; Menda, Y. Journal of Nuclear Medicine 2011, 52, 56–63.

[59] Sippel, R. S.; Kunnimalaiyaan, M.; Chen, H. Oncologist 2008, 13, 539–547.

[60] Ball, D. W. Endocrinolology and Metabolism Clinics of North America 2007, 36, 823–837.

[61] Ball, D. W. Current Opinion in Oncology 2007, 19, 18–23.

[62] Kebebew, E.; Kikuchi, S.; Duh, Q. Y.; et al. Archives of Surgery 2000, 135, 895–901.

[63] Pacini, F.; Castagna, M. G.; Cipri, C.; et al. Clinical Oncology 2010, 22, 475–485.

[64] Kloos, R. T. ; Eng, C.; Evans, D. B.; et al. Thyroid 2009, 19, 565–612.

[65] Tai, D.; Poon, D. Journal of Oncology 2010, Article ID 398564, doi: dx.doi.org/ [10]1155/2010/398564.

[66] Vainas, I.; Koussis, Ch.; Pazaitou-Panayiotou, K.; et al. Journal of Experimental and Clinical Cancer Research 2004, 23, 549–559.

[67] Rufini, V.; Salvatori, M.; Garganese, M. C.; et al. Rays 2000, 25, 273–282.

[68] Ambrosini, V.; Fani, M.; Fanti, S.; et al. Journal of Nuclear Medicine 2011, 52, 42–55.

[69] Sowa-Staszczak, A.; Pach, D.; Kunikowska, J.; et al. Endokrynologia Polska 2011, 62, 392–400.

[70] Béhé, M.; Behr, T. M. Biopolymers 2002, 66, 399–418.

[71] Laverman, P.; Sosabowski, J. K; Boerman, O. C.; et al. European Journal of Nuclear Medicine and Molecular Imaging 2012, 39, 78–92.

7

SOMATOSTATIN ANALOGS: FUTURE PERSPECTIVES AND PRECLINICAL STUDIES—PANSOMATOSTATINS

AIKATERINI TATSI[1], BERTHOLD A. NOCK[1], THEODOSIA MAINA[1], AND MARION DE JONG[2]

[1] *Molecular Radiopharmacy, INRASTES, NCSR "Demokritos," Athens, Greece*
[2] *Department of Nuclear Medicine and Radiology, Erasmus MC, University Medical Center, Rotterdam, The Netherlands*

ABBREVIATIONS

ACTH	adrenocorticotropic hormone
ATP	adenosine triphosphate
cDNA	complementary deoxyribonucleic acid
cGMP	cyclic guanosine monophosphate
DOTA	1,4,7,10-tetraazacyclododecane-1,4,7,10-tetra-acetic acid
DOTANOC	[DOTA^0-1-Nal^3]octreotide
DTPA	diethylene triamine penta-acetic acid
ECL	extracellular loop
GDP	guanosine diphosphate
GH	growth hormone
GHRH	growth hormone-releasing hormone
GPCRs	G-protein-coupled receptors
GTP	guanosine triphosphate
IGF-1	insulin growth factor type 1

Somatostatin Analogues: From Research to Clinical Practice, First Edition. Edited by Alicja Hubalewska-Dydejczyk, Alberto Signore, Marion de Jong, Rudi A. Dierckx, John Buscombe, and Christophe Van de Wiele.

KE108   pansomatostatin synthetic nonapeptide
MAPK   mitogen-activated protein kinase
mRNA   messenger ribonucleic acid
NET   neuroendocrine tumor
NHEs Na^+/H^+ exchangers
PET/CT   positron emission tomography/computed tomography
PPSST   preprosomatostatin
PTPases   phosphotyrosine phosphatases
RNA   ribonucleic acid
RT-PCR   reverse transcriptase-polymerase chain reaction
SS   somatostatin
TMS   transmembrane segments
TSH   thyroid-stimulating hormone
VIP   vasoactive intestinal polypeptide

NATIVE HORMONES

Somatostatin (SS) was first isolated from ovine hypothalamic extracts and was characterized as a tetradecapeptide. It was identified as part of the releasing hormone family for its property to inhibit the secretion of growth hormone (GH) from pituitary cells by Brazeau and colleagues in 1973 [1]. There are two naturally bioactive SS products, somatostatin-28 (SS-28) and somatostatin-14 (SS-14). SS-28 contains the entire sequence of SS-14 at its carboxy terminal, immediately preceded by a double pair of basic amino acids [2]. SS-28 was first isolated from ovine [3] and then from porcine [4] and rat [5] brain tissues. Both forms of mammalian SS are derived from a larger inactive precursor molecule, preprosomatostatin (PPSST), that is processed by posttranslational enzymatic cleavage to yield the active polypeptides (Fig. 7.1). Isolation and cloning of human and rat cDNAs encoding PPSST revealed the sequence and structure of a polypeptide consisting of 116 amino acids [6–9]. SS-14 is generated by dibasic cleavage at an Arg^{101}-Lys^{102} residue, whereas endoproteolysis of a monobasic Arg^{88} site produces SS-28 [10, 11]. A secondary monobasic site was determined in PSST, the cleavage of which results in the generation of a 10-amino-acid peptide termed antrin ($PSST_{1-10}$) [12–14]. Various mixtures of SS-14 and SS-28 are produced in mammalian tissues [15] because of the differential processing of PSST. SS-14 and SS-28 are the only known biologically active forms of PSST. Other products have also been identified in circulation following processing whose biological function remains uncertain as they are lacking of any known activity [10, 15].

SS-producing cells are typically neurons or endocrine-like cells and are found in high density in the central and peripheral nervous systems and in the endocrine pancreas and in the gut and in small numbers in the thyroid, adrenals, submandibular glands, kidneys, prostate, placenta blood vessel walls, and immune cells [16–22]. The major role of hypothalamic SS is the inhibition of GH from anterior pituitary somatotrophs [23]. SS inhibits the secretion of GH via a direct interaction on the

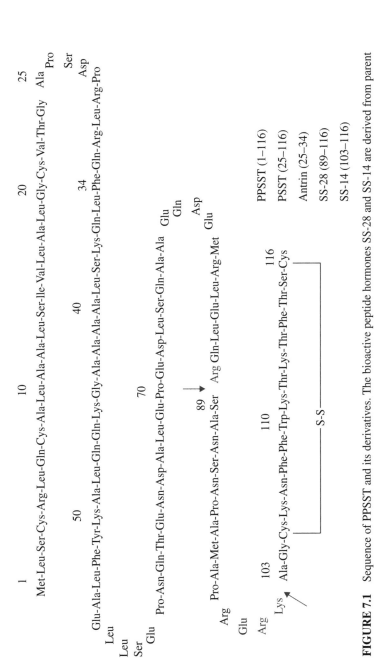

FIGURE 7.1 Sequence of PPSST and its derivatives. The bioactive peptide hormones SS-28 and SS-14 are derived from parent PPSST (92 aa, 25–116), which is generated from a larger inactive precursor molecule PPSST (1–116). Endoproteolysis of monobasic Arg[88] produces SS-28, and cleavage at a dibasic Arg[101]-Lys[102] produces SS-14.

pituitary and indirectly through suppression of growth hormone-releasing hormone (GHRH) release [24, 25]. These pathways both interact with each other at their point of convergence at the level of the pituitary and through direct neural connections within the hypothalamus [26]. There are two secretory feedback loops that modulate SS release—the short loop, where SS is negatively regulated by GHRH [24] and subject to positive feedback by GH [27], and the long loop, where insulin growth factor type 1 (IGF-I), produced by GH acting on the liver, provides a positive influence for SS release [28] (Fig. 7.2). In addition, secretion of hypothalamic SS can be further promoted by dopamine, substance P, neurotensin, glucagon, hypoglycemia, various amino acids, acetylcholine, α_2-adrenergic agonists, vasoactive intestinal polypeptide (VIP), and cholecystokinin and can be inhibited by glucose [29, 30]. Similar mechanisms also exist in the hypothalamic control of thyroid-stimulating hormone (TSH) secretion [31–35].

Moreover, SS functions as a neurotransmitter in the brain with effects on cognitive, locomotor, sensory, and autonomic functions [20, 32–34, 36]. SS inhibits the release of dopamine from the midbrain and the secretions of norepinephrine, thyroid-releasing hormone, and corticotrophin-releasing hormone including its own secretion from the hypothalamus. In contrast, SS has no effects on the release of

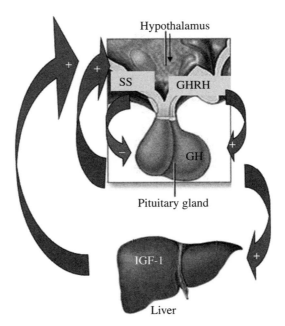

FIGURE 7.2   GH regulation by SS. Growth hormone-releasing hormone (GHRH) and somatostatin (SS) are produced in the hypothalamus. GHRH stimulates synthesis and secretion of growth hormone (GH) (right upper arrow) and SS inhibits GH (left upper descending arrow). GH feeds back positively for SS (left upper ascending arrow). There is a second positive feedback for hypothalamic SS from liver IGF-1 (left large ascending arrow). IGF-1 is produced in the liver after stimulation of pituitary GH (right lower descending arrow).

luteinizing hormone, follicle-stimulating hormone, prolactin, or adrenal corticotrophin hormone under normal physiological conditions. SS has direct effects on the thyroid by inhibiting the release of T4, T3, and calcitonin from thyroid parafollicular cells stimulated by TSH.

Within the gastrointestinal tract, virtually every gut hormone has been shown to be inhibited by SS, including gastric acid, pepsin, bile, and colonic fluid. In the pancreas, SS is an endogenous islet hormone. Its actions on the pancreas were first noted within the year of its discovery by two groups, following infusion in humans and baboons [37].

SS regulates the secretion of hormones from several tissues, including neurotransmission. When synthesized and released from δ cells of pancreatic islets, SS causes suppression of the synthesis and secretion of both insulin and glucagon, including the inhibition of pancreatic polypeptide [38–42]. Moreover, a downregulation in SS and sst expression has been associated with Alzheimer's disease [43]. Although the role of SS in Huntington's disease is controversial, it is believed that SS-positive neurons are selectively spared from disease [44].

Regulation of SS can be achieved by a broad array of secretagogues—from ions and nutrients to neuropeptides, neurotransmitters, hormones, growth factors, and cytokines—as it interacts with various bodily systems [20, 32, 33, 36]. Almost every neurotransmitter or neuropeptide tested has been shown to exert some sort of effect on SS secretion with a certain degree of tissue specificity. Potent stimulators of SS secretion are glucagon, GHRH, neurotensin, corticotrophin-releasing hormone, calcitonin gene-related peptide, and bombesin [20, 22, 45]. Thyroid, GH, IGF-I, and insulin trigger SS release from the hypothalamus [20, 22, 36], while insulin, leptin, and epinephrine inhibit its release from the pancreas and hypothalamus, respectively [20, 22, 36].

SS RECEPTOR SUBTYPES

The physiological actions of SS-14 are mediated by high-affinity plasma membrane receptors. Five SS receptor subtypes (sst_{1-5}) have been identified by gene cloning techniques, and they belong to the superfamily of G-protein-coupled receptors (GPCRs). Yamada and coworkers in Bell's laboratory cloned the first 2 ssts in early 1992, the human (h) and mouse (m) homologs of sst_1 and sst_2 [46]. This success was followed by the cloning of rat sst_1 [47]; mouse, rat, and human sst_3 [48–51]; rat and human sst_4 [52–54]; and rat and human sst_5 [55, 56]. Sst_1 and sst_{3-5} lack classical introns in the coding segment. One of the receptors (sst_2) is expressed in two alternatively spliced forms [57], a long sst_{2A} and a short sst_{2B}. The sst_{2A} and sst_{2B} differ only in the length of the cytoplasmic C-tail [21]. So, there are six putative sst subtypes of closely related size, each displaying seven helical transmembrane segments (TMS) typical of GPCRs. The five human sst genes segregate on separate chromosomes [22]. The five human sst sequences range in size from 356- to 391-amino-acid residues and show the greatest similarity in the putative TMS (55–70% sequence identity). The highest degree of TM sequence identity occurs between $hsst_1$ and $hsst_4$ (70%)

followed by $hsst_3$ and $hsst_5$ (69%) [21, 58, 59]. All sst receptor isoforms that have been cloned so far not only from humans but also from other species have a highly conserved sequence motif, YANSCANPI/VLY, in the seventh TMS, which is a signature sequence for this receptor family. The $hsst_{1-5}$ display one or multiple sites for N-linked glycosylation within the amino-terminal segment and second extracellular loop (ECL). All of the $hsst_{1-5}$ have three to eight putative recognition motifs for protein phosphorylation by protein kinase A, protein kinase C, and calmodulin kinase II in the cytoplasmic C-terminal segment and within the second and third intracellular loops. $Hsst_1$, $hsst_2$, $hsst_4$, and $hsst_5$ display a conserved cysteine residue 12 amino acids downstream from the seventh TMS, which may be the site of a potential palmitoyl membrane anchor as observed in several other members of the GPCR superfamily [22]. On the other hand, $hsst_3$ lacks the cysteine palmitoylation membrane anchor and displays a longer C-tail than the other hssts. Interestingly, sst_{1-5} subtypes exhibit a remarkable degree of structural conservation across species [21, 59]. Particularly, there is 94–99% sequence identity among the human, rat, and mouse isoforms of sst_1; 93–96% sequence identity among the human, rat, mouse, porcine, and bovine isoforms of sst_2; and 88% sequence identity between the rat and the human isoforms of sst_4, whereas sst_3 and sst_5 are somewhat less conserved, showing 82–83% sequence identity between the human and the rodent homologs [22]. All sst_{1-5} bind SS-14 with high and comparable affinity. Moreover, human sst_1, sst_2, sst_3, and sst_4 bind SS-14 with a 2–3-fold higher affinity than SS-28 [46, 51, 53, 54, 60, 61]. In contrast, $hsst_5$ binds SS-28 with a 10–30-fold higher affinity than SS-14 and can be considered SS-28 selective [56, 60, 61]. Based on these sequence homologies, the receptor subtypes can be placed into two subgroups. The first receptor group comprises sst_2, sst_3, and sst_5, whereas sst_1 and sst_4 belong to the second group [62]. The five sst_{1-5} display differences in the cell internalization capabilities [19, 63] upon ligand binding. Sst_3 and sst_2 internalize much better than sst_1 [63]. Sst_5 internalizes after ligand binding but can also assemble the sst_5 receptors from intracellular stores to the membrane [64]. The activation of G-proteins actuates the signal transduction pathway of ssts. The G-protein consists of three subunits: the α (G_α), the β, and the γ($G_{\beta\gamma}$) subunits. After the activation of the G-protein group, the nucleotide exchange of GDP for GTP results in the dissociation of the complex, and the G_α and $G_{\beta\gamma}$ are free to spread their signal. Binding of SS ligands to the ssts is transduced to several key enzymes, such as adenylate cyclase, phosphotyrosine phosphatases (PTPases), and mitogen-activated protein kinase (MAPK), along with changes in the intracellular levels of calcium and potassium ions [19, 21, 36, 58, 59, 62, 65, 66]. Sst subtype, signaling elements, sst internalization, desensitization, and receptor crosstalk are a few of the factors that will determine which signal will eventually prevail.

The first effector enzyme to be identified and regulated by ssts was adenylate cyclase, and all five sst subtypes negatively couple to the enzyme [60]. Ssts are coupled to several types of potassium channels such as the delayed rectifier, inward rectifier, ATP-sensitive potassium channels, and large conductance calcium-activated BK channels [67–71]. Ssts have also been shown to directly modulate high-voltage-dependent calcium channels [72] and may also inhibit calcium currents by activation of cGMP protein kinase [73]. Moreover, ssts have also been shown to couple to Na^+/

H$^+$ exchangers (NHEs) [74–76] to modulate such features as cell adhesion, migration, and proliferation [77]. Sst$_1$ was the first subtype to specifically regulate NHE-1 [74]. It was later determined that sst$_3$ and sst$_4$ also conduce, but not sst$_2$ and sst$_5$ [76]. Ssts activate a number of phosphatases that have been implicated in cell growth. Tyrosine phosphatase and also serine and threonine phosphatase activation has been demonstrated to be recruited by ssts [19, 62, 65, 66]. According to recent studies, ssts can function as dimers and/or oligomers [78]. Homo- and heterodimerization was first demonstrated for the hsst$_5$ subtype [79, 80]. Hsst$_5$ can form heterodimeric association with hsst$_1$ but not with hsst$_4$, indicating that SS receptor heterodimerization is restricted to specific subtype combinations [80]. Western blot and coimmunoprecipitation experiments showed that sst$_{2A}$ and sst$_3$ can form homodimers when expressed alone and heterodimers when coexpressed in HEK293 cells [81].

EXPRESSION OF sst$_{1-5}$ IN NORMAL TISSUES AND TUMORS

The expression of sst subtypes has been examined in human and rodent tissues using different procedures including Northern blot, reverse transcriptase-polymerase chain reaction (RT-PCR) amplification of cellular RNA, ribonuclease protection assay, and *in situ* hybridization [21, 22, 53, 58, 59, 66, 82]. In rat, mRNA for sst$_{1-5}$ has been localized in the cerebral cortex, striatum, hippocampus, amygdale, olfactory bulb, and preoptic area [82]. In particular, sst$_1$ mRNA in rat is expressed in the brain, pituitary, islets, jejunum, stomach, heart, spleen, and adrenals [46]; rsst$_2$ mRNA is abundantly expressed in the brain as well as in the pituitary, pancreatic islets, and adrenals [57, 83]; and rsst$_3$ mRNA is densely expressed in the cerebellum and in moderate amounts in the amygdala, cortex, and striatum and in the spleen, liver, and pituitary [50]. Rsst$_4$ mRNA is poorly expressed in the brain, cerebral cortex, hippocampus, and olfactory bulb [53], and rsst$_5$ mRNA is expressed in the brain, in high levels in the hypothalamus and preoptic area and in a moderate extent in cortical and subcortical regions [56, 84].

In human, ssts are expressed in the brain [65] as well as in numerous peripheral tissues, including the pituitary, pancreas, gut, thyroid, adrenal, and kidney, and the immune system. Most studies are based on mRNA measurements. In contrast to rat brain, human brain is devoid of hsst$_1$ expression. Hsst$_2$ mRNA is expressed in high levels in the brain and kidney and in moderate levels in the jejunum, colon, and liver [46, 57]. Hsst$_3$ mRNA has been detected in brain pituitary and islets [50], and hsst$_4$ mRNA has been detected in moderate levels in the brain, stomach, and lung and other tissues in lesser amounts [54]. Hsst$_5$ mRNA shows limited expression in man including the pituitary gland in adults and the pituitary and hypothalamus in the fetus [56, 84]. Interestingly, the liver, kidneys, cerebellum, and lungs, which are atypical target organs with negative or low SS binding, show high expression of sst mRNAs [21]. Sst$_{1-5}$ are often coexpressed in various amounts and combinations in different tissues and cell types [19, 21, 36, 58, 59, 66, 85]. Coexpression of sst$_1$ and sst$_2$ is found in GHRH-producing arcuate

TABLE 7.1 sst_{1-5} expression in human tumors x/y.

Tumors	sst_1	sst_2	sst_3	sst_4	sst_5
Neuroblastomas	ne	hd	ne	ne	ne
Meningiomas	ne	hd	ne	ne	ne
Medulloblastomas	ne	hd	ne	ne	ne
Breast carcinomas	1/31 hd	29/31 hd 1/31 md	ne	ne	ne
Lymphomas	ne	7/7 hd	ne	ne	3/7 ld
Renal cell carcinomas	ne	5/7 hd 2/7 md	ne	ne	ne
Paragangliomas	2/10 hd	8/10 hd	2/10 ld	ne	ne
Small cell lung cancers	ne	2/2 hd	ne	ne	ne
Hepatomas	ne	2/2 hd	ne	ne	ne
Prostate carcinomas	4/6 hd 2/6 md	ne	ne	ne	1/6 ld
Sarcomas	6/9hd 2/9md	1/9md	ne	1/9md	ne
Inactive pituitary adenomas	1/30hd	8/30hd 4/30md 4/30ld	16/30hd 1/30md	ne	1/30md 3/30ld
GH pituitary adenomas	ne	9/24hd 12/24md 1/24ld	2/24ld	ne	2/24hd 8/24md
Gastroenteropancreatic tumors	3/22hd 3/22md	1/22hd 12/22md 2/22ld	ne	ne	3/22md 1/22ld
Pheochromocytomas	3/19hd 2/19md	13/19hd 2/19md	ne	ne	1/19ld
Gastric carcinomas	2/5hd 1/5ld	1/5hd	ne	ne	2/5hd
Ependymomas	1/2hd	ne	ne	ne	1/2hd

x = cases per pattern, y = total cases, hd = high density, md = medium density, ld = low density, and ne = no expression.

neurons [86]. Hsst$_1$, hsst$_2$, hsst$_3$, and hsst$_5$ are expressed in the adult human pituitary, whereas in the pituitary of rats, all five rssts are expressed [82, 87–89].

Of particular interest is the fact that sst$_{1-5}$ are expressed alone or in various combinations in many human tumors (Table 7.1) where they show higher densities than in normal tissues. Ssts are often highly expressed in neuroendocrine tumors (NETs) including GH-secreting pituitary adenomas and gastroenteropancreatic tumors. Furthermore, ssts are expressed as well in many other tumors including neoplasias of the brain, breast carcinomas, lymphomas, renal cell cancers, mesenchymal tumors, and prostatic, ovarian, gastric, hepatocellular, and nasopharyngeal carcinomas. The most frequently detected subtype is sst$_2$ and is expressed in neuroblastomas, medulloblastomas, breast carcinomas, meningiomas, paragangliomas, lymphomas, small cell lung cancers, hepatomas, and renal cell carcinomas. Moreover, sst$_2$ seems to play an important role in tumor growth inhibition in agreement with reports by Buscail et al., suggesting that the absence of sst$_2$ subtype may be responsible for the rapid growth of pancreatic and colonic cancers [90]. A few tumors with clear predominance of sst$_1$ have been identified such as prostate carcinomas and sarcoma. Nevertheless, several other tumors may also have moderate expression of sst$_1$, including gastroenteropancreatic tumors, pheochromocytomas, gastric carcinomas, and ependymomas. The same tumors may also alternatively express sst$_2$ and sst$_5$ [91]. No tumor expression for sst$_3$ was identified except for inactive pituitary adenomas where sst$_3$ are often present. A possible explanation that sst$_3$ are not often detected may be that this receptor is expressed intracellularly in some tumors and not on the cell membrane. Moreover, sst$_4$ proteins and also sst$_4$ mRNA were found to be frequently expressed in tumors. Sst$_4$ expression in moderate levels was documented only in one sarcoma, suggesting that membrane-bound sst$_4$ does not play a significant role in human cancer [91]. Sst$_5$ are often expressed in pituitary adenomas and in moderate levels in sarcomas, prostate cancer, gastroenteropancreatic tumors, gastric carcinomas, and ependymomas. Several cancer cells coexpress ssts with different combinations. Sst$_2$ and sst$_5$ are frequently expressed, often together, in GH-secreting pituitary adenomas [84, 91]. A mixture of sst$_1$ and sst$_2$ is expressed in neuroendocrine gastroenteropancreatic tumors and pheochromocytomas [91].

Thus, despite the predominance of sst$_2$ expression in many human tumors, coexpression of sst$_2$ with other sst$_{1-5}$ subtypes is frequent enough. Furthermore, tumors devoid of sst$_2$ may express one of the other sst$_{1-5}$. These two findings have clinical implications for sst-targeted diagnosis and therapy, so far attempted mainly with sst$_2$-preferring peptide analogs (*vide infra*). It is reasonable to assume that pansomatostatin-like peptides, that is, analogs binding with high affinity to all five human sst$_{1-5}$, will broaden the clinical indications of their sst$_2$-selective counterparts, as they will localize and treat a wider range of human tumors. Furthermore, the use of pansomatostatin-like analogs will enhance the diagnostic sensitivity and/or therapeutic efficacy in such cases where sst$_{1-5}$ coexpression occurs given that localization at the tumor is expected to increase. Due to the poor metabolic stability of native SS-14, synthetic and sst$_2$-preferring analogs have been clinically applied so far in diagnosis and therapy. However, interest in the development of new metabolically stable pansomatostatins for clinical use has been increasing recently due to their inherent potential to fully exploit the molecular (co)existence of more than one sst$_{1-5}$ in human cancers. Furthermore, such analogs will provide excellent molecular tools to deeper explore and understand the pharmacology and function of sst$_{1-5}$ *in vitro* and *in vivo*.

FROM sst_2 PREFERRING TO PANSOMATOSTATIN RADIOLIGANDS

New SS derivatives have been synthesized with receptor subtype selectivity, higher stability, and improved pharmacokinetics compared to the native peptide. Octreotide, lanreotide, vapreotide, and the hexapeptide MK678 are such analogs (Fig. 7.3). However, these reduced-size SS analogs do not bind with high affinity to all five presently known SS receptor subtypes [19]. They showed high-affinity binding to sst_2, moderate binding to sst_3 and sst_5, and no binding to sst_1 and sst_4 [92]. Although these analogs are available for clinical use, none of them can detect the tumors expressing preferentially sst_1, such as sarcomas and prostate cancer, or sst_3, such as inactive pituitary adenomas [91].

Moreover, many neuroendocrine tumors are resistant to SS sst_2-selective analog therapy because they often express several SS receptor subtypes other than sst_2. As a result, drug development has recently been focused on stable SS analogs with high affinity to all five receptors subtypes (pansomatostatins) to imitate the broad actions of the native hormones. Some SS analogs, such as In- and Y-DOTA-lanreotide, have been claimed to bind with high affinity to all five sst_{1-5}. However, binding studies with sst_{1-5}-transfected cells revealed that Y-DOTA-lanreotide has low affinity for the sst_1 and sst_4 receptor subtypes [93].

The multi-SS analog SOM230 (pasireotide) and the true pansomatostatin synthetic nonapeptide KE108 are both reduced-size analogs of SS-14 (Fig. 7.3). SOM230 has affinity for the four of the five SS subtypes, with IC_{50} values of 9.3 nM (sst_1), 1.0 nM (sst_2), 1.5 nM (sst_3), >1000 nM (sst_4), and 0.16 nM (sst_5) [94]. Although SOM230 has comparable sst_2 affinity with octreotide, it shows incapacity to stimulate sst_2 internalization *in vitro* and *in vivo* [95]. In addition, SOM230 is more potent than octreotide in inducing internalization and signaling of the sst_3 and the sst_5. Moreover, sst_2 receptors recycle faster to the plasma membrane in SOM230 than in octreotide-treated cells and may counteract homologous desensitization in sst_2-expressing cells [96]. SOM230 has also potent inhibitory effects on GH and IGF-I release and very high metabolic stability *in vivo*. SOM230 primarily targets the sst_5, and accordingly, it is a promising candidate for the treatment of ACTH adenomas, octreotide-resistant GH adenomas, and octreotide-resistant carcinoid tumors. Early phase I clinical trials have started with SOM230 to fully explore its therapeutic potential in man [97].

KE108 has high affinity to all five SS receptors, with IC_{50} values of 2.6 ± 0.4 nM (sst_1), 0.9 ± 0.1 nM (sst_2), 1.5 ± 0.2 nM (sst_3), 1.6 ± 0.1 nM (sst_4), and 0.65 ± 0.1 nM (sst_5) [98]. Both KE108 and SOM230 are full agonists for adenylyl cyclase inhibition but antagonize the actions of SS-14 to stimulate calcium and ERK phosphorylation [99]. KE119 is the 3-I-Tyr^0 analog of KE108 and has also pansomatostatin properties and could be used as radiotracer *in vitro* and *in vivo* to identify tissues expressing all five SS receptor subtypes [98].

KE88 is an analog of KE108 modified by coupling DOTA to the N-terminus, and it retains a high affinity to all five sst_{1-5}. The KE88-based radioligands may be interesting and suitable for imaging sst_2-expressing tumors at early time points (^{68}Ga) and sst_3-expressing tumors at later imaging time points using a longer-lived radionuclide (^{64}Cu, ^{86}Y) [100]. Coupling of acyclic tetraamines to KE108 and *des*-Tyr^0-KE108 resulted in Demopan 1 and Demopan 2, respectively, both able to bind ^{99m}Tc in good

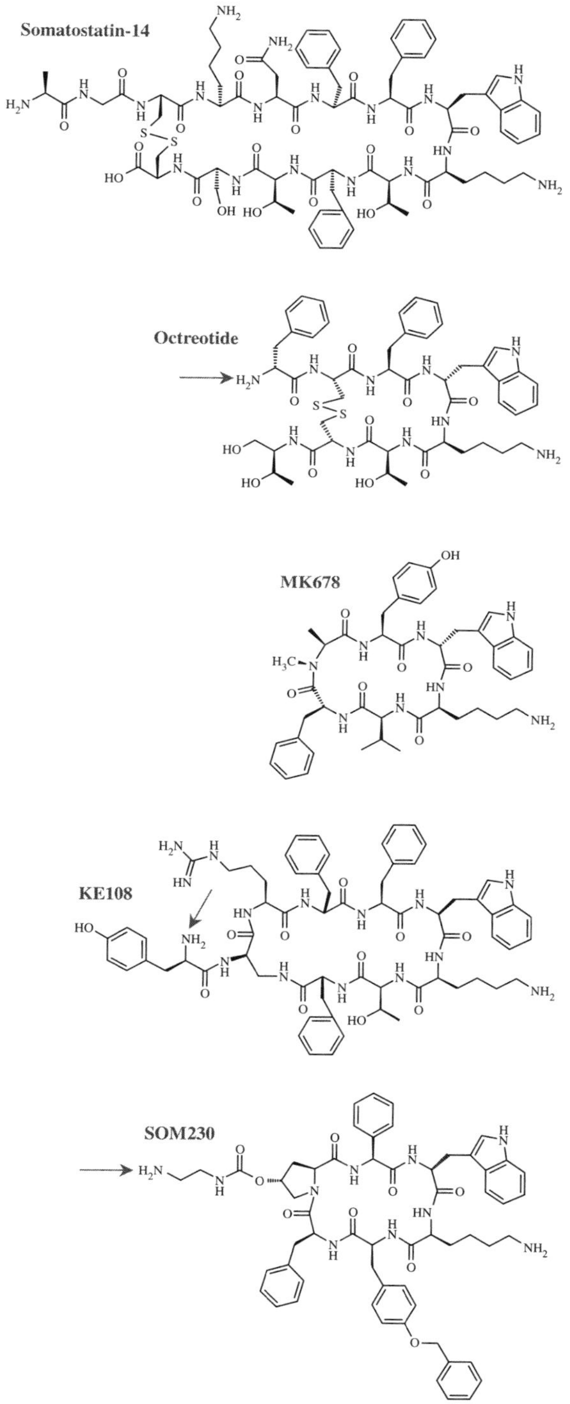

FIGURE 7.3 Molecular structures of SS-14 and smaller-ring synthetic analogs. Molecular structure of native 12-AA ring tetradecapeptide SS-14 (top), followed by the cyclic octapeptide octreotide, the cyclic hexapeptide MK678 and the nonapeptide KE108; the lower structure represents SOM230. Coupling of DOTA, NOTA, N4 chelators and their derivatives for stable labeling with medically useful radiometals has been performed at the positions indicated by the arrow, either directly or via suitable spacers.

yields. These analogs, despite their true *in vitro* pansomatostatin profile, exhibited suboptimal pharmacokinetics in animal models [101].

In the late 1970s, the Merck group has synthesized the first bicyclic peptides to increase the metabolic stability of the native hormones. Some years later, Veber et al. [102–104] performed structure–activity relationships studies with these analogs to elucidate the structural parameters implicated in SS function. Following these results, Maecke et al. synthesized AM3 [DOTA-Tyr-*cyclo*(Dab-Arg-*cyclo*(Cys-Phe-D-Trp-Lys-Thr-Cys)] wherein the DOTA chelator is coupled for trivalent radiometal binding [105]. AM3 shows a pansomatostatin profile with higher affinity for sst_2 and sst_3. The conjugate is an agonist on sst_2 and sst_3, as shown by immunofluorescence-based internalization assays and by Ca^{2+} flux studies. AM3 seems to be comparable with similar broad sst binding analogs, such as DOTANOC [106, 107].[68]Ga-AM3showed pharmacokinetics similar to ^{177}Lu-AM3, including clear tumor delineation and receptor-mediated uptake. This radiopeptide may be good a candidate for PET/CT studies of sst-expressing tumors [105].

REFERENCES

[1]   Brazeau, P.; Vale, W. W.; Burgus, R.; et al. Science 1973, 179, 77–79.

[2]   Pradayrol, L.; Jornvall, H.; Mutt, V.; et al. FEBS Letters 1980, 109, 55–58.

[3]   Esch, F.; Böhlen, P.; Ling, N.; et al. Proceedings of the National Academy of Sciences of United States of America 1980, 77, 6827–6831.

[4]   Schally, A.; Huang, W.; Chang, R.; et al. Proceedings of the National Academy of Sciences of United States of America 1980, 77, 4489–4493.

[5]   Böhlen, P.; Brazeau, P.; Esch, F.; et al. Regulatory Peptides 1981, 2, 359–369.

[6]   Goodman, R. H.; Lund, P. K.; Jacobs, J. W. Journal of Biological Chemistry 1980, 255, 6549–6552.

[7]   Zingg, H. H.; Patel, Y. C. Journal of Clinical Investigation 1982, 70, 1101–1109.

[8]   Oyama, H.; O'Connell, K.; Permutt, A. Endocrinology 1980, 107, 845–847.

[9]   Shields, D. Journal of Biological Chemistry 1980, 255, 11625–11628.

[10] Patel, Y. C.; O'Neil, W. Journal of Biological Chemistry 1988, 263, 745–751.

[11] Bersani, M.; Thim, L.; Baldissera, F. G. Journal of Biological Chemistry 1989, 264, 10633–10636.

[12] Benoit, R.; Ling, N.; Esch, F. Science 1987, 238, 1126–1129.

[13] Rabbani, S. N.; Patel, Y. C. Endocrinology 1990, 126, 2054–2061.

[14] Ravazzola, M.; Benoit, R.; Ling, N.; et al. Journal of Clinical Investigation 1989, 83, 362–366.

[15] Patel, Y. C.; Wheatley, T.; Ning, C. Endocrinology 1981, 109, 1943–1949.

[16] Patel, Y. C.; Reichlin, S. Endocrinology 1978, 102, 523–530.

[17] Aguila, M. C.; Dees, W. L.; Haensly, W. E.; et al. Proceedings of the National Academy of Sciences of United States of America 1991, 88, 11485–11489.

[18] Arimura, A.; Sato, H.; Dupont, A.; et al. Science 1975, 189, 1007–1009.

[19] Patel, Y. C. Frontiers in Neuroendocrinology 1999, 20, 157–198.

[20] Weil, C.; Muller, E. E.; Thorner, M. O.; ed., *Basic and Clinical Aspects of Neuroscience.* Berlin: Springer-Verlag, Vol. 4, 1992, 1–16.

[21] Patel, Y. C.; Greenwood, M. T.; Panetta, R.; et al. Life Sciences 1995, 57, 1249–1265.

[22] Jefferson, L. S.; Cherrington, A. D.; ed., *The Handbook of Physiology, The Endocrine Pancreas and Regulation of Metabolism.* New York: Oxford University Press, 2001.

[23] Barinaga, M.; Bilezikjian, L. M.; Vale, W. W.; et al. Nature 1985, 314, 279–281.

[24] Katakami, H.; Downs, T. R.; Frohman, L. A. Endocrinology 1988, 123, 1103–1109.

[25] Tannenbaum, G. S.; McCarthy, G. F.; Zeitler, P.; et al. Endocrinology 1990, 127, 2551–2560.

[26] Horvath, S.; Palkovits, M.; Gorcs, T.; et al. Brain Research 1989, 481, 8–15.

[27] Berelowitz, M.; Firestone, S. L.; Frohman, L. A. Endocrinology 1981, 109, 714–719.

[28] Berelowitz, M.; Szabo, M.; Frohman, L. A.; et al. Science 1981, 212, 1279–1281.

[29] Berelowitz, M.; Dudlak, D.; Frohman, L. A. Journal of Clinical Investgation 1982, 69, 1293–1301.

[30] Chihara, K.; Arimura, A.; Schally, A. V. Endocrinology 1979, 104, 1656–1662.

[31] Arimura, A.; Schally, A. V. Endocrinology 1976, 98, 1069–1072.

[32] Reichlin, S. New Engand Journal of Medicine 1983, 309, 1495–1501.

[33] Reichlin, S. New England Journal of Medicine 1983, 309, 1556–1563.

[34] Vale, W.; Brazeau, P.; Rivier, C.; et al. Recent Progress in Hormone Research 1975, 31, 365–397.

[35] Siler, T. M.; Yen, S. C.; Vale, W.; et al. Journal of Clinical Endocrinology and Metabolism 1974, 38, 742–745.

[36] Barnett, P. Endocrine 2003, 20, 255–264.

[37] Koerker, D. J.; Ruch, W.; Chideckel, E.; et al. Science 1974, 184, 482–484.

[38] German, M. S.; Moss, L. G.; Rutter, W. J. Journal of Biological Chemistry 1990, 265, 22063–22066.

[39] Philippe J. Diabetes 1993, 42, 244–249.

[40] Redmon, J. B.; Towle, H. C.; Robertson, R. P. Diabetes 1994, 43, 546–551.

[41] Kleinman, R.; Gingerich, R.; Ohning, G.; et al. International Journal of Pancreatology 1995, 18, 51–57.

[42] Ballian, N.; Brunicardi, F. C.; Wang, X. P. Pancreas 2006, 33, 1–12.

[43] Kumar U. Neuroscience 2005, 134, 525–538.

[44] Kumar, U.; Asotra, K.; Patel, S. C.; et al. Experimental Neurology 1997, 145, 412–424.

[45] Epelbaum, J.; Dournaud, P.; Fodor, M.; et al. Critical Reviews in Neurobiology 1994, 8, 25–44.

[46] Yamada, Y.; Post, S. R.; Wang, K.; et al. Proceedings of the National Academy of Sciences of United States of America 1992, 89, 251–255.

[47] Li, X. J.; Forte, M.; North, R. A.; et al. Journal of Biological Chemistry 1992, 267, 21307–21312.

[48] Yasuda, K.; Rens-Domiano, S.; Breder, C. D.; et al. Journal of Biological Chemistry 1992, 267, 20422–20428.

[49] Meyerhof, W.; Wulfsen, I.; Schoenrock, C.; et al. Proceedings of the National Academy of Sciences of United States of America 1992, 89, 10267–10271.

[50] Yamada, Y.; Reisine, T.; Law, S. F.; et al. Molecular Endocrinology 1993, 6, 2136–2142.

[51] Corness, J. D.; Demchyshyn, L. L.; Seeman, P.; et al. FEBS Letters 1993, 321, 279–284.

[52] Bruno, J. F.; Xu, Y.; Song, J.; et al. Proceedings of the National Academy of Sciences of United States of America 1992, 89, 11151–11155.

[53] Rohrer, L.; Raulf, F.; Bruns, C.; et al. Proceedings of the National Academy of Sciences of United States of America 1993, 90, 4196–4200.

[54] Demchyshyn, L. L.; Srikant, C. B.; Sunahara, R. K.; et al. Molecular Pharmacology 1993, 43, 894–901.

[55] O'Carroll, A. M.; Lolait, S. J.; Konig, M.; et al. Molecular Pharmacology 1992, 42, 939–946.

[56] Panetta, R.; Greenwood, M. T.; Warszynska, A.; et al. Molecular Pharmacology 1994, 45, 417–427.

[57] Patel, Y. C.; Greenwood, M. T.; Kent, G.; et al. Biochemical and Biophysical Research Communications 1993, 192, 288–294.

[58] Patel, Y. C.; Srikant, C. B. Trends in Endocrinology and Metabolism 1997, 8, 398–405.

[59] Reisine, T.; Bell, G. I. Endocrine Review 1995, 16, 427–442.

[60] Patel, Y. C.; Greenwood, M. T.; Warszynska, A.; et al. Biochemical and Biophysical Research Communications 1994, 198, 605–612.

[61] Patel, Y. C.; Srikant, C. B. Endocrinology 1994, 135, 2814–2817.

[62] Olias, G.; Viollet, C.; Kusserow, H.; et al. Journal of Neurochemistry 2004, 89, 1057–1091.

[63] Nouel, D.; Gaudriault, G.; Houle, M.; et al. Endocrinology 1997, 138, 296–306.

[64] Stroh, T.; Jackson, A. C.; Sarret, P.; et al. Endocrinology 2000, 141, 354–365.

[65] Csaba, Z.; Dournaud, P. Neuropeptides 2001, 35, 1–23.

[66] Moller, L. N.; Stidsen, C. E.; Hartmann, B.; et al. Biochimica et Biophysica Acta 2003, 1616, 1–84.

[67] De Weille, J. R.; Schmid-Antomarchi, H.; Fosset, M.; et al. Proceedings of the National Academy of Sciences of United States of America 1989, 86, 2971–2975.

[68] Wang, H.; Bogen, C.; Reisine, T.; et al. Proceedings of the National Academy of Sciences of United States of America 1989, 86, 9616–9620.

[69] White, R. E.; Schonbrunn, A.; Armstrong, D. L. Nature 1991, 351, 570–573.

[70] Sims, S. M.; Lussier, B. T.; Kraicer, J. Journal of Physiology 1991, 441, 615–637.

[71] Akopian, A.; Johnson, J.; Gabriel, R.; et al. Journal of Neuroscience 2000, 20, 929–936.

[72] Kleuss, C.; Hescheler, J.; Ewel, C.; et al. Nature 1991, 353, 43–48.

[73] Meriney, S. D.; Gray, D. B.; Pilar, G. R. Nature 1994, 369, 336–339.

[74] Hou, C.; Gilbert, R. L.; Barber, D. L. Journal of Biological Chemistry 1994, 269, 10357–10362.

[75] Ye, W. Z.; Mathieu, S.; Marteau, C. Journal of Cell and Molecular Biology 1999, 45, 1183–1189.

[76] Lin, C. Y.; Varma, M. G.; Joubel, A.; et al. Journal of Biological Chemistry 2003, 278, 15128–15135.

[77] Putney, L. K.; Denker, S. P.; Barber, D. L. The Annual Review of Pharmacology and Toxicology 2002, 42, 527–552.

[78] Devi, L. A. Trends in Pharmacological Sciences 2000, 21, 324–326.

[79] Rocheville, M.; Lange, D. C.; Kumar, U.; et al. Science 2000, 288, 154–157.

[80] Rocheville, M.; Lange, D. C.; Kumar, U.; et al. Journal of Biological Chemistry 2000, 275, 7862–7869.

[81] Pfeiffer, M.; Koch, T.; Schroder, H.; et al. Journal of Biological Chemistry 2000, 276, 14027–14036.

[82] Bruno, J.-F.; Xu, Y.; Song, J.; et al. Endocrinology 1993, 133, 2561–2567.

[83] Kong, H.; DePaoli, A. M.; Breder, C. D.; et al. Neuroscience 1994, 59, 175–184.

[84] Panetta, R.; Patel, Y. C. Life Sciences 1994, 56, 333–342.

[85] Srikant, C. B., ed., *Somatostatin*. Norwell: Kluwer, 2004, 123–142.

[86] Tannenbaum, G. S.; Zhang, W. H.; Lapointe, M.; et al. Endocrinology 1998, 139, 1450–1453.

[87] Panetta, R.; Patel, Y. C. Life Sciences 1995, 56, 333–342.

[88] O'Carroll, A. M.; Krempels, K. Endocrinology 1995, 136, 5224–5227.

[89] Day, R.; Dong, W.; Panetta, R.; et al. Endocrinology 1995, 136, 5232–5235.

[90] Buscail, L.; Saint-Laurent, N.; Chastre, E.; et al. Cancer Research 1996, 56, 1823–1827.

[91] Reubi, J. C.; Waser, B.; Schaer, J. C.; et al. European Journal of Nuclear Medicine 2001, 28, 836–846.

[92] Hoyer, D.; Bell, G. I.; Berelowitz, M.; et al. Trends in Pharmacological Sciences 1995, 16, 86–88.

[93] Reubi, J. C.; Schar, J. C.; Waser, B.; et al. European Journal of Nuclear Medicine 2000, 27, 273–282.

[94] Lewis, I.; Bauer, W.; Albert, R.; et al. Journal of Medicinal Chemistry 2003, 46, 2334–2344.

[95] Waser, B.; Cescato, R.; Tamma, ML.; et al. European Journal of Pharmacology 2010, 644, 257–262.

[96] Lesche, S.; Lehmann, D.; Nagel, F.; et al. Journal of Clinical Endocrinology and Metabolism 2009, 94, 654–661.

[97] Bruns, C.; Lewis, I.; Briner, U.; et al. European Journal of Endocrinology 2002, 146, 707–716.

[98] Reubi, J. C.; Eisenwiener, K. P.; Rink, H.; et al. European Journal of Pharmacology 2002, 456, 45–49.

[99] Cescato, R.; Loesch, K. A.; Waser, B.; et al. Molecular Endocrinology 2009, 24, 240–249.

[100] Ginj, M.; Zhang, H.; Eisenwiener, K. P.; et al. Clinical Cancer Research 2008, 14, 2019–2027.

[101] Charalambidis, D.; Cescato, R.; Marsouvanidis, P. J. et al. European Journal of Nuclear Medicine and Molecular Imaging 2008, 35, 491, S209.

[102] Veber, D. F.; Holly, F. W.; Paleveda, W. J.; et al. Proceedings of the National Academy of Sciences of United States of America 1978, 75, 2636–2640.

[103] Veber, D. F.; Holly, F. W.; Nutt, R. F.; et al. Nature 1979, 280, 512–514.

[104] Veber, D. F.; Freidlinger, R. M.; Perlow, D. S.; et al. Nature 1981, 292, 55–58.

[105] Fani, M.; Mueller, A.; Tamma, M. L.; et al. Journal of Nuclear Medicine 2010, 51, 1771–1779.

[106] Antunes, P.; Ginj, M.; Zhang, H.; et al. European Journal of Nuclear Medicine and Molecular Imaging 2007, 34, 982–993.

[107] Wild, D.; Schmitt, J. S.; Ginj, M.; et al. European Journal of Nuclear Medicine and Molecular Imaging 2003, 30, 1338–1347.

8

RADIOLABELED SOMATOSTATIN RECEPTOR ANTAGONISTS

MELPOMENI FANI[1] AND HELMUT R. MAECKE[2]

[1]Clinic of Radiology and Nuclear Medicine, University of Basel Hospital, Basel, Switzerland
[2]Department of Nuclear Medicine, University Hospital Freiburg, Freiburg, Germany

ABBREVIATIONS

7TMR	seven-transmembrane receptors
CB-TE2A	4,11-bis(carboxymethyl)-1,4,8,11-tetraazabicyclo[6.6.2]hexadecane
CNS	central nervous system
DOTA	1,4,7,10-tetraazacyclododecane-1,4,7,10-tetraacetic acid
DTPA	diethylenetriaminepentaacetic acid
GPCR	G-protein-coupled receptors
HEK	human embryonic kidney
hsst	human sst-expressing tumors
HYNIC	hydrazinonicotinamide
EDDA	ethylenediamine-N,N'-diacetic acid
NOC	[1-Nal^3]octreotide
NODAGA	1,4,7-triazacyclononane,1-glutaric acid-4,7-acetic acid
Pal	3-pyridyl-alanine
PET	positron emission tomography

Somatostatin Analogues: From Research to Clinical Practice, First Edition. Edited by Alicja Hubalewska-Dydejczyk, Alberto Signore, Marion de Jong, Rudi A. Dierckx, John Buscombe, and Christophe Van de Wiele.

rsst   rat sst-expressing tumors
SPECT   single-photon emission computed tomography
sst   somatostatin receptor subtype

INTRODUCTION

Somatostatin receptors belong to the large family of G-protein-coupled receptors (GPCRs) also called seven-transmembrane receptors (7TMR). This receptor class is the largest and most ubiquitous plasma membrane receptors and encompasses the most successfully used druggable targets in the market. The receptors transfer extracellular stimuli—exerted by light, odor, or agonistic ligands—into intracellular signals [1, 2].

Typically, GPCR activation by agonists results in activating many signaling pathways, which for most receptors are mediated by G proteins and β-arrestins. Distinct biological responses are often linked to these pathways. Usually, full agonists activate the full range of receptors signaling network (full spectrum of signaling pathways), whereas antagonists inactivate the signaling network, and they block the signaling action of agonists. However, biased ligands selectively activate some signaling pathways while inactivating others mediated by the same receptor. This enhanced functional selectivity allows developing agents with increased efficacy (e.g., effective internalization and retention) and/or decreased adverse effects. Ligands developed in this field are mainly focused on the cardiovascular and central nervous system (CNS) diseases.

Aberrant expression and activity of GPCRs are also considered important in tumorigenesis [3, 4]. In a deep sequencing study, O'Hayre et al. found that about 20% of human tumors harbor mutations in GPCRs [3]. Long before these studies, Reubi et al. showed that GPCRs are overexpressed on a variety of human tumors and therefore found to be important targets in cancer imaging and targeted radionuclide therapy [5]. The prototypes of these cancers are neuroendocrine tumors overexpressing somatostatin receptors as shown by autoradiography [6]. Five somatostatin receptor subtypes (sst1–sst5) are known, and they are all overexpressed to some extent in these tumors; the most important one is sst2 for reasons not known at present. As discussed in Chapters 8 and 10, radiolabeled somatostatin-based peptides were developed for imaging (SPECT and PET) and for targeted radionuclide therapy [6–8]. They are successfully used in the clinic and are the only radiolabeled peptides having an impact in patient care. These radioligands are all potent agonists that internalize into tumor cells upon binding to the receptor. This mechanism was considered to be essential (mandatory) for active and high accumulation of the radioligand in the tumor along with long tumor retention. This paradigm was supported by data from the Anderson and our own group. Lewis et al. showed that the rate of internalization correlated with the tumor uptake in a group of four ^{64}Cu-labeled somatostatin-based octapeptides [9]. In a series of ^{99m}Tc-labeled HYNIC/EDDA-octreotide (HYNIC/EDDA-TOC, where HYNIC refers hydrazinonicotinamide and EDDA to ethylenediamine-N,N'-diacetic acid), compared with ^{111}In-DOTA-TOC (DOTA: 1,4,7,10-tetraazacyclododecane-1,4,7,10-tetraacetic acid) and OctreoScan, Storch et al. have shown a strong linear correlation between the uptake in the tumor and in the pancreas (sst2-positive organ) and the rate of internalization in cell culture [10]. In addition, Ginj et al. [11] described a family of carbocyclic octapeptides with pansomatostatin properties (high affinity for all five receptor subtypes). When these peptides were tested for their agonist properties using a cAMP assay (inhibition of forskolin-stimulated cAMP production), they showed potent agonistic properties at all five receptor subtypes. On the contrary, they did not internalize in sst2 nor did they show any Ca^{2+} release when incubated with human embryonic kidney (HEK)293-human sst2 (hsst2) cells. They antagonize somatostatin stimulation of intracellular Ca^{2+} flux. Interestingly, they showed rapid internalization in HEK293-hsst3 cells. Consequently, in a dual tumor model xenografted with HEK293-hsst2 and HEK293-hsst3, a fast washout was found from the sst2 tumor, but high tumor uptake and long retention were found in the hsst3 tumor. These peptides were later shown to be biased ligands on sst2 and true agonists on sst3 [12]. These data argued for the use of radiolabeled somatostatin-based agonists as internalization indeed appears to be of high importance for high and long-lasting tumor uptake.

However, a number of recent observations have challenged this strategy. Antagonists may have characteristics other than those related to internalization that may make their radiolabeled derivatives suitable tools for *in vivo* receptor targeting. Most relevant is the *in vitro* evidence that, in certain circumstances, antagonist radioligands may recognize a higher number of receptor-binding sites than agonist radioligands [13, 14]. A higher number of receptor-binding sites may result in higher uptake than agonist radioligands, and the lack of internalization may compensate for a faster washout, which can nicely be combined with short-lived radionuclides such as ^{68}Ga and ^{18}F for PET or [213]Bi for therapy. This was the hypothesis to start a project on somatostatin-based radiolabeled antagonists along with the University of Berne (Jean Claude Reubi) and the Salk Institute (Jean Rivier).

RADIOLABELED ANTAGONISTS

Chemistry

The first somatostatin-based antagonists were published by Bass et al. in 1996 [15]. They found that in the octapeptide series (disulfide cyclized, hexapeptide core), the inversion of chirality at positions 1 and 2 of the octapeptide (octreotide family) converted an agonist into a potent antagonist. Later structure–activity relationship studies by the Coy group afforded new structures. The most potent of these antagonists were H-Cpa-cyclo[DCys-Tyr-DTrp-Lys-Thr-Cys]-Nal-NH$_2$ and H-Cpa-cyclo[DCys-Pal-DTrp-Lys-Thr-Cys]-2Nal-NH$_2$ (Pal = 3-pyridyl-alanine) [16]. Based on these findings, the antagonists were used as leads for further modifications as well as DOTA coupling for radiometal labeling by Cescato et al. [17] and Ginj et al. [18].

Preclinical Studies

The first preclinical studies with radiolabeled somatostatin antagonists were performed by Ginj et al. targeting rat sst2- and sst3-overexpressing transfected HEK293 [18] cells. It was shown that the potent sst2 and sst3 antagonists are superior to radiolabeled agonists of similar or even higher receptor affinity with regard to *in vivo* tumor uptake as well as tumor-to-normal tissue ratios.

An impressively high uptake of the antagonist sst3-ODN-8 radioligand (Table 8.1) in HEK293-sst3 tumors (60%IA/g, 1 h p.i.) has been reported, which actually has indeed never been achieved by any somatostatin receptor agonist ligand. Not only the uptake at the peak time point was very high, but also the long-lasting accumulation of the antagonist radioligand up to 72 h after injection was a remarkable result and represents a considerable advantage over targeting with established agonists. The same observation was obtained in HEK293-sst2 tumors with the sst2 antagonist ^{111}In-DOTA-sst2-ANT (Table 8.1). In this study, ^{111}In-DOTA-sst2-ANT has been compared with ^{111}In-DTPA-TATE (DTPA: diethylenetriaminepentaacetic acid), and knowing the outstanding targeting abilities of ^{111}In-DTPA-TATE [19], it is striking to see that the *in vivo* uptake at 4 and 24 h for the sst2 antagonist is twice as high, despite the fact that the antagonist is not internalized into the tumor cells and that its sst2-binding affinity is lower than for the agonist [18].

Explanations for these excellent *in vivo* targeting properties of antagonists may be found, at least in part, in the higher number of binding sites recognized by antagonists, compared to agonists, as shown using the HEK293-sst2 and HEK293-sst3

TABLE 8.1 **Somatostatin-based antagonists that have been developed as potential radiotracers**

Code	Chemical structure	Chelator	(Radio) metal	References
sst3-ODN-8	NH_2-CO-c(D-Cys-Phe-Tyr-D-Agl[8](Me,2-naphthoyl)-Lys-Thr-Phe-Cys)-OH	DOTA	[111]In	[18]
sst2-ANT (BASS)	p-NO_2-Phe-cyclo(D-Cys-Tyr-D-Trp-Lys-Thr-Cys)-D-Tyr-NH_2	DOTA CB-TE2A	[111]In [177]Lu [64]Cu	[18, 20, 31, 32, 35, 36]
LM3	p-Cl-Phe-cyclo(D-Cys-Tyr-D-Aph(Cbm)-Lys-Thr-Cys)-D-Tyr-NH_2	DOTA NODAGA CB-TE2A	[68]Ga [64]Cu [nat]In	[21, 30]
JR10	p-NO_2-Phe-c[D-Cys-Tyr- D-Aph(Cbm)-Lys-Thr-Cys]-D-Tyr-NH_2	DOTA NODAGA	[nat/68]Ga [nat]In [nat]Lu	[30, 39]
JR11	Cpa-c[D-Cys-Aph(Hor)-D-Aph(Cbm)-Lys-Thr-Cys]-D-Tyr-NH_2	DOTA NODAGA	[nat/68]Ga [nat]In [nat/177]Lu [nat]Cu [nat]Y	[30]

transfected cell lines [18]. It appears that in an *in vivo* situation, an agonist that triggers a strong internalization but binds to a limited number of high-affinity receptors is a less efficient targeting agent than an antagonist lacking internalization capabilities but binding to a larger variety of receptor conformations.

In another study, the antagonist sst2-ANT was compared with the powerful agonist TATE, both conjugated to 4,11-bis(carboxymethyl)-1,4,8,11-tetraazabicyclo[6.6.2]hexadecane (CB-TE2A) and labeled with ^{64}Cu, employing the frequently used rat pancreatic cell line AR42J [20]. Interestingly, this study did not show the superiority of the antagonist *in vivo*, despite the fact that a 14-fold higher number of binding sites for the antagonist were found, similar to what was found with the transfected cell line HEK293-sst2 for ^{111}In-DOTA-sst2-ANT compared to TATE [18]. These data indicated that the *in vitro* and *in vivo* properties of radiolabeled GPCR antagonists are not understood well yet and need further studies.

Relatively little is known in regard to structural parameters determining the pharmacological properties of radiolabeled somatostatin-based antagonists. A recent study demonstrates that the chelate makes the difference on the affinity and also pharmacokinetics of these antagonists [21]. In this study, the sst2 antagonist LM3 (Table 8.1) was coupled to three different macrocyclic chelators, namely, CB-TE2A, 1,4,7-triazacyclononane,1-glutaric acid-4,7-acetic acid (NODAGA), and DOTA. The NODAGA and CB-TE2A conjugates were labeled with ^{64}Cu, while the NODAGA and DOTA conjugates with ^{68}Ga. *In vitro* and *in vivo* studies showed the strong dependence of the affinity and pharmacokinetics on the chelator and radiometal. For instance, ^{68}Ga-NODAGA-LM3 has a 10-fold higher sst2 affinity than the corresponding DOTA conjugate (^{68}Ga-DOTA-LM3) and a 5-fold higher sst2 affinity than the corresponding ^{64}Cu complex (^{64}Cu-NODAGA-LM3), indicating that there is a much stronger influence of the appended chelate than found in somatostatin-based agonists [22]. Significant differences were observed in the *in vivo* behavior of the two ^{64}Cu-labeled peptides, namely, ^{64}Cu-NODAGA-LM3 and ^{64}Cu-CB-TE2A-LM3. Even though ^{64}Cu-CB-TE2A-LM3 has 1.6-fold better affinity than ^{64}Cu-NODAGA-LM3, this is not reflected in a higher tumor uptake. On the contrary, ^{64}Cu-NODAGA-LM3 shows a distinctly higher uptake than ^{64}Cu-CB-TE2A-LM3 at 1 h (35.5 vs. 19.3 %IA/g) and 4 h (37.9 vs. 26.9 %IA/g). Significant amount of radioactivity is accumulating in the kidneys for both radiolabeled antagonists but 2- to 9-fold higher for ^{64}Cu-CB-TE2A-LM3, depending on the time point, probably due to the positive charge of ^{64}Cu-CB-TE2A-LM3 versus the neutral charge of ^{64}Cu-NODAGA-LM3.

An important issue in tumor targeting, especially concerning therapeutic applications, is the retention of the radioligand in the tumor. As mentioned previously, studies indicated that low or absent internalization correlates with very short tumor retention [11, 23]. Concerning ^{64}Cu-NODAGA-LM3 and ^{64}Cu-CB-TE2A-LM3, both showed low internalization. Therefore, it is remarkable that ^{64}Cu-NODAGA-LM3 has a (relatively) slow but distinctly faster washout from the tumor (~15 %IA/g, 24 h p.i.) compared to ^{64}Cu-CB-TE2A-LM3, which shows almost no washout between 4 h (26.9 %IA/g) and 24 h p.i. (21.6 %IA/g). This may be of high importance if therapy is considered with the beta emitter ^{67}Cu ($t_{1/2} = 64$ h) as the area under the curve may be larger for the $^{64/67}$Cu-CB-TE2A conjugate. Despite the negligible washout of ^{64}Cu-CB-TE2A-LM3 from the tumor, the tumor-to-normal tissue ratios are remarkably higher for ^{64}Cu-NODAGA-LM3 compared to ^{64}Cu-CB-TE2A-LM3 (Fig. 8.1a; e.g., tumor-to-kidney, 12.8 vs. 1.7, and tumor-to-muscles, 1342 vs. 75.2, at 24 h), an important parameter for good image contrast. These differences were especially impressive for the tumor-to-kidney ratios, which strongly increase with time and are much higher than usually seen with somatostatin-based radiolabeled agonists. MicroPET imaging shows clear tumor localization and very high image contrast, especially for ^{64}Cu-NODAGA-LM3 (Fig. 8.1b).

It has been shown that the chelated somatostatin agonists are sensitive to N-terminal radiometal modifications, with Ga-DOTA agonists having significantly higher binding affinity than their Lu-, In-, and Y-DOTA correlates [24]. For instance, Ga(III)-DOTA-OC has three times higher affinity for sst2 than Y(III)-DOTA-OC, Ga(III)-DOTA-TOC has six times higher affinity than Y(III)-DOTA-TOC, and Ga(III)-DOTA-TATE has eight times higher affinity than Y(III)-DOTA-TATE [24]. These improved binding affinities also translated into improved internalization rates and concomitantly higher tumor uptake [25]. In those studies, it appeared adequate to generalize that $^{67/68}$Ga is a radiometal that systematically improved the sst2 affinity of somatostatin agonists as well as their pharmacokinetics. It was further confirmed *in vivo* in patients with neuroendocrine tumors that ^{68}Ga-DOTA-NOC (NOC: [1-Nal^3]octreotide), ^{68}Ga-DOTA-TOC, and ^{68}Ga-DOTA-TATE were better imaging agents than the ^{111}In-DOTA congeners [25–29].

Recently, a study was published on whether chelated somatostatin antagonists are also sensitive to radiometal modifications. In this study, three different somatostatin antagonists, namely, JR10, JR11, and LM3 (Table 8.1), were conjugated to the chelators DOTA and NODAGA, and various (radio)metals including In(III), Y(III), Lu(III), Cu(II), and Ga(III) were added [30]. Surprisingly, in all three resulting antagonists, the Ga-DOTA analogs were the least affine radioligands, with an sst2-binding affinity up to 60 times lower than the respective Y(III)-DOTA, Lu(III)-DOTA, and In(III)-DOTA compounds (Table 8.2). Interestingly, however, substitution of DOTA by the NODAGA chelator in the antagonist conjugate was able to increase massively its binding affinity in contrast to the Ga(III)-DOTA analog.

An important finding in this study is the 8-fold difference between In(III)-DOTA-JR11 (IC$_{50}$=3.8±0.7 nM) and Y(III)-DOTA-JR11 (IC$_{50}$=0.47±0.05 nM). ^{111}In is commonly used as a surrogate for ^{90}Y for imaging the biodistribution of ^{90}Y-labeled tracers and for dosimetry (e.g., ^{111}In/^{90}Y-DOTA-TOC). The recent results indicate that ^{111}In may not be a suitable surrogate of ^{90}Y and therefore not very reliable for dosimetric studies of the therapeutic radiolabeled peptide based on somatostatin antagonists. Moreover, this may be the case for other peptide families and has to be investigated.

A striking finding from these studies is the comparison of the antagonists with clinically used agonists. The side-by-side *in vivo* comparison between ^{68}Ga-DOTA-TATE and ^{68}Ga-DOTA-JR11 illustrates the great potential of the antagonists. ^{68}Ga-DOTA-JR11, having a dramatically lower affinity for the sst2 (~150-fold) compared to ^{68}Ga-DOTA-TATE (Table 8.2), showed a 1.3-fold higher tumor uptake (23.8 vs. 17.8 %IA/g, 1 h p.i.), while ^{68}Ga-NODAGA-JR11, with a

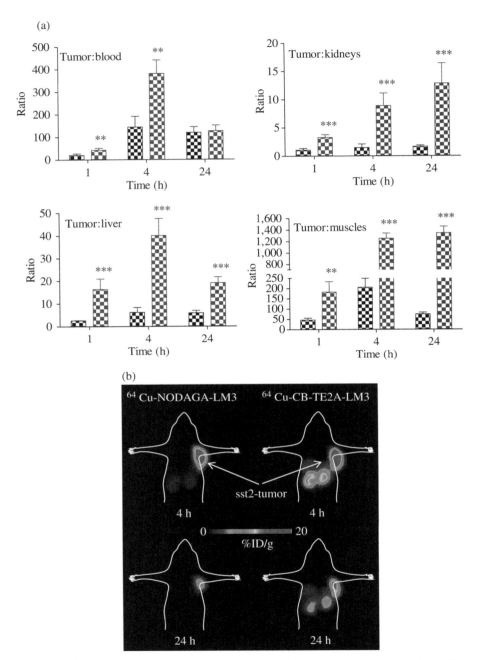

FIGURE 8.1 (a) Tumor-to-normal tissue ratios of the radiolabeled antagonists ^{64}Cu-NODAGA-LM3 (grey bars) and ^{64}Cu-CB-TE2A-LM3 (black bars) in biodistribution studies in HEK293-sst2 tumor xenografts at 1, 4, and 24 h p.i. are significantly higher for the NODAGA conjugate than for the CB-TE2A conjugate ([**]$P<0.01$ and [***]$P<0.001$). (b) MicroPET images of ^{64}Cu-NODAGA-LM3 and ^{64}Cu-CB-TE2A-LM3 at 4 and 24 h p.i. (coronal sections) showed the potential of these radiopeptides for *in vivo* imaging of sst2-expressing tumors and highlighted the improved tumor-to-background contrast of ^{64}Cu-NODAGA-LM3, especially tumor to kidney [21].

TABLE 8.2 Chemical structures of three sst_2 antagonists based on the JR10, JR11, and LM3 family and their metallated conjugates and IC_{50} values for sst2 (mean ± SEM, $n \geq 3$) [30]

Code	Chemical structure	IC_{50} (nM)
DOTA-JR11	DOTA-Cpa-c[D-Cys-Aph(Hor)-D-Aph(Cbm)-Lys-Thr-Cys]-D-Tyr-NH$_2$	0.72 ± 0.12
Ga-DOTA-JR11	Ga-DOTA-Cpa-c[D-Cys-Aph(Hor)-D-Aph(Cbm)-Lys-Thr-Cys]-D-Tyr-NH$_2$	29 ± 2.7
Y-DOTA-JR11	Y-DOTA-Cpa-c[D-Cys-Aph(Hor)-D-Aph(Cbm)-Lys-Thr-Cys]-D-Tyr-NH$_2$	0.47 ± 0.05
Lu-DOTA-JR11	Lu-DOTA-Cpa-c[D-Cys-Aph(Hor)-D-Aph(Cbm)-Lys-Thr-Cys]-D-Tyr-NH$_2$	0.73 ± 0.15
Cu-DOTA-JR11	Cu-DOTA-Cpa-c[D-Cys-Aph(Hor)-D-Aph(Cbm)-Lys-Thr-Cys]-D-Tyr-NH$_2$	16 ± 1.2
In-DOTA-JR11	In-DOTA-Cpa-c[D-Cys-Aph(Hor)-D-Aph(Cbm)-Lys-Thr-Cys]-D-Tyr-NH$_2$	3.8 ± 0.7
NODAGA-JR11	NODAGA-Cpa-c[D-Cys-Aph(Hor)-D-Aph(Cbm)-Lys-Thr-Cys]-D-Tyr-NH$_2$	4.1 ± 0.2
Ga-NODAGA-JR11	Ga-NODAGA-Cpa-c[D-Cys-Aph(Hor)-D-Aph(Cbm)-Lys-Thr-Cys]-D-Tyr-NH$_2$	1.2 ± 0.2
DOTA-JR10	DOTA-p-NO$_2$-Phe-c[D-Cys-Tyr- D-Aph(Cbm)-Lys-Thr-Cys]-D-Tyr-NH$_2$	0.62 ± 0.21
Ga-DOTA-JR10	Ga-DOTA-p-NO$_2$-Phe-c[D-Cys-Tyr-D-Aph(Cbm)-Lys-Thr-Cys]-D-Tyr-NH$_2$	8.9 ± 2.2
In-DOTA-JR10	In-DOTA-p-NO$_2$-Phe-c[D-Cys-Tyr-D-Aph(Cbm)-Lys-Thr-Cys]-D-Tyr-NH$_2$	2.3 ± 0.5
Lu-DOTA-JR10	Lu-DOTA-p-NO$_2$-Phe-c[D-Cys-Tyr-D-Aph(Cbm)-Lys-Thr-Cys]-D-Tyr-NH$_2$	1.2 ± 0.2
NODAGA-JR10	NODAGA-p-NO$_2$-Phe-c[D-Cys-Tyr-D-Aph(Cbm)-Lys-Thr-Cys]-D-Tyr-NH$_2$	23.0 ± 1.5
Ga-NODAGA-JR10	Ga-NODAGA-p-NO$_2$-Phe-c[D-Cys-Tyr-D-Aph(Cbm)-Lys-Thr-Cys]-D-Tyr-NH$_2$	6.5 ± 0.5
DOTA-LM3	DOTA-p-Cl-Phe-c[D-Cys-Tyr-D-Aph(Cbm)-Lys-Thr-Cys]-D-Tyr-NH$_2$	0.39 ± 0.05
Ga-DOTA-LM3	Ga-DOTA-p-Cl-Phe-c[D-Cys-Tyr-D-Aph(Cbm)-Lys-Thr-Cys]-D-Tyr-NH$_2$	12.5 ± 4.3
In-DOTA-LM3	In-DOTA-p-Cl-Phe-c[D-Cys-Tyr-D-Aph(Cbm)-Lys-Thr-Cys]-D-Tyr-NH$_2$	1.3 ± 0.1
Ga-NODAGA-LM3	Ga-NODAGA-p-Cl-Phe-c[D-Cys-Tyr-D-Aph(Cbm)-Lys-Thr-Cys]-D-Tyr-NH$_2$	1.3 ± 0.2
Reference agonist		
Ga-DOTA-TATE	Ga-DOTA-D-Phe-c[Cys-Tyr-D-Trp-Lys-Thr-Cys]-Thr	0.2 ± 0.04

6-fold lower affinity (Table 8.2), showed an up to 1.7-fold higher tumor uptake (30.7 vs. 17.8 %IA/g, 1 h p.i.) [30]. The low-affinity antagonist is slightly superior, which again may be explained by the higher number of binding sites for antagonists versus agonists [18], outweighing the affinity differences. This was also impressively demonstrated with *in vitro* autoradiographic studies of human tumor specimens [31], discussed afterward. PET images highlighted the higher uptake of the antagonists compared to agonist (Fig. 8.2) and the excellent background clearance already at 1 and 2 h p.i.

There are already two studies in animal tumor models where a side-by-side comparison of somatostatin receptor antagonist versus agonist showed that the antagonists bind to sst2- and sst3-expressing tumors *in vivo* better than agonists with comparable or even higher affinity [18, 30]. This was proven also to be the case in a side-by-side *in vitro* binding studies on several different, well-characterized human tumor samples of the antagonist ^{177}Lu-DOTA-BASS versus the agonist ^{177}Lu-DOTA-TATE [31].

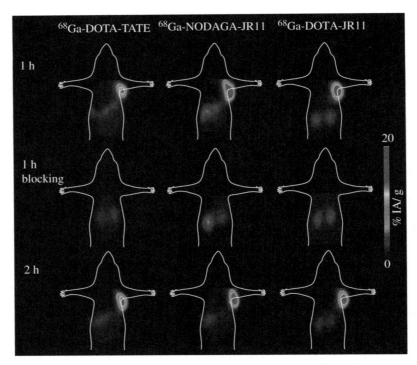

FIGURE 8.2   Small-animal PET images (coronal sections) of HEK-hsst2 tumor-bearing mice injected with ^{68}Ga-DOTA-TATE, ^{68}Ga-DOTA-JR11, and ^{68}Ga-NODAGA-JR11, 1 and 2 h p.i., show the potentiality of the radiolabeled antagonists ^{68}Ga-DOTA-JR11 and ^{68}Ga-NODAGA-JR11 to image sst2-expressing tumors *in vivo*. They also illustrate the higher tumor uptake of the radiolabeled antagonists compared to agonist, and they confirm the specificity of all radiotracers as no tumor is visualized in the blocking experiments [30].

About 50 sst2-positive human tumor tissue samples were analyzed by *in vitro* receptor autoradiography for the expression of sst2, comparing the binding capacity of ^{177}Lu-DOTA-BASS and ^{177}Lu-DOTA-TATE in successive tissue sections. Both radioligands efficiently bind to the sst2 receptor on the different tumor tissues. However, in all cases, the radiolabeled antagonist ^{177}Lu-DOTA-BASS bound to more sst2-receptor sites in the tumors than did the agonist ^{177}Lu-DOTA-TATE. Autoradiograms of representative examples for all tumor types analyzed are shown in Figure 8.3a. The mean ratios of the antagonist ^{177}Lu-DOTA-BASS to the agonist ^{177}Lu-DOTA-TATE ranged from 4.2 up to 12.3 (Fig. 8.3b). This significantly increased binding may increase the localization accuracy for tumors and metastases but also increase the efficacy of radiotherapeutic intervention with such antagonist radiotracers. Of particular interest is the fact that tumors other than neuroendocrine, such as tumors that normally express a low density of receptors, for example, breast carcinomas, renal cell cancers, or non-Hodgkin lymphomas, showed the same results. Those tumors are not currently among those being routinely investigated with somatostatin receptor imaging. It is therefore impressive to see that by using the antagonist radioligand, it is possible to increase, at least *in vitro*, the number of binding sites in breast carcinomas and renal cell cancers 11.4- and 5.1-fold, respectively (Fig. 8.3b). As well, the case of non-Hodgkin lymphomas normally expressing a low amount of sst2 showed a 4.8-fold increase in receptor number with the antagonist radioligand, reaching levels that may be detected more easily *in vivo* than with current agonists. These *in vitro* human data, together with the *in vivo* animal tumor data, are strong arguments indicating that sst2 antagonists may be worth testing in patients in a wide range of tumors including nonneuroendocrine tumors. Such sst2 antagonist radioligands should be useful not only for diagnostic tumor imaging but also for targeted tumor radiotherapy.

Clinical Studies

The first pilot study of radiolabeled somatostatin-based antagonist demonstrates their feasibility of imaging sst2-expressing tumors in patients [32]. This pilot study provides the first clinical evidence that radiolabeled sst2 antagonists not only detect sst2-expressing neuroendocrine tumors but also may even be superior to agonists for imaging and therapy of neuroendocrine tumors. The studied antagonist ^{111}In-DOTA-BASS had a favorable biodistribution profile (higher tumor uptake and lower organ uptake) than the agonist ^{111}In-DTPA-OC, resulting in a higher tumor detection rate.

The two radiotracers, ^{111}In-DOTA-BASS and ^{111}In-DTPA-OC, were compared in 5 patients with metastatic thyroid carcinoma or neuroendocrine tumors. ^{111}In-DOTA-BASS detected 25 of 28 lesions, whereas ^{111}In-DTPA-OC detected only 17 of 28 lesions. All lesions visible on ^{111}In-DTPA-OC scans were also detected by ^{111}In-DOTA-BASS, whereas there were 8 lesions on ^{111}In-DOTA-BASS that were not visible on ^{111}In-DTPA-OC. There were additionally 3 bone lesions that were negative on both ^{111}In-DOTA-BASS and ^{111}In-DTPA-OC scans, but bone metastases were confirmed by serial CT. Figure 8.4a shows representative images of the same patient

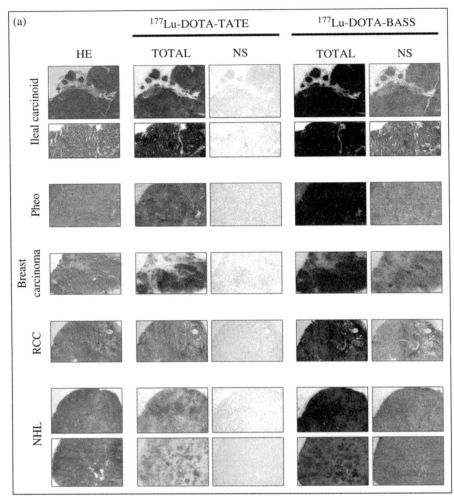

FIGURE 8.3   (a) Comparison of ¹⁷⁷Lu-DOTA-TATE and ¹⁷⁷Lu-DOTA-BASS receptor auto-radiographic binding in successive sections of various types of human cancers (ileal carcinoid, pheochromocytoma, breast carcinoma, renal cell carcinoma, and non-Hodgkin lymphoma) expressing sst2. Columns from left to right represent hematoxylin and eosin staining, total and nonspecific binding of ¹⁷⁷Lu-DOTA-TATE, and total and nonspecific binding of ¹⁷⁷Lu-DOTA-BASS. Binding is markedly stronger with the antagonist ¹⁷⁷Lu-DOTA-BASS. HE, hematoxylin and eosin; NHL, non-Hodgkin lymphoma; NS, nonspecific; Pheo, pheochromocytoma; RCC, renal cell carcinoma. (*See insert for color representation of the figure.*) (b) Quantitation of *in vitro* receptor autoradiography experiments with various types of human cancers (ileal carcinoid, pheochromocytoma, breast carcinoma, renal cell carcinoma, and non-Hodgkin lymphoma) using ¹⁷⁷Lu-DOTA-TATE and ¹⁷⁷Lu-DOTA-BASS as radioligands. Shown are bar graphs of specific binding (counts/h) of radioligands to tumor sections after quantitation using InstantImager. For all tested human tumor types, the antagonist ¹⁷⁷Lu-DOTA-BASS exhibited markedly better binding behavior. NHL, non-Hodgkin lymphoma; Pheo, pheochromocytoma; RCC, renal cell carcinoma; Spec., specific [31].

(b)

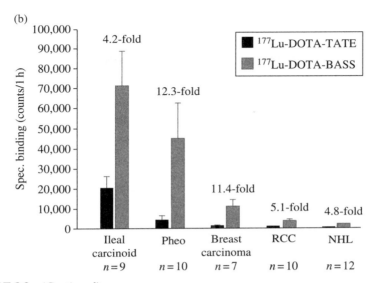

FIGURE 8.3   (*Continued*)

(a)   (b)

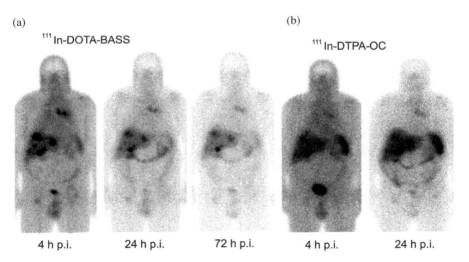

FIGURE 8.4   (a) ^{111}In-DOTA-BASS (4, 24, and 72 h p.i. left) and ^{111}In-DTPA-OC (4 and 24 h p.i. right) planar anterior whole-body scans from patient with neuroendocrine carcinoma. The images illustrate the better tumor-to-background uptake ratio of ^{111}In-DOTA-BASS (e.g., tumor-to-liver uptake ratio of 6.7 at 4 h p.i. and 3.5 at 24 h p.i.) compared to ^{111}In-DTPA-OC (1.2 at both time points). In this patient, ^{111}In-DOTA-BASS detected 16 metastases, whereas ^{111}In-DTPA-OC detected only 11 metastases. (b) Comparison of ^{111}In-DOTA-BASS and ^{111}In-DTPA-octreotide uptake in the kidneys, spleen, and tumors. Values are expressed as mean $\pm$ SD of 5 patients. OC = octreotide; p.i. = after injection [32].

scanned with both radiotracers. The longest residence times of ^{111}In-DOTA-BASS were observed in the tumor and kidneys.

The quantitative analysis showed relevant differences in the biodistribution between the antagonist ^{111}In-DOTA-BASS and agonist ^{111}In-DTPA-OC (Fig. 8.4b). The antagonist showed up to 4.1 times higher uptake in the tumor (3.5 ± 2.8 %IA vs. 1.0 ± 0.99 %IA) 4 h after injection. At the same time, the uptake in the kidneys was lower for the antagonist (1.5 vs. 2.3 %IA), resulting in up to 5.2 times higher tumor-to-kidney uptake ratio in favor of the antagonist. The antagonist also showed lower uptake than the agonist in organs such as the liver (1.4 ± 0.2 vs. 1.7 ± 0.6 %IA) and spleen (1.0 ± 0.2 vs. 2.3 ± 1.0 %IA) 4 h after injection.

With respect to radionuclide therapy, it is particularly encouraging that tumor-to-kidney ratios were up to 5.2 times higher for the antagonist than for the agonist as the kidneys are the major dose-limiting organ in peptide receptor radionuclide therapy. The several-fold higher tumor-to-kidney uptake ratio could significantly improve the efficacy and toxicity profile of radionuclide therapy. In an attempt to further evaluate the potential of somatostatin receptor antagonists for radionuclide therapy, the organ doses of ^{90}Y-DOTA-BASS were calculated based on the assumption that ^{90}Y-DOTA-BASS shows the same biodistribution as ^{111}In-DOTA-BASS. Compared with literature data for ^{90}Y-DOTA-TOC [33, 34], ^{90}Y-DOTA-BASS showed lower renal, hepatic, and splenic radiation doses [32].

Some Interesting Aspects

The pharmacokinetics of the first radiolabeled antagonist in humans, ^{111}In-DOTA-BASS, has been studied more thoroughly in order to investigate the potency of these new radiolabeled antagonists as therapeutic radiopharmaceuticals. The same antagonist was studied with ^{177}Lu to clarify if the radiometal has an influence on the pharmacokinetics [35]. The tumor uptake was almost identical for the two radiolabeled peptides ^{111}In-DOTA-BASS and ^{177}Lu-DOTA-BASS (30–35%IA/g, 4 h p.i.), and the washout is relatively slow for both; about 50% still remains in the tumor 48 h p.i. This indicates no radiometal influence of this DOTA-conjugated antagonist, rendering ^{177}Lu-DOTA-BASS a promising radiotherapeutic agent. This result is somewhat contrary to what was shown with the somatostatin-based radiolabeled antagonist LM3 [21] as discussed previously, where distinct differences in pharmacology depending on radiometal and chelator were found. Attempts to block the tubular cell-mediated kidney uptake of antagonistic radiopeptides with agents, which are successful in the clinical use of corresponding radiolabeled agonists, showed that lysine and Gelofusine were both effective in kidney blocking of ^{111}In-DOTA-BASS. Both agents did not block tumor uptake but blocked about 50% of kidney uptake, increasing the tumor-to-kidney ratio about 2-fold. Last but not least, this study excluded species differences as the biodistribution results of ^{111}In-DOTA-BASS in mice bearing hsst2-expressing tumors (hsst2) in comparison to mice bearing rat sst2-expressing tumors (rsst2) were found to be very similar. This study indicated the relevance of radiolabeled somatostatin-based antagonist not only in diagnostic imaging but also in targeted radionuclide therapy of somatostatin receptor-positive tumors.

All the previously mentioned metallated peptides demonstrated antagonist potency in immunofluorescence microscopy-based internalization experiments [18, 21, 30, 35]. They did not trigger sst2 receptor internalization and antagonized the receptor internalization mediated by an agonist. However, in the radioligand internalization experiments, the corresponding radiotracers, for example, ^{68}Ga-NODAGA-LM3, ^{68}Ga-DOTA-LM3, or ^{111}In-DOTA-BASS, but also other radiolabeled antagonists, such as ^{64}Cu-CB-TE2A-BASS [20, 36], showed significant levels of internalization in sst2-positive cell lines, albeit much lower than the internalization of corresponding radiolabeled agonists. Internalization studies of ^{111}In-DOTA-BASS in cells not expressing sst2 showed no internalization of the radioligand, while pretreatment of the sst2-expressing cells with hypertonic sucrose, known to prevent receptor endocytosis by preventing clathrin-coated pit formation [37], distinctly reduced radioligand internalization. These findings are strong indications of a specific receptor-mediated internalization process of the radioligands. It is still an open question why these antagonists internalize in a radioligand experiment but are not able to stimulate sst2 internalization. It should be considered that in the immunofluorescence microscopy experiments, the sst2 receptor trafficking itself is monitored where ligand trafficking is monitored in the radioligand experiment. Moreover, the ligand is used in high excess in the immunofluorescence microscopy experiments, while in the radioligand internalization experiments, the concentration of the radioligand is far below tumor saturation. As far as *in vivo* concerns, the first *in vivo* data indicate receptor internalization triggered by agonists but apparently not by antagonists [38].

CONCLUSION

The recent developments indicate the change of paradigm for *in vivo* targeting of somatostatin receptors using radiolabeled somatostatin-based antagonists than agonists, as this may considerably improve the sensitivity of diagnostic procedures and the efficacy of targeted radionuclide therapy of somatostatin receptor-positive tumors.

REFERENCES

[1] Moore, C. A.; Milano, S. K.; Benovic, J. L. Annual Review of Physiology 2007, 69, 451–482.

[2] Lefkowitz, R. J. Angewandte Chemie International Edition English 2013, 52, 6366–6378.

[3] O'Hayre, M.; Vazquez-Prado, J.; Kufareva, I.; et al. Nature Reviews. Cancer 2013, 13, 412–424.

[4] Dorsam, R. T.; Gutkind, J. S. Nature Reviews. Cancer 2007, 7, 79–94.

[5] Reubi, J. C. Endocrine Reviews 2003, 24, 389–427.

[6] Maecke, H. R.; Reubi, J. C. Journal of Nuclear Medicine 2011, 52, 841–844.

[7] de Jong, M.; Breeman, W. A.; Kwekkeboom, D. J.; Valkema, R.; Krenning, E. P. Accounts of Chemical Research 2009, 42, 873–880.

[8] Ambrosini, V.; Fani, M.; Fanti, S.; Forrer, F.; Maecke, H. R. Journal of Nuclear Medicine 2011, 52 (Suppl 2), 42S–55S.

[9] Lewis, J. S.; Lewis, M. R.; Srinivasan, A.; Schmidt, M. A.; Wang, J.; Anderson, C. J. Journal of Medicinal Chemistry 1999, 42, 1341–1347.

[10] Storch, D.; Behe, M.; Walter, M. A.; Chen, J.; Powell, P.; Mikolajczak, R.; Macke, H. R. Journal of Nuclear Medicine 2005, 46, 1561–1569.

[11] Ginj, M.; Zhang, H.; Eisenwiener, K. P.; et al. Clinical Cancer Research 2008, 14, 2019–2027.

[12] Cescato, R.; Loesch, K. A.; Waser, B.; et al. Molecular Endocrinolology 2010, 24, 240–249.

[13] Perrin, M. H.; Sutton, S. W.; Cervini, L. A.; Rivier, J. E.; Vale, W. W. The Journal of Pharmacology and Experimental Therapeutics 1999, 288, 729–734.

[14] Sleight, A. J.; Stam, N. J.; Mutel, V.; Vanderheyden, P. M. Biochemical Pharmacology 1996, 51, 71–76.

[15] Bass, R. T.; Buckwalter, B. L.; Patel, B. P.; et al. Molecular Pharmacology 1996, 50, 709–715.

[16] Hocart, S. J.; Jain, R.; Murphy, W. A.; Taylor, J. E.; Coy, D. H. Journal of Medicinal Chemistry 1999, 42, 1863–1871.

[17] Cescato, R.; Erchegyi, J.; Waser, B.; et al. Journal of Medicinal Chemistry 2008, 51, 4030–4037.

[18] Ginj, M.; Zhang, H.; Waser, B.; et al. Proceedings of the National Academy of Sciences of the United States of America 2006, 103, 16436–16441.

[19] de Jong, M.; Breeman, W. A.; Bakker, W. H.; et al. Cancer Research 1998, 58, 437–441.

[20] Wadas, T. J.; Eiblmaier, M.; Zheleznyak, A.; et al. Journal of Nuclear Medicine 2008, 49, 1819–1827.

[21] Fani, M.; Del Pozzo, L.; Abiraj, K.; et al. Journal of Nuclear Medicine 2011, 52, 1110–1118.

[22] Eisenwiener, K. P.; Prata, M. I.; Buschmann, I.; et al. Bioconjugate Chemistry 2002, 13, 530–541.

[23] Fani, M.; Mueller, A.; Tamma, M. L.; et al. Journal of Nuclear Medicine 2010, 51, 1771–1779.

[24] Reubi, J. C.; Schar, J. C.; Waser, B.; et al. European Journal of Nuclear Medicine 2000, 27, 273–282.

[25] Antunes, P.; Ginj, M.; Zhang, H.; et al. European Journal of Nuclear Medicine and Molecular Imaging 2007, 34, 982–993.

[26] Gabriel, M.; Decristoforo, C.; Kendler, D.; et al. Journal of Nuclear Medicine 2007, 48, 508–518.

[27] Srirajaskanthan, R.; Kayani, I.; Quigley, A. M.; Soh, J.; Caplin, M. E.; Bomanji, J. Journal of Nuclear Medicine 2010, 51, 875–882.

[28] Campana, D.; Ambrosini, V.; Pezzilli, R.; et al. Journal of Nuclear Medicine 2010, 51, 353–359.

[29] Ambrosini, V.; Campana, D.; Bodei, L.; et al. Journal of Nuclear Medicine 2010, 51, 669–673.

[30] Fani, M.; Braun, F.; Waser, B.; et al. Journal of Nuclear Medicine 2012, 53, 1481–1489.

[31] Cescato, R.; Waser, B.; Fani, M.; Reubi, J. C. Journal of Nuclear Medicine 2011, 52, 1886–1890.

[32] Wild, D.; Fani, M.; Behe, M.; et al. Journal of Nuclear Medicine 2011, 52, 1412–1417.

[33] Cremonesi, M.; Ferrari, M.; Zoboli, S.; et al. European Journal of Nuclear Medicine 1999, 26, 877–886.

[34] Forrer, F.; Uusijarvi, H.; Waldherr, C.; et al. European Journal of Nuclear Medicine and Molecular Imaging 2004, 31, 1257–1262.

[35] Wang, X.; Fani, M.; Schulz, S.; Rivier, J.; Reubi, J. C.; Maecke, H. R. European Journal of Nuclear Medicine and Molecular Imaging 2012, 39, 1876–1885.

[36] Nguyen, K.; Parry, J. J.; Rogers, B. E.; Anderson, C. J. Nuclear Medicine and Biology 2012, 39, 187–197.

[37] Heuser, J. E.; Anderson, R. G. The Journal of Cell Biology 1989, 108, 389–400.

[38] Waser, B.; Tamma, M. L.; Cescato, R.; Maecke, H. R.; Reubi, J. C. Journal of Nuclear Medicine 2009, 50, 936–941.

[39] Baum, R. P.; Kulkarni, H. R. Theranostics 2012, 2, 437–447.

9

CORTISTATINS AND DOPASTATINS

Manuela Albertelli and Diego Ferone

Endocrinology Unit, Department of Internal Medicine and Center of Excellence for Biomedical Research, University of Genova, Genova, Italy

ABBREVIATIONS

CST	cortistatin
DA	dopamine
DRs	dopamine receptors
GPCRs	G-protein-coupled receptors
IL	interleukin
MAPK	mitogen-activated protein kinase
NETs	neuroendocrine tumors
NFPAs	clinically nonfunctioning pituitary adenomas
PI3K	phosphatidylinositol 3-kinase
SRIF	somatostatin
SSTRs	somatostatin receptors
sst1–5	somatostatin receptor subtypes 1–5
TEC	thymic epithelial cells
TMDs	transmembrane domains

Somatostatin (SRIF) is a widespread neuropeptide with mainly inhibitory function on hormone release in the anterior pituitary and the gastrointestinal tract, as well as with antiangiogenic, antiproliferative, and analgesic properties [1–3]. Within the nervous system, SRIF acts as a neuromodulator with physiological effects on neuroendocrine,

Somatostatin Analogues: From Research to Clinical Practice, First Edition. Edited by Alicja Hubalewska-Dydejczyk, Alberto Signore, Marion de Jong, Rudi A. Dierckx, John Buscombe, and Christophe Van de Wiele.

motor, and cognitive functions [4]. SRIF displays antiproliferative activities triggering different intracellular pathways, at least *in vitro*, involved in the control of cell growth as well as apoptosis [1–4]. The effects of SRIF are mediated by five different SRIF receptors (SSTRs), heterogeneously distributed on target cells [1–4].

Recently, the complexity of the somatostatinergic system has been extended by two new important findings: the discovery of cortistatin (CST) as well as the evidence of cross interaction between the same SSTRs and between these and other membrane receptors, in particular dopamine receptors (DRs). These new challenging discoveries have led the research toward finding potential new drugs and to the opportunity to capitalize on these new pathophysiological acquisitions.

CST

The cDNA of this new neuropeptide, displaying strong structural similarities with SRIF, has been discovered and cloned in 1996 by De Lecea et al., who identified for the first time CST in rat cortex [5].

CST is a cyclic peptide derived from a prepropeptide of 114 residues, and comparable to SRIF-14 and SRIF-28, proteolytic processing of procortistatin results in the production of two isoforms in rat, CST-14 and CST-29, and in human, CST-17 and CST-29 [5–7]. In analogy with SRIF, CST contains the four amino acids Phe7, Trp8, Lys9, and Thr10 [5], which are essential for binding to the five SSTR subtypes (sst) [8]. Because of its high structural resemblance to SRIF, CST binds with high affinity to the five known sst; however, despite these analogies, these neuropeptides are products of different genes [7].

CST shares several functional properties with SRIF; however, some biological activities of CST are unique. Indeed, many differences in the direct effects of SRIF and CST have been described. Among these are the induction of slow-wave sleep, the reduction of locomotor activity, and the activation of cation-selective currents not responsive to SRIF [9]. Most of the actions of CST have been designated to the regulation of behavior, sleep, and memory mechanisms localized in the brain. Both compounds seem to depress neuronal activity in the hippocampus [5] and reduce development of seizures [10] and can deteriorate memory consolidation [11], whereas CST-treated rats showed a clear hypoactive behavior, and the electroencephalogram (EEG) showed a dramatic increase in cortical slow waves [5]. Subsequently, after administration of CST, rats spent more time in slow-wave sleep and less time in rapid eye movement (REM) sleep. On the other hand, the administration of SRIF results in sleep periods dominated by REM sleep, without significantly affecting the other phases of sleep [5, 7, 12]. Moreover, it has been demonstrated that CST expression is upregulated during sleep deprivation in rats. In this phase, also upregulation of SRIF has been demonstrated in the brain [13, 14]. This suggests that both SRIF and CST may have sleep regulatory functions and probably interplay in the regulation of sleep in their own specific way. Summarizing, in behavior and sleep regulation, either SRIF or CST play important roles, whereas the clinical significance of these peptides in regulating these processes needs to be more extensively investigated [15].

As already mentioned, CST was initially found to be selectively expressed in the brain cortex and hippocampus; however, further studies revealed its expression in the human kidney, stomach, testes, as well as leukocytes [16]. The functional significance of CST in peripheral tissues is currently under investigation. Since SRIF plays important roles in regulating various processes in humans through its SSTRs and CST may act via these G-protein-coupled receptors as well, we could expect that CST might have comparable effects in the body with respect to regulation of cell proliferation and secretion. However, as previously stated, in the human brain, differences in effects of CST and SRIF have been described [9]. Indeed, these differences may be explained by different postreceptor signaling pathways or by the existence of a specific CST receptor. In fact, recently, a novel MrgX2 receptor has been cloned. CST binds this receptor with a relative higher affinity of over 2500 peptides tested, including SRIF [17]. These findings suggest that CST acts via either the SSTRs or MrgX2, which may explain some differences in effects between SRIF and CST. Although further studies are warranted to support this hypothesis, MrgX2 could represent the putative CST receptor; however, CST has been more recently shown to bind, with lower affinity, the type 1a growth hormone secretagogue receptor (GHS-R1a) as well [18].

Focus on the Immune System

A significant anti-inflammatory effect of SRIF has been demonstrated in different experimental animal models displaying an inflammatory state. CST has also been shown to inhibit the production of proinflammatory cytokines and nitric oxide by activated macrophages and to stimulate the secretion of the anti-inflammatory cytokine interleukin (IL)-10, affecting circulating levels of these molecules and strikingly improving survival and clinical outcomes in mouse models of sepsis, Crohn's disease, and rheumatoid arthritis [19–21]. In fact, very recently, the role of CST in immune cells during inflammatory processes in rodent models has been deeply investigated. Firstly, Gonzalez-Rey et al. demonstrated in murine models of lethal endotoxemia that CST prevented the septic shock-associated histopathology, including inflammatory cell infiltration and multiorgan intravascular disseminated coagulation [19]. It was suggested that the main action of CST in regulation of septic shock was the deactivation of resident and infiltrating macrophages. Similarly, in a mouse model of inflammatory bowel disease, it was found that CST was able to reduce disease severity, also by decreasing the levels of proinflammatory/cytotoxic cytokines, chemokines, and acute-phase proteins during inflammation [20]. Finally, the effects of CST were investigated in a murine model of rheumatoid arthritis [21]. In this setting, CST treatment significantly reduced the grade of collagen-induced arthritis, completely abrogating swelling of joints and destruction of cartilage and bone [21]. These recent findings in murine models imply an important regulatory role for CST in immune-mediated diseases and potential implications for future clinical use. As SSTRs are also widely expressed in affected tissues in patients suffering from immune disease, such as rheumatoid arthritis, as well as granulomatous disease and inflammatory bowel disease, CST may have comparable inhibitory effects in humans [22–24]. It may therefore be hypothesized that CST may have future clinical implications in controlling immune cell-mediated disease. However, it should be noted that many differences in sst expression between rodents and humans have been described [25, 26]. Differences in sst expression may account for differences in effects between both organisms. However, future *in vitro* and *in vivo* studies will be necessary in order to evaluate the functional significance of CST in human immune disease [15].

Therefore, two different ligands (SRIF and CST) can activate SSTRs to dampen inflammation, but CST anti-inflammatory action appears to be stronger than that of SRIF, possibly because of CST binding to GHS-R1a and, hypothetically, to MrgX2 as well. In animals, cyclosomatostatin, an SSTR-antagonist, fully blocked the anti-inflammatory effect of SRIF, but not of CST, while a GHS-R antagonist partially antagonized CST activity [19–21, 27]. In addition, only CST inhibited basal and IL-1β-stimulated release of prostaglandin E2 by primary rat microglia, a population of CNS-resident macrophage-like cells [28]: thus, CST action on innate immunity cells might also be broader than that of SRIF.

As is known, the wide expression of ssts in the human immune system has suggested a potential functional significance of SRIF and its receptors in immune cell function. The thymus plays a pivotal role in the control of the immune system and for the establishment of immunocompetence. Apart from the local production of SRIF, three receptor subtypes, sst1, sst2A, and sst3, have been found in the thymic tissue [29]. However, SSTR expression in the human thymus is age dependent, and the expression of the three SSTR subtypes is heterogeneously distributed within the different cell subsets forming the complex architecture of this organ [29, 30]. A selective expression of sst1 and sst2A has been detected in cultured thymic epithelial cells (TEC), whereas the whole population of isolated thymocytes express sst2A and sst3 [31, 32]. In fact, sst2A seems selectively expressed in very early thymocytes; however, while this latter SSTR is downregulated during thymocyte maturation, sst3 is upregulated and appears predominantly, or even selectively, expressed in the mature subsets [32]. Conversely, thymic macrophages and dendritic cells maintain the selective expression of sst2A [31–33]. Interestingly, in other human lymphoid tissues, such as the spleen, SSTRs are heterogeneously expressed, again maintaining a preferential localization of sst2 in cells of the monocyte–macrophage lineage and of sst3 in resting and activated lymphocytes [4, 25, 34]. In support of this concept, it is intriguing that peripheral human T lymphocytes (derived from thymocytes) seem to selectively express sst3 [34].

In an interesting paper of 2003, the authors investigated mRNA expression of SRIF and CST (isoforms) by RT-PCR in different immune tissues, such as the thymus, spleen, and bone marrow [35]. SRIF was only detected in the whole thymic tissue and TEC, but not in thymocytes, whereas no SRIF mRNA was recorded in the human spleen and bone marrow. CST mRNA, however, was clearly expressed in all tissues tested. Moreover, by ligand-binding studies, it has been also demonstrated that CST can displace [125I-Tyr3]octreotide binding with relatively high affinity on human thymic tissue and sst2-expressing cells [35]. Then, in preliminary studies, it has been demonstrated that CST inhibits proliferation of isolated human thymocytes, suggesting that also CST may contribute to the development of mature T lymphocytes [36].

Interestingly, differences in CST mRNA expression levels were observed between monocytes and their functionally derived cells, that is, macrophages and dendritic cells: in fact, when monocytes differentiated into mature macrophages *in vitro*, CST mRNA levels increased approximately 60-fold [35], pointing to an upregulatory mechanism for CST mRNA expression during differentiation and maturation of monocytes into both macrophages and dendritic cells and, possibly, a more important role for CST in the mature immune system.

Moreover, because no SRIF mRNA but only CST mRNA was detected in the samples tested, in view of the presence of ssts in these tissues, and on the basis of both the expression of CST and the observation that CST is able to bind to human sst2 receptors, it has been hypothesized that CST, rather than SRIF, may act via SSTRs in the human immune system in an autocrine and/or paracrine manner [35]. Noteworthy, CST, but not SRIF, mRNA can be detected in human monocytes–macrophages and can be upregulated upon activation in parallel with sst2 mRNA, suggesting the existence of an autocrine CST–sst2 loop [27].

Focus on the Pituitary

In a number of recent either *in vivo* or *in vitro* studies, it has been found that SRIF and CST may exert inhibitory effects on prolactin secretion by prolactinomas [37, 38]. Indeed, in a series of cultured prolactinomas, CST significantly inhibited prolactin secretion, similarly to the effects induced by the sst5-selective agonist BIM23206, while the sst2-selective agonist BIM23210 only exerted significant effects in one case [38]. In an *in vivo* study, a significant, comparable reduction in prolactin levels was obtained after infusion with either SRIF or CST [37]. These effects may be explained by their high binding affinities to both sst2 and sst5, mainly by acting via the predominantly expressed sst5, while on the other hand, sst2-selective agonists were found to have limited effects on prolactin levels [39]. The clinical importance of sst analog therapy in prolactinomas remains uncertain as their effects seem not to be additive to the effects of the widely used dopamine (DA) agonist treatment regimens. CST has been tested also in clinically nonfunctioning pituitary adenomas (NFPAs), where there is effectively reduced cell viability in 6 out of 13 NFPAs. However, CST resulted as less potent in reducing cell viability in these tumors compared with either SRIF or sst-selective analogs [40]. Interestingly, recent evidence shows a growing number of spliced variants of SSTRs, particularly sst5, in other species, such as pig, as well as in humans. In particular, truncated isoforms of five and four transmembrane domains (TMDs), named sst5TMD5 and sst5TMD4, respectively, have been recorded in pituitary adenomas of diverse etiology, including NFPAs, corticotropinomas, somatotropinomas, and prolactinomas [41]. In contrast to the predominant membrane localization of full-length sst5, both sst5TMD5 and sst5TMD4 show a preferentially intracellular localization. Despite their truncated nature, both receptors seem to be functional. However, whereas sst5TMD5 is selectivity activated by SRIF compared with CST, cells transfected with sst5TMD4 almost exclusively respond to CST and not to SRIF [41]. In summary, the available data suggests that CST substantially may partially reproduce the *in vitro* effects of SRIF on pituitary secretions of human and animal models. Conversely, the functions of CST in the majority of peripheral endocrine (and nonendocrine) tissues are almost unknown. Despite this last observation, the differential tissue expression of SRIF, CST, and their receptors suggests that CST may act as a sort of natural SRIF analog in a number of tissues, whereas in specific endocrine tissues, it may play a predominant, unique regulatory role, although the challenge is now to find the differences between these neuroendocrine peptides [42].

DOPASTATIN

In addition to the discovery of the CST, in recent years, the somatostatinergic system has been studied for its interactions with another complex but similar system, the dopaminergic one.

DA is the predominant catecholaminergic neuropeptide in human central nervous system where it controls a variety of functions including cognition, emotion, locomotor activity, and regulation of the endocrine system. DA plays multiple roles also in the periphery, as a modulator of cardiovascular and renal function, gastrointestinal motility, as well as hormone synthesis and secretion.

The various actions of DA in the endocrine system are mediated by specific receptors (DRs) that can be differentially expressed on both endocrine and neuroendocrine cells. SSTRs and DRs share some similarities: they are both G-protein-coupled receptors (GPCRs) and belong to two distinct receptor superfamilies, each consisting of 5 subtypes. Two isoforms of the D2 have been also found and characterized, the long (D2long) and short (D2short) isoforms. These two forms are generated via alternative splicing and differ only for a small amino acidic fraction at intracellular level. However, the D2short seems more important and deeply involved in the control of cell activities, at least in neuroendocrine cells [43, 44].

Similarly to SSTRs, also DRs are linked to different intracellular pathways, leading mainly to the negative control of hormonal secretion and/or cell cycle or to induction of apoptosis through different signaling transduction mechanisms [45]. Indeed, particular interest is rising in the potential role of these regulatory neuropeptides in the control of cell cycle. In pituitary tumor cells, SRIF and its analogs exert an antiproliferative effect by acting on the phosphatidylinositol 3-kinase (PI3K)/AKT signaling pathway, whereas apoptosis has been observed upon binding of SRIF and SRIF analogs to sst3 and possibly to sst2 as well [46]. Also, both isoforms of D2 receptor play a relevant role in the signaling pathways involved in the proliferation and cell death of pituitary tumor cells, possibly through p38 mitogen-activated protein kinase (MAPK) and ERK activation [47].

Until recently, it was believed that a single dominant SSTR or DR subtype could control a single biologic function. Consequently, ligands with high affinity for specific receptor subtypes were developed and introduced in the clinical practice, including the SRIF analogs octreotide and lanreotide and their slow-release (SR) depot formulations, octreotide LAR and SR lanreotide first and Autogel later, all of which bind preferentially sst2. On the other side, DA agonists such as bromocriptine, quinagolide, and cabergoline, which bind predominantly D2 receptor, have been largely used in the treatment of endocrine and neurological disorders. Indeed, several studies have demonstrated a close positive correlation between the presence of each receptor and the clinical response to the analog targeting that specific receptor [48, 49]. However, a lack of clinical response to SRIF and DA analogs has been observed in a rather high percentage of patients despite the presence of functional sst2 or D2 receptors. To overcome the resistance to single-agent treatment, the use of a combined SRIF analog and DA agonist treatment has also been explored with modest success [50].

More recently, further studies on the characterization of the receptor profile have definitively shown that the concept of a single dominant SSTR or DR subtype controlling a specific biological function is too simplistic and does not account for the lack of efficacy expected for the corresponding medical therapy.

In the past years, the knowledge about SSTRs physiopathology has been extended by the discovery that SSTRs can interact on cell membrane forming receptor dimers [51]. A series of studies, carried out on transfected cell lines, have shown that dimers can consist of two identical SSTR subtypes (homodimers) or two different subtypes (heterodimers), with a range of possible combinations depending on the specific subtype and, probably, on the specific SSTR-expressing cell population. SSTR dimerization can be a ligand-dependent phenomenon, since natural SRIF and subtype-specific SRIF agonists have different effects on dimerization, leading to formation or, on the contrary, dissociation of SSTR dimers [51–56]. SSTR homo- and heterodimerization also involve cellular events beyond the membrane, since it modifies SSTR internalization and trafficking, as well as signal transduction [57, 58]. More interestingly, SSTRs can also form functional dimers interacting at membrane level with DRs [51]. Indeed, in a recent study, the dimerization between SSTRs and between SSTRs and D2 receptor has been finally observed also on prostate and non-small cell lung cancer lines that, in contrast to previous studies, were not transfected with the specific GPCRs, but expressed constitutively both families of receptor subtypes [59].

However, few studies have been conducted so far to assess the final consequences of receptor dimerization on cellular function [43, 56], and the clinical relevance of heterodimerization between SSTRs and DR must be proven [59].

The combination of all these basic research observations, concerning the interactions of SSTRs and DRs, and the clinical reports of enhanced efficacy of combined SRIF and DA analog treatment lead to the concept of creating chimeric molecules combining structural features of both compound classes [60, 61]. These hybrid products indeed combine structural elements of both SRIF and DA in the same molecule and simultaneously recognize, with high-affinity binding activity, and activate both SSTRs and DRs. Generally, these new intriguing products may display greatly enhanced potency and efficacy, as compared with that of individual SSTR or DR agonists [60]. Among the new classes of chimeric compounds, two have been more largely explored, BIM-23A387, targeting simultaneously sst2 and D2 receptors, and BIM-23A760, targeting sst2, sst5, and D2, and have been named dopastatins.

Preclinical Experience

In primary cultures of human GH-secreting pituitary adenoma cells, it was observed that individual SRIF and DA analogs interacting with either sst2 or D2 receptors both suppress GH secretion in a dose-dependent manner. The combination of individual sst2 and D2 agonists achieves no greater suppression of GH than the single agonist alone. In contrast, when the dopastatin sst2–D2 was tested, a greater than 100-fold increase in potency was observed in terms of suppressing GH, and similar potency was observed in suppressing prolactin levels from mixed GH-/prolactin-secreting human pituitary adenoma cells [60, 62]. In a recent paper, a small series of thyrotropin-secreting pituitary adenomas were evaluated for SSTR and DR expression and the relationships between receptor expression, *in vitro* antiproliferative response, and clinical data [63]. In this study, tumor cell proliferation was tested *in vitro* using octreotide, cabergoline, and two chimeric compounds, BIM-23A760 and BIM-23A387. BIM-23A760 caused the highest antiproliferative effect among all the compounds tested, while combined treatment with octreotide and cabergoline displayed an additive effect of magnitude comparable to that of the other chimeric compound, BIM-23A387. Moreover, octreotide resistance in one case could be overcome by treatment with the chimeric compounds. From this study, it can be extrapolated that a high sst5/sst2 ratio might be predictive of a positive outcome to long-term treatment with SRIF analogs in this type of pituitary adenoma. However, combined SSTR and D2 receptor targeting might be considered as a potential tool to improve the response rate in SRIF analog-resistant tumors [63].

Even more intriguing is the evidence that these chimeric compounds appear to be more effective than traditional molecules also in inhibiting cell proliferation. Data derived from a preclinical study on tumor cells, in which the GPCR profile of a well-established human non-small lung cancer cell line, CALU-6, was characterized, displayed a more potent inhibitory effects on cell proliferation of two new chimeras, BIM-23A387 and BIM-23A370, compared with monomeric receptor activators and also with combination of specific SRIF analogs and DA agonists [43]. This evidence was lately confirmed in a multicenter study by Florio et al. in NFPAs unresponsive to conventional analogs [64]. In this setting, the chimeric compound BIM-23A760 was able to achieve a better control of cell growth, measured by 3H-thymidine incorporation in cell cultures, when compared with the individual DA and/or SRIF analogs, alone or in combination [64]. In the study by Ferone et al., both BIM-23A387 and BIM-23A370 were significantly more potent in inhibiting CALU-6 cell proliferation compared with classical and new experimental SRIF analogs and DA agonists, tested either alone or in combination [43]. Moreover, in another study, an increased inhibition of proliferation in either prostate cancer or non-small lung cell lines has been observed, which was more evident with those ligands stimulating the formation of receptor dimers [59].

The most effective antisecretory and antiproliferative activity of chimeric compounds is not yet fully elucidated, and currently, a definitive explanation for the peculiar properties of these new hybrid molecules has not been found yet. Definitively, possible explanations for the increased efficacy in suppressing hormone production include their ability to bind and activate multiple receptors, resulting in the activation of multiple signal transduction pathways or even alternative pathways, and/or the possibility that the interaction between receptor ligands allows a stabilization of the receptor in its active conformation or reduces the rate of internalization and, finally, that the chimeras make rapid passage of the ligand, released from the receptor, to another receptor [65].

Moreover, the receptor distribution on tumor cells is certainly one of the key elements in regulating the cellular response to both SRIF analogs and DA agonists [45]. Recent evidence indicates that the chimeric molecules can act differently in various tissues tested and that their effect may vary depending on the cell type. In fact, a different degree of cytotoxicity of chimeric compounds, for example, in neuroendocrine bronchopulmonary cell lines and in cells lines resembling the neuro-endocrine tumors (NETs) of the small intestine, was observed [58]. The responses of each cell line suggest that the NETs of the various districts, resulting from different neuroendocrine cells, may require cell-specific antiproliferative agents on the basis of the uniqueness of receptor profile of individual lesions [58].

Having molecules with multiple receptor affinity and being able to modulate the activity on the various receptors, obtaining a product with maximum effectiveness and minimal side effects was attempted. Interestingly, when the activity was strong either on sst2, sst5, or D2, the compound has no greater efficacy than one of currently available clinical analogs of ssts [65]. Since it is known that the sst5 directly suppresses insulin secretion from pancreatic beta cells, with potential side effect on glycemic balance, the ideal activity ratio for a chimeric molecule to be used clinically would seem to be potent sst2 and D2 activity, with moderate activity at sst5 level [60, 66]. This condition has been developed with the chimeric compound BIM-23A760.

In Vivo Experience

BIM-23A760 has also been tested *in vivo* in normal cynomolgus monkeys and found to induce a significant dose-related suppression of both GH and prolactin. A consensual and expected reduction in the levels of IGF-I was observed after 24 hours of treatment with this dopastatin. No effect was observed on either insulin secretion or circulating glucose at any time or dose of BIM-23A760 during the whole study [60]. In a phase I study in normal individuals, BIM-23A760 was observed to dramatically suppress prolactin levels for prolonged periods of time (>8 days) from single subcutaneous injections without significant side effects [67]. Further, in a phase II study, single subcutaneous administration of BIM-23A760 to acromegalic patients resulted in a significant, prolonged suppression of circulating GH levels [68]. Unexpectedly, however, when tested in a phase IIb study examining chronic administration of BIM-23A760 in acromegalic patients, the suppressive effect on GH was unimpressive, while prolactin was profoundly suppressed. Follow-up study of the clinical blood sample and supporting basic experimentation revealed that, in human, BIM-23A760 produces a metabolite with potent dopaminergic activity that interferes with the activity of the parent molecule. However, a new experimental phase is still ongoing to modify the compound structure to avoid production of interfering metabolites while maintaining the potent, enhanced activity [60, 61]. The new products will be soon available for exploring their clinical properties in different sectors of endocrine and nonendocrine diseases.

CONCLUSIONS

In conclusion, due to the strong evidence that SSTR and D2 receptors are widely coexpressed in pituitary adenomas, as well as in a large variety of well-differentiated NET, originating from different tissues such as the gastrointestinal tract, pancreas, lung, adrenal, and thymus, the perspective for the use of dopastatins is wide and challenging. Indeed, the new findings in the pathophysiology of SSTRs and DRs, together with the availability of new universal and subtype-specific SRIF and D2-selective analogs, may offer in the near future new weapons to clinicians facing these complex endocrine disorders and to patients for improving their survival and reducing the high morbidity associated with these diseases.

REFERENCES

[1] Weckbecker, G.; Lewis, I.; Albert, R.; et al. Nature Reviews Drug Discovery 2003, 2, 999–1017.

[2] Møller, L. N.; Stidsen, C. E.; Hartmann, B.; et al. Biochimica et Biophysica Acta 2003, 1616, 1–84.

[3] Olias, G.; Viollet, C.; Kusserow, H.; et al. Journal of Neurochemistry 2004, 89, 1057–1091.

[4] Ferone, D.; Boschetti, M.; Resmini, E.; et al. New York Academy of Sciences 2006, 1069, 129–144.

[5] de Lecea, L.; Criado, J. R.; Prospero-Garcia, O.; et al. Nature 1996, 381, 242–245.

[6] de Lecea, L.; Ruiz-Lozano, P.; Danielson, P. E.; et al. Genomics 1997, 42,499–506.

[7] Fukusumi, S.; Kitada, C.; Takekawa, S.; et al. Biochemical and Biophysical Research Communications 1997, 232, 157–163.

[8] Veber, D. F.; Holly, F. W.; Nutt, R. F.; et al. Nature 1979, 280, 512–514.

[9] Spier, A. D.; de Lecea, L. Brain Research Reviews 2000, 33, 228–241.

[10] Braun, H.; Schulz, S.; Becker, A.; et al. Brain Research 1998, 803, 54–60.

[11] Sanchez-Alavez, M.; Gomez-Chavarin, M.; Navarro, L.; et al. Brain Research 2000, 858, 78–83.

[12] Feige, J. J.; Cochet, C.; Rainey, W. E.; et al. Journal of Biological Chemistry 1987, 262, 13491–13495.

[13] Toppila, J.; Alanko, L.; Asikainen, M.; et al. Journal of Sleep Research 1997, 6, 171–178.

[14] Toppila, J.; Asikainen, M.; Alanko, L.; et al. Journal of Sleep Research 1996, 5, 115–122.

[15] Dalm, V. A.; Hofland, L. J.; Lamberts, S. W. Molecular and Cellular Endocrinology 2008, 286, 262–277.

[16] Ejeskar, K.; Abel, F.; Sjoberg, R.; et al. Cytogenetics and Cell Genetics 2000, 89, 62–66.

[17] Robas, N.; Mead, E.; Fidock, M. Journal of Biological Chemistry 2003, 278, 44400–44404.

[18] Siehler, S.; Nunn, C.; Hannon, J.; et al. Molecular and Cellular Endocrinology 2008, 286, 26–34.

[19] Gonzalez-Rey, E.; Chorny, A.; Robledo, G.; et al. Journal of Experimental Medicine 2006, 203, 563–571.

[20] Gonzalez-Rey, E.; Varela, N.; Sheibanie, A. F.; et al. Proceedings of the National Academy of Sciences 2006, 103, 4228–4233.

[21] Gonzalez-Rey, E.; Chorny, A.; Del Moral, R. G.; et al. Annals of the Rheumatic Diseases 2007, 66, 582–588.

[22] Van Hagen, P. M.; Krenning, E. P.; Reubi, J. C.; et al. European Journal of Nuclear Medicine 1994, 21, 497–502.

[23] Van Hagen, P. M.; Markusse, H. M.; Lamberts, S. W.; et al. Arthritis & Rheumtism 1994, 37, 1521–1527.

[24] Reubi, J. C.; Mazzucchelli, L.; Laissue, J. A. Gastroenterology 1994, 106, 951–959.

[25] Dalm, V. A.; Hofland, L. J.; Ferone, D.; et al. Journal of Endocrinological Investigation 2003, 26, 94–102.

[26] Lichtenauer-Kaligis, E. G.; Van Hagen, P. M.; Lamberts, S. W.; et al. European Journal of Endocrinology 2000, 143, S21–S25.

[27] Ameri, P.; Ferone, D. Neuroendocrinology 2012, 95, 267–276.

[28] Dello Russo, C.; Lisi, L.; Navarra, P.; et al. Journal of Neuroimmunology 2009, 213, 78–83.

[29] Ferone, D.; Van Hagen, P. M.; Van Koetsveld, P. M.; et al. Endocrinology 1999, 140, 373–380.

[30] Ferone, D.; Pivonello, R.; Van Hagen, P. M.; et al. American Journal of Physiology—Endocrinology and Metabolism 2000, 279, E791–E798.

[31] Ferone, D.; van Hagen, P. M.; Colao, A.; et al. Annals of Medicine 1999, 31, 28–33.

[32] Ferone, D.; Pivonello, R.; Van Hagen, P. M.; et al. American Journal of Physiology—Endocrinology and Metabolism 2002, 283, E1056–E1066.

[33] Ferone, D.; van Hagen, P. M.; Pivonello, R.; et al. European Journal of Endocrinology 2000, 143, S27–S34.

[34] Lichtenauer-Kaligis, E. G.; Dalm, V. A.; Oomen, S. P.; et al. European Journal of Endocrinology 2004, 150, 565–577.

[35] Dalm, V. A.;Van Hagen, P. M.; van Koetsveld, P. M.; et al. Journal of Clinical Endocrinology and Metabolism 2003, 88, 270–276.

[36] Dalm, V. A.; Van Hagen, P. M.; de Krijger, R. R.; et al. Clinical Endocrinology 2004, 60, 625–629.

[37] Grottoli, S.; Gasco, V.; Broglio, F.; et al. Journal of Clinical Endocrinology & Metabolism 2006, 91, 1595–1599.

[38] Rubinfeld, H.; Hadani, M.; Barkai, G.; et al. Journal of Clinical Endocrinology & Metabolism 2006,91, 2257–2563.

[39] Lamberts, S. W.; Zweens, M.; Klijn, J. G.; et al. Clinical Endocrinology 1986, 25, 201–212.

[40] Padova, H.; Rubinfeld, H.; Hadani, M.; et al. Molecular and Cellular Endocrinology 2008, 286, 214–218.

[41] Durán-Prado, M.; Gahete, M. D.; Martínez-Fuentes, A. J.; et al. Journal of Clinical Endocrinology & Metabolism 2009, 94, 2634–2643.

[42] Gahete, M. D.; Durán-Prado, M.; Luque, R. M.; et al. Molecular and Cellular Endocrinology 2008, 286, 128–134.

[43] Ferone, D.; Arvigo, M.; Semino, C.; et al. American Journal of Physiology: Endocrinology & Metabolism 2005, 289, E1044–E1050.

[44] Missale, C.; Nash, S. R.; Robinson, S. W.; et al. Physiological Review 1998, 78, 189–225.

[45] Ferone, D.; Gatto, F.; Arvigo, M.; et al. Journal of Molecular Endocrinology 2009, 42, 361–370.

[46] Theodoropoulou, M.; Zhang, J.; Laupheimer, S.; et al. Cancer Research 2006, 66, 1576–1582.

[47] Ferone, D.; Pivonello, R.; Resmini, E.; et al. European Journal of Endocrinology 2007, 156, S37–S43.

[48] Ferone, D.; de Herder, W. W.; Pivonello, R.; et al. Journal of Clinical Endocrinology & Metabolism 2008, 93, 1412–1417.

[49] Pivonello, R.; Matrone, C.; Filippella, M.; et al. Journal of Clinical Endocrinology & Metabolism 2004, 89, 1674–1683.

[50] Pivonello, R.; Ferone, D.; Lamberts, S. W.; et al. New England Journal of Medicine 2005, 352, 2457–2458.

[51] Rocheville, M.; Lange, D. C.; Kumar, U.; et al. Journal of Biological Chemistry 2000, 275, 7862–7869.

[52] Pfeiffer, M.; Koch, T.; Schroder, H.; et al. Journal of Biological Chemistry 2001, 276, 14027–14036.

[53] Patel, R. C.; Kumar, U.; Lamb, D. C.; et al. Proceedings of the National Academy of Sciences 2002, 99, 3294–3299.

[54] Grant, M.; Collier, B.; Kumar, U. Journal of Biological Chemistry 2004, 279, 36179–36183.

[55] Grant, M.; Patel, R. C.; Kumar, U. Journal of Biological Chemistry 2004, 279, 38636–38643.

[56] Grant, M.; Alturaihi, H.; Jaquet, P.; et al. Molecular Endocrinology 2008, 22, 2278–2292.

[57] Baragli, A.; Alturaihi, H.; Watt, H. L.; et al. Cell Signalling 2007, 19, 2304–2316.

[58] Kidd, M.; Drozdov, I.; Joseph, R.; et al. Cancer 2008, 113, 690–700.

[59] Arvigo, M.; Gatto, F.; Ruscica, M.; et al. Journal of Endocrinology 2010, 207, 309–317.

[60] Culler, M. D. Hormone and Metabolic Research 2011, 43, 854–857.

[61] Gatto, F.; Hofland, L. J. Endocrine Related Cancer 2011, 18, R233–R251.

[62] Saveanu, A.; Lavaque, E.; Gunz, G.; et al. Journal of Clinical Endocrinology & Metabolism 2002, 87, 5545–5552.

[63] Gatto, F.; Barbieri, F.; Gatti, M.; et al. Clinical Endocrinology (Oxford) 2012, 76, 407–414.

[64] Florio, T.; Barbieri, F.; Spaziante, R.; et al. Endocrine Related Cancer 2008, 15, 583–596.

[65] Jaquet, P.; Gunz, G.; Saveanu, A.; et al. European Journal of Endocrinology 2005, 153, 135–141.

[66] Zambre, Y.; Ling, Z.; Chen, M. C.; et al. Biochemical Pharmacology 1999, 57, 1159–1164.

[67] Froehlich, J.; Ramis, J.; Lesage, C.; et al. 91st Annual Meeting of the Endocrine Society 2009, P3-685, June 10–13, 2009, Washington, DC.

[68] Lesage, C.; Seymour, C.; Urbanavicius, V.; et al. 91st Annual Meeting of the Endocrine Society 2009, P3–673, June 10–13, 2009, Washington, DC.

INDEX
